T0396409

Operationalising e-Democracy through a System Engineering Approach in Mauritius and Australia

Soobhiraj Bungsraz

Operationalising e-Democracy through a System Engineering Approach in Mauritius and Australia

palgrave
macmillan

Soobhiraj Bungsraz
University of Newcastle Australia
Newcastle, NSW, Australia

ISBN 978-981-15-1779-2 ISBN 978-981-15-1777-8 (eBook)
https://doi.org/10.1007/978-981-15-1777-8

This Palgrave Macmillan imprint is published by the registered company Springer Nature Singapore Pte Ltd.
The registered company address is: 152 Beach Road, #21-01/04 Gateway East, Singapore 189721, Singapore

This book is dedicated to, scholars, honest politicians, citizens, and the younger generations like my son Karan Bungsraz, may it help them shape the now and the coming future.

Preface

The development of the Internet in the 1990s and the subsequent proliferation of Information and Communication Technologies (ICTs) prompted considerable optimism about new possibilities for addressing some of the perceived problems with contemporary representative democracy. ICTs held out the potential to create a particular form of democracy, namely e-democracy, in which people could be empowered to engage more actively with the processes of governing. In particular, e-democracy heralded the possibility for citizens to become actively engaged in policy making and decision making such that the ideal democracy (in the sense of a self-governing *demos*) would finally become a reality. However, e-democracy in practice has fallen somewhat short of expectations. The ideal of a self-governing *demos*, or as it is termed in the book, a democracy that democratises, seems as elusive as ever.

Asking why these developments in e-democracy have fallen short of expectations constitutes the central research problematic of the book. Therefore, the first part of the book looks at some ongoing problems in contemporary representative democracy. It also examines how various attempts at implementing an e-democracy appeared to be an appropriate means of solving them. In reviewing these developments and the debates in the scholarly literature, it emerges that what has been regarded as e-democracy is mostly a variation on the theme of simply digitising currently existing practices. These types of approach contribute to the failure of these ventures into e-democracy because they end up reproducing the very problems that are trying to be solved. That is, such approaches focus on digitising already existing political processes–processes that were never meant to deliver a self-governing *demos*. A democracy that democratises cannot eventuate.

This provides the context for thinking about what might need to be done to enable an ICT-driven e-democracy to be successful. At the very least, an e-democracy worthy of the name needs a democratic design solution (or more appropriately a people-driven design solution) that builds in democratic practices from the start. Hence, what is needed is a means to theorise about how to use the ICTs' capability to democratise. To that end, the book explores a novel approach to such problems of digitisation. It looks beyond the field of political science (though at the same time drawing on the rich literature of political theory and democratic theory) to that of engineering and the systems-based approach of Systems Engineering (SE). Originating in the US Air Force, SE has been adapted for use in many diverse multidisciplinary contexts.

SE is understood in this book as a systems-based technique that facilitates the development of effective solutions to meet a specified need or objective. It is a proven methodology that uses systems thinking to design solutions based on an identified need. In this case, the identified need is one of the empowering citizens to be able to engage in meaningful self-rule, to be truly self-governing. Or in the context of creating an e-democracy, the task is to demonstrate how the digitisation process might be developed in such a way as to overcome some of the problems that have beset past attempts to create e-democracies.

The core research question for the book then turned on explaining how a SE approach might address the currently perceived problems with e-democracy. To answer that central question the book engaged in what amounts to a thought experiment. It considered how SE could be used to develop and deliver a workable e-democracy. Hence, the book provided an exposition of the principles of SE, its key concepts, and their inter-relationships. It then turns to showing how they might be developed and applied successfully to create a workable e-democracy.

The book demonstrates, in theory at least, that SE has something to offer political scientists interested in exploring the potential of ICTs for creating an e-democracy, but not just an e-democracy as previously understood. The analysis pursued in the book offers an approach that moves beyond past ways of thinking about e-democracy. This approach develops an innovative theoretical agenda that makes suggestions about its feasibility for creating an e-democracy that is capable of empowering its citizens to be truly self-governing. In so doing, the book makes a major contribution to the theory and practice of e-democracy.

Newcastle, Australia

Dr. Soobhiraj Bungsraz
CPENG, MIEAust, NER, APEC (Eng),
IntlPE(Aus), MRAeS, Ph.D. (Politics),
MBA, MIT, Grad. Dip. Mgt.- UoN,
BE (Hons) (Aero)- PEC

Acknowledgements

My foray in politics is the culmination of a gracious acceptance of two key players, Professor Dr. Jim Jose and Dr. Robert Imre, who through their mentorship made it feasible. An especial thanks to Dr. Jose for improving my understanding of the field of politics as without those hours of patience in guiding my fledging steps I would not have reached to this crucial stage in my endeavour. The book has made me realise that the journey for change has just started, for that I am always indebted to you, Jim.

I wish to thank many friends I made during the journey in the cohort of Ph.D. with whom I shared some fascinating moments about Ph.D. life. I would like to thank the Faculty of Business and Law and the Discipline of Politics, Policy and International Relations for their material and academic support. I benefited enormously from the spirited debates ranging from the epistemology of colonialism to creativity, from strategic management to the concept of love in marketing, and from entrepreneurship to innovation. A special acknowledgement goes to Dr. Raj Yadav and my friends Benji and Rohit who participated in those spirited debates with a generosity of spirit so essential to creative research. Thanks are also due to my previous lecturers in the Information and Communication Technology (ICT) department who helped me with deepening my interest in that field.

This book is a bridge for other engineers, scholars and politicians to engage in developing systemic solutions that are integrative from a politics lens rather than purely Information Technology (IT) or engineering.

Without values like the promise of democracy and its past insights that allow us to imagine a better future for all, the next evolutionary breakthrough may be missed as the lessons of the past are not acknowledged to guide the future foresights awaiting to be created. Years of practice in a systems environment where I guided junior engineers to embrace holistic system thinking in the fast jet environment found its way through this endeavour. Thus, a big thanks to those, many in my previous workplace, who contributed to my knowledge and experience, and also encouraged my risky decision to travel this new path, especially Air Vice-Marshal Kim Osley and my friends, Group Captain Dave Langlois, and Air Vice-Marshal Cathy Roberts.

My special thanks to my family, my son Karan Bungsraz and my wife Neena Sharma, for their patience and support during my Ph.D. journey which culminated in this book. Also, a gratitude to my brother, Mr Hemraj Bungsraz, for his support in this new field. Without you, my interviews of the Ministers, Members of Parliament specially Honourable Francisco Francois for his vision for improving Rodrigues' government through technology, and the senior public servants in Mauritius would not have been possible. Your long hours seeking access to those to be interviewed allowed me to get some unique insight into what truly ails the political system in Mauritius and to some degree elsewhere. I also wish to thank those who agreed for interviews and spared the time and effort to contribute. If our conversations and this book help you create a better system for Mauritius, then the effort has been purposeful.

To many friends and new acquaintances, which I made during the three field trips to Mauritius, and during my Ph.D. journey thank you for your support in trying to improve that political system. I dedicate this book to my son, policy makers, those Mauritius Ministers past and present, and importantly the people and others elsewhere like in Australia wishing to improve their political system. Let the future be not just for the few, but rather as a family and community a dream for all realised through the systematic efforts of the many. Dare to empower the many, so they have a say to contribute and be heard, and then the means for creating and achieving their dreams. This book is an acknowledgement of the work you are all putting in to achieve an as yet unfulfilled promise—the possibility of a democracy for all and a world where no one will be less equal. As an afterthought, the chapter on Australia is included to show this potentiality.

Without you all, this book and my journey in politics may have remained a conceptual endeavour. This project has enriched me in bringing a whole new perspective to what could be possible if we but dare it to be. This e-democracy operationalisation is but the beginning as it is anticipated that each addendum (nation-state) with variants customised from its applied context will be generated of *Operationalising e-democracy a system engineering approach* so it helps to continuously improve, add and enrich with a pluralism that will be a creating of unique political systems designs that evolve to realise the ultimate goal of people's rule. May the system approach here mentioned assist you in that unique adventure and beyond!

Good luck.

Contents

About the Author

Soobhiraj Bungsraz started his career as a graduate engineer in the field of Aeronautical engineering at the Punjab Engineering College (PEC), he then became a member of Royal Aeronautical Society (RAeS) and Engineers Australia (EA). He has a long career in aviation especially fast jets ranging from training aircraft like the Macchi to the Hornet in the 1990s, earlier it also included civilian jets like B767, B747, A330s, ATR42 and then the JSF finally in the 2010s along the way with roles as Project Engineering Manager and CEO of an airline. This created a deep experience of aviation in differing contexts and first-hand exposure to the impact from introduction of new technology that is shaping that field. He also trained many engineers in that period in the use of engineering managements systems which he developed as part of the organisations he worked for. His career spanned from a junior engineer learning about applied airworthiness and its regulations at the Civil Aviation in Mauritius to developing engineering management system for the Royal Australian Air Force namely early prototypes like the Engineering Management Database in the early 1990s for the Hornet and Macchi fleet technical management. In parallel, these work activities were reinforced by further academic forays in management, administration and Information Technology at the University of Newcastle (UoN), that culminated in a Graduate Diploma in Management, followed by Master's in Business Administration and then Master's in Information Technology and finally a Doctor in Philosophy in Politics.

A quest to improve and find better ways is what drove these academic activities for Bungsraz, but inherently there was also a need to reform his home country Mauritius where massive corruption existed and is still reported. This need to change the system found an expression in his seeking to understand the political system in depth and a venture into the field of politics where he sought to bring in engineering experience both of a practical nature (his long aviation work experience) and also academic nature (politics, management and IT). However, this quest to an improvement initiative was prompted by a short stint as CEO of the national airline in Mauritius where the PM of the country had invited him to fix the airline's performance which was suffering from significant losses due to bad decision making. Armed with his knowledge and experience he ventured to fix the system but the system resisted him and his reforms, changes which he had started were undermined even though a result from his tenure's reforms influenced the record profit for the airline.

This event made him reflect on the nature of the problem, and there came a realisation that systems are politically wasteful and driven often by the incompetence of a powerful few which the system generated and unfortunately protected. Using his aeronautical experience where technology drove reforms to continuously improve the system's performance, the idea of a technical input to politics emerged in 2013 for reforming of political systems through an upgrade. This meant further research and the Ph.D. which culminated led to this book. Experience showed that it was not in the politician's interest to bring change that adversely affected their entrepreneurial activities as political parties which the existing systems under guise of democracy facilitated. Technology could undermine democracy as tolerated by the people and threaten representatives 'democratic reforms' that still protects a privileged few.

So, a new system was needed using a different paradigm, a systems approach, which the book proposes suggestive in that it eliminates the waste endemic in the system of government that is followed under so-called guise of democracies around the world. This book allows honest politicians and the people critical of the performance of their political systems to bring their political system to align. It provides a means to create a system e-democracy. One where a new breed of politicians deliver what the people want and most importantly it is a system accountable to transparently deliver what the people have mandated. The book is aimed at both the policy and practice of democracy. These key ideas can be adapted to suit based on the political system's context. This upgrade depends on both the people and the country where the e-democracy system is to be created, maintained and further improved upon.

Current technology should allow reforms to build better systems and importantly continuously improve the quality of politics and politicians who get elected in a democracy. Hence, through engineering's furthering the quality of democracy in countries that adopt this system thinking, Systems Engineering is recommended by the book for exploration of the Mauritian system's upgrade and it includes the Australian political system as a *what if* experiment. It is hoped that others will build on this work to create in future better political systems around the world for the benefit of all. This book provides a roadmap for those citizens willing to reform their current political system. It prompts conversation to air why their current system still allows incompetence to rule and critically keep ruling to continue undermining the potential for better systems to emerge like an e-democracy that democratises. The book argues that technology can be harnessed to upgrade democracy in any political system. It provides a framework for this upgrade with two cases as examples of engineering application to politics. It is hoped that more cases (countries) will be explored.

Acronyms

ABC	Australian Broadcasting Corporation
ALP	Australian Labor Party
APEC (Eng)	Asia Pacific Economic Co-operation (Engineer)
BCE	Before Common Era
CAB	Citizens Advice Bureau
CCB	Configuration Control Board
CI	Configuration Item
CM	Configuration Management
CMP	Configuration Management Plan
CoRE	Consortium Requirements Engineering
CPENG	Chartered Professional Engineer
CRL	Concept Requirements List
DOORS	Telelogic Requirements Management Tool
EU	European Union
FCI	Functional Configuration Item
FPP	First-Past-the-Post
GDP	Gross Domestic Product
ICT	Information and Communication Technology
INCOSE	International Council on Systems Engineering
IntPE(Aus)	International Professional Engineer (Australia)
JEDMICS	Joint Engineering Data Management and Information Control System
MIEAust	Member of Institution of Engineers Australia
Mil-Std	Military Standard
MP	Member of Parliament
MRAeS	Member of Royal Aeronautical Society

MSM	Militant Socialist Movement
NDU	National Development Unit
OECD	Organisation for Economic and Commercial Development
OPR	Organisation du Peuple de Rodrigues
PEC	Punjab Engineering College
PPP	Purchasing Power Parity
PPS	Private Parliamentary Secretary or junior Minister
PR	Proportional Representation
PTr	Partie Travailliste (Labor Party)
PV	Preferential Voting
Q&A	Question and Answer
SE	Systems Engineering
SEAM	Systems Engineering Assessment Model
SoS	System of Systems
UN	United Nations
UoN	University of Newcastle
USAF	United States Air Force
WTO	World Trade Organization

List of Figures

Chapter 4

Chapter 5

Chapter 6

Chapter 7

Chapter 9

List of Terms

Collage: A Mauritian word to mean ‘glued together’. The book uses this term to describe temporary association for bloc formation so as to form two supra-parties as an electoral winning combination, which is announced a few weeks before an election event.

Configuration Item: The construct used to define the smallest building block of a system in engineering, a political system in this book.

Faucher: A Mauritian term to mean destroyed, wronged or corrupted, democracy is ‘faucher’ means democracy is destroyed.

Functional Configuration Item: The construct used to define a function when more than one configuration items (CIs) are made to interact in such a way so as to produce a specific outcome. It allows in this book to differentiate between simple inputs and complex outcomes, which is obtained when a process transforms the input from CIs that are aggregated into specific configurations.

Reflexivity: A term to describe the premium which is placed on speed of action when digital technology is used.

System: A concept used to describe a tangible or intangible product. A system is a construct that allows the engineer to develop something (solution) to solve a requirement (need).

Technical system: System that uses technology. It could be conceptual or physical (occupy space).

CHAPTER 1

Introduction: Representation as a Case for Upgrade

The question of a democratic deficit in Mauritius and reflections on how it might be addressed were the prompt for undertaking this book. In Mauritius, the seeming regularity of corruption scandals suggests that there is a growing democratic deficit (Transparency Mauritius 2013, 2015, 2017). This is not uncommon in many contemporary democracies even Australia, where concerns about a democratic deficit have been noted for some time (Zweifel 2002; White and Nevitte 2012; Ward 2002; Norris 1997, 2011, 2012; Beetham 2012). For present purposes, the idea of a democratic deficit can be understood as a perceived gap between what democratic institutions purport to be about and what they actually do. Put another way, 'the liberal democracies are failing to fulfil their normative ideals such as maintaining liberal values and practices and providing democratic channels for people to have a say in their own collective destiny' (Stokes and Carter 2001: 3). However, the main research and analysis undertaken in this book is not going to be concerned with solving the specific problems of Mauritius's democratic deficit. Rather this book offers what could best be described as a thought experiment about how to operationalise an effective form of e-democracy.

The advent of Information Communication Technologies (ICTs) via the development of the Internet in the 1990s was seen by many as providing a means to connect government with citizens. The communicative capability of ICTs initiated a new scholarly debate about the prospect for

S. Bungsraz, *Operationalising e-Democracy through a System Engineering Approach in Mauritius and Australia*,
https://doi.org/10.1007/978-981-15-1777-8_1

technologically enhanced forms of democracy to emerge in which people might be involved more actively in the decision-making processes of democratic government and governing (Rios Insua and French 2010; Qvortrup 2007; Päivärinta and Sæbø 2006; OECD 2003; Mulder and Hartog 2013; Meier 2012; Kneuer 2016; Hilbert 2009). The innovative and positive promises of ICTs also prompted some caution about their negative and even anti-democratic potential in terms of abrogating civil liberties and various taken for granted rights as well as concerns about privacy issues. However, as Chen et al. (2006: 9) suggested, 'ICTs can be seen as yielding considerable pluses against … some likely negatives'. Many scholars saw great promise in the development of ICTs and thought that ICTs' capability could deliver a new form of democracy, one that came to be known as e-democracy (Rios Insua and French 2010; Qvortrup 2007; Päivärinta and Sæbø 2006; OECD 2003; Mulder and Hartog 2013; Hague and Loader 1999; Grönlund 2001; Dahlberg and Siapera 2007; Moss and Coleman 2014).

The initial claims for improvement to democracy created some conceptual confusion. It was not just e-democracy that was developed through the use of ICTs. There also emerged other forms of ICT-driven applications to governing, and hence new terms emerged: e-government, e-governance, and the like (along with a proliferation of other terms not necessarily concerned with governing such as e-business, e-commerce, and e-society) (Crespo et al. 2013: 2). The proliferation of e-terms was treated as an infancy issue for a new emergent field, loosely termed e-governance (Grönlund 2004). However, it was more than an infancy issue as the terminological confusion failed to disappear as the capacity and sophistication of ICTs grew, making it difficult to operationalise relevant research findings (Grönlund 2008). Of particular interest here are three key terms: e-democracy, e-government, and e-governance. These warrant some initial clarification so that we do not lose sight of the core focus of the book, namely e-democracy and the development of a possible means to implement it successfully.

There is often considerable confusion between the idea of e-democracy and e-government. While both concepts have arisen as a result of the development of sophisticated ICTs, they capture different activities and have different purposes (Lee et al. 2011: 444). E-democracy refers to some form of electronic democracy, also called 'teledemocracy or digital democracy', whereas ICTs are used 'to connect politicians and citizens by means of information, voting, polling, or discussion' (Grönlund

2001: 23). The emphasis here is on the use of ICTs to connect ordinary citizens to political debates and decision-making processes (Päivärinta and Sæbø 2006: 3) with a goal according to Backus (2001) to move 'citizens from passive information access to active citizen participation in the governing process' (Lee et al. 2011: 444). As we shall see, it is the prospect of engaging citizens actively in the processes of government decision making and policy making that gives e-democracy its appeal. It is this specific potential that differentiates e-democracy, in theory and practice, from e-government. It is also this potential that leads some to resist the idea of e-democracy since the emphasis on the potential for engaging in direct democracy gives rise to familiar arguments about 'the risk of populism, lower-level political discourse, loss of deliberation, creating an unclear role of politicians, and instability of democratic institutions, among others', as well as concerns about 'security and privacy' (Grönlund 2011: 23–24). Some of these concerns will be taken up later in the book but here it can be agreed with Grönlund's earlier view that ICTs provide a means to improve democracy in representative systems (Grönlund 2001). To involve citizens in e-democracy's political process, two requirements broadly identified as digital participation and engagement or e-participation and e-engagement are suggested as key processes (Lee et al. 2011: 445). As we will discuss from the Organisation for Economic Co-operation and Development (OECD) study of e-democracy conducted in twelve countries, digitising these key processes are challenging in practice (OECD 2003).

'E-government', on the other hand, is much more diffuse in its scope. The United Nations (2005: 14) has defined it as the 'use of [information and communication technology] ICT and its application by the government for the provision of information and public services to the people'. It thus describes the provision of 'digital government, one-stop government, and online government' (Grönlund and Horan 2004: 713). These services are about delivering internal government operations to their citizens (who in the process come to be redefined as clients) through various interactive forms and payment facilities apparently under guise to improve electronic engagement of stakeholders. The aim would appear to be more about simplifying government operations in effecting efficiencies in the delivery of services than about empowering citizens. While there may be interactivity between citizens and the particular service or form of information being provided there is no necessary interactivity directly involved in the creation of that service or the provision of that information. The role of the citizen is essentially passive. The actual process of governing continues behind the scenes with little or no involvement of citizens.

Some scholars have suggested that e-democracy has the capacity to overcome these problems, and indeed of overcoming the problem of the democratic deficit (Rios Insua and French 2010; Dahlberg and Siapera 2007). The OECD suggests that better government, though not necessarily more democratic government, would result from digitisation for developing countries (Grönlund and Horan 2004). Variations on this view seem to be repeated with the announcement of every novel technology (Weinberger 1988). On the other hand, there is considerable evidence that e-democracy, despite its promises to improve existing democratic systems and practices, more often than not fails to deliver effectively on those promises (Scholl 2006; Norris and Reddick 2013). Thus, Päivärinta and Sæbø (2006) describe some examples of e-democracy as fragmented experiments, and Moss and Coleman (2014) suggest that the reason for the failures of e-democracy is due to a lack in both strategic direction and coherent policy at national (macro) level, whereas others seem to attribute the failure of this initiative to an over-confidence in the capacity for digital technology to improve the existing system automatically in a more democratic direction (Wilhelm 2000; Hague and Loader 1999; Hoff et al. 2003; Hindman 2009). The continuing problems with designing and implementing a viable political system organised as an e-democracy provided this book with its central idea that perhaps a solution to these problems might lie in adapting the insights from Systems Engineering (SE).

SE was developed by the US Department of Defence to meet its need to design and use complex technological systems. SE is extensively applied by the US Air Force (USAF) to develop and operationalise new technological solutions, for example complex aircrafts from concept to operation to their retirement when obsolete. Although originating in the aerospace sector, SE has begun to be taken up in areas like medicine, emergency services and information technology (Eisner 2011; Nielsen et al. 2015). The success of SE in these fields suggests that it might have potential for use in political contexts understood as a system of processes. In essence, this is the core of the thought experiment undertaken in the book: to explore how SE might be adapted to address a systemic issue in political science, namely improving, if not solving, the limitations to the realisation of a genuinely democratic political system of government.

A thought experiment is a means to think through ideas and go beyond the text (Sainsbury 2017: 153). In political theory, thought experiments make it possible to communicate what may be feasible if it were to be implemented in some way or another and can be used as means to explore concepts (McCall and Widerquist 2017). A thought experiment allows 'the event of thinking [that is] reconstituted as a narrative' to conceptualise reality (Lambert 2012: 46). There were experiments of e-democracy carried as described in the book that failed to deliver an operational e-democracy. E-democracy for some scholars has generated theoretical discussions but 'very little in the way of concrete success' (Chadwick 2006: 84). This lack of success needs further investigation. The thought experiment in this book develops an approach to e-democracy that can be actualised using the design methodology of SE.

While there are many ways that this experiment could be approached, for this book the driving idea is SE. It poses a *what if* experiment to be conducted using a system (e-democracy) as a construct to be realised through design for the people and by the people (citizens). The system is designed to be as dynamic as it is evolutionary. As will be discussed later in the book, the SE system approach emerged from a military context. However, SE was not developed to reinforce centralisation of the traditional command and control model of the military, but rather it emerged from a need to develop flexibility in allowing cutting edge, evolutionary technology to emerge. Decentralised decision making was the norm, to enable routine decisions at the lower level to be automated. The thought experiment culminates with an *interpretative* exercise concerning the Republic of Mauritius. Mauritius serves as a means to consider how a SE-informed e-democracy might be implemented as some of Mauritius's democratic institutions readily lend themselves to adaptation to the key features of a SE approach. Australia as described differs from the Mauritian context and so does the SE system to be developed.

Put somewhat simplistically, SE might provide a solution to the problem of the democratic deficit. It will be argued in this book that SE can be adapted and applied to address the problem of implementing e-democracy. That is, an understanding of SE will be used to explore how it might enable new approaches to implementing e-democracy, once e-democracy is understood as a system of processes. Thus, the book explores as a core question: *how might a Systems Engineering approach address the currently perceived problems with e-democracy*? This question sits inside a broader problematic, namely (as noted above) the promise of ICTs to provide workable solutions to the problem of the democratic deficit by delivering a genuinely democratic form of e-democracy. Implicit here is a distinction between e-democracy and the idea of a genuinely democratic form of e-democracy. The point, to be taken up later in the book, is that a genuinely democratic e-democracy involves a radical transformation about how we think about e-democracy and the digital processes and ICT solutions that it might require.

The first part of this book investigates the nature of e-democracy, the problems that have arisen in attempts to implement it, and the causes that have thus far been identified with its lack of success. In so doing, it will offer an understanding of e-democracy and its limitations, identify relevant implementation issues, and determine if there are common themes in terms of its failure to actualise at a national level. This, together with a preliminary discussion of the nature of representative democracy, provides the assumed background material for the thought experiment. Hence, the book begins with an overview of various understandings of the idea of democracy, in particular representative democracy, the dominant institutional form that emerged in the West during the nineteenth and twentieth centuries.

Considered in the abstract, as distinct from actually existing representative democracies, a representative democracy operates according to some basic principles where citizens are able to make cyclical changes to those who govern via free and fair elections. Citizens also have the freedom to make their views known through various forms of political participation, such as joining political parties, holding political meetings, lobbying their representatives and engaging in protests and civil disobedience. The electoral process and freedom of expression are normally protected in the representative democracy's Constitution. The elected representatives make policy on behalf of all citizens, and every adult citizen has one vote used at an election to select their representatives. The elected representatives are also accountable to their constituents for their actions and can be replaced at the next election cycle (Mezey 2008). Some of the e-democracy experiments to be covered in this book tend to make the representative system more efficient with the use of digital technology but these reforms are within an existing paradigm (e-Voting) and therefore limited in their use of the technology to create a democratic system. A technology which is capable and growing more capable as we discuss in the book is lagging due to an appropriate framework, which can consider ICTs' role in designing an e-democracy (OECD 2003: 19). This book argues that SE can address this through its innovative use of digital technology to develop technical systems (Blanchard and Fabrycky 1998).

For many scholars of democracy, one of the key problems of the contemporary era is that in most representative democracies, there has been a growing disconnection between citizens and those empowered through the electoral system to govern. This disconnection also involves a degree of alienation on the part of citizens and in turn produces a systemic problem known as the 'democratic deficit'. The idea of a democratic deficit is not simply the difference between some idealised view of representative government and its operation in practice, or even a fundamental discrepancy between democratic norms and institutional practice. A democratic deficit comes to the fore when people are dissatisfied with the performance of democracy itself. Concerns about a perceived lack of fairness in the system of representative government, accountability issues (concerning elected representatives, non-elected officials, and governing executives), and a need to have a say seem to be central in shaping this dissatisfaction (Norris 2012: 30). A perception gap emerges between what people desire from their government and what actually results. The political system becomes increasingly stressed exhibiting symptoms ranging from citizens' protests to revolutions.

However, the emergence of a democratic deficit is not an overnight issue; it is a latent flaw built into the system as has been argued by Hindess (2002), insofar as the majority really only has a say when allowed to vote to choose their political representatives, and in some cases leaders (Zweifel 2002; White and Nevitte 2012; Ward 2002; Norris 1997, 2011, 2012; Beetham 2012). A representative democracy is a deficit by design (Hindess 2002). It will be argued in this book that a SE approach provides a means to fix this design issue.

Norris (1997) has investigated the idea of 'democratic deficit' with regard to legitimacy for the European Union (EU) merger. In her view, the perceptions of a democratic deficit over the workings of the EU emerged both from a lack of a sense of transparency and the difficulties faced by citizens in linking complex policies to outcomes (Norris 1997). This legitimacy issue arises as the powers delegated to EU Ministers seem unaccountable to the Parliaments of the member states and thus creates a perceived lack of transparency. The critical citizens of individual states within the EU questioned the legitimacy of the EU's decisions process. Now for democratic deficit, Norris mentions that the 'original idea was to judge the legitimacy of decision making processes within the EU against democratic standards of European nation-states' (Norris 2012: 24). Each state in the EU may have different standards of democratisation, and thus, accepting the EU's decisions becomes problematic for a particular nation-state. Also, complexity and opacity of EU governing processes rendered citizens at the level of individual member states suspicious of decisions made by the larger EU entity.

The EU appears to emulate a representative system at a supranational level, and this system recreates an imperfect democracy on a larger scale. Ward points out that the EU merger of nation-states creates an entity that from the 'loss of national sovereignty to powers that are unwieldy and unaccountable' also end up overriding the voice of the people (Ward 2002: 1). Though size of the nation-state arguably makes the representative system a feasible alternative (Dahl and Tufte 1973), for some citizens this larger EU system is not democratic. These critical citizens view the EU decisions process as complex and incomprehensible at their level. Using a communication lens, Ward suggests that people's lack of ability to have a say in the public sphere is a root cause. He argues that the EU entity 'has created a democratic dilemma for nation[-]states in that it poses the problem of fracturing of the relationship between output and input processes' with repercussions on the member state's welfare programme (Ward 2002: 5). Economic integration in creating a larger EU market was meant to improve life; however, it yielded some side effects for the social agenda of its members, which are hard to explain to those affected.

The solutions for resolving the democratic deficit 'usually focus on institutional reform' rather than 'some form of communication model, with citizens discursively interacting with one another, as well as with the institutions elected to govern, [which] is the central problem' (Ward 2002: 1). In Ward's estimation, the democratic deficit is a problem that has been existing in every form of representative government system since its inception (Ward 2002). Ward suggests this can be addressed using a communication model and thus argues for 'a public sphere coterminous with the institutional scale of decision making structure of the EU in order that policies are governed, and steered by the authentic expression of the citizenry' (Ward 2002: 1). ICTs potentially broaden the public sphere for decision making for an e-democracy. With ICTs, the EU can build a new public sphere to engage not only media but also its citizens in democracy. The EU remit extends beyond the geographic boundaries of the member states within which individual representative democracies operate. The EU representative democratic system, if it may be described as such, operates outside those national boundaries. At the EU level, this magnifies the imperfections of its system of representation. In appearance, if not in fact, this makes the EU's representative decision-making process seem largely unaccountable to citizens of nation-states.

In a comparative study of the USA, Switzerland and the EU political systems, Zweifel (2002: 12) suggests that a democratic deficit involves the lack of five key elements: legitimacy, transparency, consensus, accountability and protection. The low voter turnout as measured through Eurobarometer data shows a lack of legitimacy, while the EU Council of Ministers secretive decisions make EU institutions opaque. From the voting patterns, a lack of consensus exists he claims. This is due to an EU voting pattern showing a tyranny of the majority exists and that he argues suppresses the will of the minority. The EU agencies like the European Commission, EU standardisation bodies and the European Court of Justice are unaccountable to the EU electorate. Their unelected staff powers grow unchecked by the weak elected Parliament. In pursuit of competitiveness, member states have reduced welfare protection. However, when comparing the EU with the USA and Switzerland against those elements, he claims that the EU is not less democratic than the USA and Switzerland in its decision-making process (Zweifel 2002). Zweifel developed his analysis based on a Schumpeterian view of democracy in which elections were understood as the systemic means for changing one group of rulers or decision makers for another. That is, power was delegated through the election of representatives (Zweifel 2002: 6–7). Zweifel compares what he alleges are comparative representative democracies, the USA and Switzerland, to suggest that in both countries there is a lag in terms of engaging their people in decision making. In the USA and Switzerland seemingly, the people have withdrawn from governing leaving democracy to others (Zweifel 2002: 144). For Zweifel, it is a *'status quo'* situation in which people by their actions enable a democratic deficit to be a latent defect in representative democracy. Further, Zweifel argues that the USA, Switzerland and EU, '... *all three polities* suffer from a democratic deficit not because of their institutions or rules, but because "We the People" have let them' (Zweifel 2002: 144–145). An underlying issue is that the support of representative democracy may already be so low that citizens may have given up hope for seeking an improved democratic system. A lack of participation observed in elections may give this view some credence.

Another study by Norris (2012: 23) on the 'quality of contemporary democracy' in Canada and the USA found that there were the 'issues of low and eroding voter turnout, widespread disaffection with political parties, and weak linkages connecting citizens and the state', which she referred to as imperfect democracies. Her investigation of the levels of support in these representative systems revealed what she saw as symptomatic of voters' dissatisfaction with the representative system. Her approach was more inclusive as it does not assume that democracy should be understood as being limited to voting for a representative. Norris drew on Easton's systems theory approach (Easton 1965a, b, 1971), in a way that allowed the democratic performance debate to encompass various forms of representative democracies. Each representative democracy considered as a regime could be treated as an instance of a political system with specific processes to select representatives. In a country, democratic deficit is a variable to measure people's support of the country's representative system. This variable quantifies the gap between the people's aspirations for democracy and their evaluations of the performance of democratic regime in that country (Norris 2012: 24). Democratic deficit is then potentially a standard for benchmarking a perceived lack of democracy in political systems which claim to be democratic.

Another issue contributing to a growing sense of a democratic deficit is citizens' lack of trust in democratic performance of contemporary democracy (Levi and Stoker 2000). It appears to be undergoing corrosive decay, and some observers note a 'democratic recession is underway' (Norris 2011: 18). This democratic regression also translates into citizen's lack of loyalty towards political parties and, lack of trust of government, as well as, low voters turnout, and voters' anger directed towards main parties in representational democracies (Norris 2011: 3). For Norris, the demand for democracy is reflected in what people aspire for, that is, 'what people want out of life' which democracy is symbolised to supply globally (Norris 2011: 31). The supply of democracy itself, she defines as the people's level of satisfaction about democratic governance performance through their 'cognitive awareness of democratic procedures' in their own country (Norris 2011: 33). These democratic processes shape people's views, and she argues democratic deficit is the gap perceived to exist between democratic supply and democratic demand in a political system. However, Norris indicates that it is not all negative. She mentions that dissatisfied people in Latin America did not drop out of politics but rather found 'alternative political arenas' (Norris 2011: 17). Despite the obstacles, it would appear that people find a way to try and get their voices heard in a state with a democratic deficit.

The notion of solving the democratic deficit provides a key context (though it is not the only one) for bringing forth the idea of e-democracy as a possible solution that ensures citizens' participation. In principle, e-democracy equips every citizen with access to communicate and be involved in government decision making (Rios Insua and French 2010), while at the same time it promises enhanced deliberation by allowing the depth of engagement for quality (fair) democratic debates on a national scale. This is because the Internet facilitates information generation and its conveyance like no other communication technology before, whether television, radio, telephone or fax (Weinberger 1988). However, incorporating democratic principles and practices into e-democracy is complex and entails more than just digitisation of existing representational processes (Hilbert 2009), as this would in all likelihood reproduce the already known flaws that either do not meet or never fully deliver the anticipated democratic values.

Processes that allow citizens' empowerment must be created first and then be digitised using an information era paradigm. In other words, the existing digital democratic processes will need to be transformed so that in the Internet-driven era, they might lead away from the democratic models adapted for a bygone industrial era and begin to head towards realising the ideals of e-democracy. The key issue here is the transformation of digital democratic processes. This is crucial to deliver genuine democratic practices. Yet, the prospects for this sort of transformation are not self-evident as there appear to be a number of problems associated with the uptake of this technology.

Examining the Internet uptake amongst the elected representatives, Boyd (2008) found that there was little evidence of transformative adoption of Internet capabilities. The digital uptake from politicians shows a wide variance from raising hope for 'a more democratic future' to the fear that it 'would destroy the democratic process' (Boyd 2008: 1). This variance makes actualisation of an e-democracy a risky process as implementation lags even when it would appear that the technological capacity exists. In itself, an unwillingness to use technology is a significant barrier for an e-democracy. Moreover, the Internet itself shapes electronic participation and deliberation, two of the key processes for an e-democracy. Electronic participation and deliberation are largely due to the significant capacity for the Internet to deliver information in quantities and at speeds unfamiliar to earlier eras of democratic experience.

It is a commonplace that a democracy requires a free flow of information. However, in a digital environment this requirement is intensified, not as a result of a lack of information but rather of its opposite, an information overload. The sheer volume of information can be overwhelming. In turn, this makes it difficult for citizens to be able to sort fact from fiction, truth from untruth. In addition, and related to this is the issue of free and fair access to information, an area that seems neglected (Steele-Vivas 1996). This might seem incongruous given what has just been said about the volume of information made possible by ICTs. However, the issue is not so much quantity but access. Citizens should be able to access the information made available via the Internet. This might seem obvious, but it is the starting point for considering how a digitally driven democracy might work. A key requirement is investment in the means to enable people to gain access. There is a growing danger of a new technological divide being created between the information rich and those who cannot or do not have access to information to the disadvantage of the latter (Norris 2001; Steele-Vivas 1996).

The assumption by advocates of e-democracies concerning equal access to the Internet for every citizen might be misplaced considering that in many polities technology has significant costs. For example, the OECD (2010b: 7) reported that Spain's Avanza Plan encountered 'increasing fiscal constraints' leading to a reprioritisation of its digitisation initiative. The anticipated uptake of the technology did not eventuate, making the original investment something of a white elephant. The OECD study recommended a 'demand-driven, user-centred strategy that delivers visible results to society' by looking at a whole of government strategy for seeking innovative solutions (OECD 2010b: 3). The OECD position was that ICTs needed to be seen as an innovation development means rather than just a tool of communication (OECD 2010a: 13). A targeted approach which innovates with technology was suggested for Spain, as its broad diffusion strategy adopted from a government policy created a digital divide 'in certain socio-economic/demographic groups, territories or sectors' (OECD 2010b: 4). The study found that the Avanza plan could have been used to develop a roadmap which could then have been applied more precisely to reduce the digital divide and improve user uptake of the technology (OECD 2013: 7–12). There are lessons to be learnt here for polities considering an e-democracy, in particular the importance of a holistic approach within a 'methodological framework that addresses how ICT can be designed and used to effectively and efficiently support information provision, consultation and participation in policymaking' (OECD 2003: 19). A digital divide, for example, creates an information poor and an information rich, whereby entrenched inequalities are exacerbated even after significant ICTs investments have been made following a government policy to digitise.

Similarly, Norris has argued that the new communicative technologies may be contributing to a democratic divide in terms of the ability to participate in public life (Norris 2001: 4). Norris terms this a '*democratic divide*' from information resources poverty and it is a new form of inequality (Norris 2001: 4). An e-democracy that uses ICTs to digitise existing representative processes could give rise to a digital divide exacerbating the existing inequality with new information inequality. This new inequality, if not addressed, works against the possibility of an e-democracy being able to eliminate the democratic deficit. Using a time series survey evidence, Norris found that within established democracies commentators believe that 'support for the political system has gradually weakened'; she linked this to reduced trust in the legislative system's performance in the, for example, USA, Spain, Italy, France and Germany (Norris 2012: 46). Politicians' performance creates both public and scholarly concern about improvement to the political system if they lack an understanding of ICTs. An inappropriate implementation of ICTs from poor decision making has consequences for the political system.

This connects back to the point mentioned above about the lack of uptake of the technology by representatives, though increasingly this is ceasing to be an option as the demands of the digital age presuppose appropriate levels of digitisation of services. Put a little differently, information technology has the potential to make things worse in that it can just as easily fragment the public sphere as bind it together. The Internet is in effect a double-edged sword; it is just as capable of liberating as oppressing those who engage with it (Barber 1997). Through information overload or through the deliberate misleading of voters, it can consolidate the hold on power by ruling elites. The manipulation of information contributes to election outcomes. Election outcomes depend on election rules, rules that are open to exploitation by political parties.

These election rules could be linked to parties' strategic use of information to exploit social cleavages, patterns of voting behaviour which in turn influenced their candidate selection so as to optimise votes for the party.

Elections are influenced with the use of the ICTs capability to market both the candidate and the parties' agenda for vote maximising strategies based on the voting behaviour of the public. Norris (2004) found that electoral rules mattered for who gets elected. She suggests that the engineering of election rules can change the election outcomes as different rules can address some of representative democracies' weaknesses (Norris 2004). Engineered electoral rules she argues make election events less predictable for parties to manipulate the public for getting elected to govern.

Numerous scholars have pointed to the ways in which democratically deficient political systems have succeeded in propping themselves up through the use of digital technology by digitising the existing representative democracy processes to maintain a neoliberal status quo (Wilhelm 2000; Hague and Loader 1999; Hoff et al. 2003; Hindman 2009). These efficient market-oriented processes of a flawed system have enabled them to entrench themselves further in power. This is despite e-democracy alternatives that are being proposed (Rios Insua and French 2010; Dahlberg and Siapera 2007). This informational aspect requires transparent and accountable processes to enable e-democracy to emerge in a form capable of closing the democratic deficit. Digitisation without reform of existing processes serves to entrench this deficit further. These brief points about the relationship between democracy and the technology that shapes e-democracy are detailed in Chapters 2 and 3.

It is central to the arguments of this book that a new approach to thinking about e-democracy is needed. This is where the applicability of established processes used by SE might prove of value for enhancing and launching a workable and operational e-democracy. Now it is time to turn to a brief overview of systems theory and SE.

Systems are commonly used in every walk of life, and SE designs are purposive systems to meet a functional need in the field of engineering (Blanchard and Fabrycky 1998). To develop successful processes and outcomes, SE uses, as its name suggests, a systems approach (Eisner 2011; Department of Defense 1993; Blanchard and Fabrycky 1998). However, systems theory is not new in political science. David Easton, echoing the parallel work of the sociologist Talcott Parsons, was a leading advocate of a systems approach in the 1950s and 1960s (Parsons 1956; Easton 1957). The attempt to develop a conceptual framework in political science around systems theory has been described by Birch (1993: 211) as 'an ambitious attempt … to replace institutional analysis in political science'. A systems approach was part of attempts at the time to get closer to applying what was understood at the time to be the scientific method to create a grand theory in politics. A more detailed analysis of Easton's systems approach and its contribution to political science and representative democracy is developed later in Chapter 4. However, what can be noted here (and also to be discussed in more depth later in the book) is that Easton's systems approach and that of SE share some commonalities. Hence, this provides some initial support for the idea that SE might be applicable to developing e-democracy as a political system. SE treats an existing system as a legacy system that can be upgraded to become an improved (and mostly different type of) system.

Easton's main work focused on political systems (Easton 1971); the object of his study was the political system as a whole and its behaviour. A systems approach makes it immaterial whether the political system being studied is a democracy or not. The political system is conceptualised as being made up of processes with inputs and outputs that transport and translate information. In this sense, it resonates with SE techniques. It will be argued in this book that these latter techniques overcome a key weakness of Easton's systems theory in that they have the potential to allow a new democratic process in decision making to emerge that is potentially applicable to any political system.

Despite falling from favour with many political scientists, Easton's systems approach has still retained some contemporary value. For example, Norris (2011: 9) acknowledges her use of Easton's seminal contribution of 'system support' (Easton 1975), to develop her own concept of 'democratic deficit'. Norris developed an empirical study of the perceived decline in support for democracy for about 200 countries using World Value Survey data collected over decades. Treating 'democratic deficit' as an independent variable, she conducted a detailed analysis of whether citizens perceive institutions to be failing to uphold basic principles of democracy such as the legitimacy of decision making through elections or rights of protest for example. She used individual-level and national-level indicators to interpret the World Value Survey data between 1981 and 2007 to explain democratic satisfaction (Norris 2011: 247–257). At the individual level, she used demographic characteristics, and socio-economic resources and media use. While at national level, she used 'Historical index of democratization' and 'Development: GDP per capita in

PPP' (Norris 2011: 255). Using a multilevel inferential model, she measured citizen's satisfaction with democratic performance in their respective countries, from which she identified themes that she displayed graphically to show trends. Borrowing terminologies from Easton, she differentiated between diffuse and specific support, that she aligned with the supply and demand sides of democracy (Norris 2011: 6–37).

Using Easton's support concept, Norris identified five levels of a representative system's support from the most specific to the most diffuse with the specific being embedded in the more diffuse components in a nested loop (Norris 2011: 21–25). She explained these levels as nested loops (similar to Russian dolls) extending from the most specific citizen support at the centre to the most diffuse political system support at the outermost layer. The most specific or innermost core was about 'approval on incumbent officeholders', which is within 'confidence in regime institutions', and that is within 'evaluations of regime performance'. These inner cores sit inside the next outer level that is 'approval of core regime principles and values' which is then nested within the outermost core which is 'national identities', she regarded national identities as the most diffuse component of a political system's support (Norris 2011: 24–25). She drew a general model of democratic deficit with the difference in supply (as state policy outcomes) and demand (inputs from public aspirations) defining the supply and demand consequences to democratisation in the state (Norris 2011: 6). From this, she used the idea of a democratic deficit as a variable to measure democracy.

Another dimension of her analysis that can be traced back to Easton's approach is the concept of unmet needs and demands that stress the system (Easton 1965a). Easton suggested that inputs from the people to boundaries of his political system become transformed through gatekeepers into demands. Needs from citizens' inputs create demands and their satisfaction defined citizens' levels of support for the regime. Unmet demands from a state of ongoing demands may strain the political system resulting in regime failure. This strain may threaten the life of the political system and a regime change results to allow the system to survive. This is where a SE-integrated e-democracy could process needs that could be more effectively identified and ways developed to realise their satisfaction. This would assist in reducing stress to the political system.

For example, consider the model of Easton's system theory provided by Birch (1993: 221). Easton's systemic processes as viewed by Birch have inputs generated from citizens that are translated into demands by the political system. In Birch's interpretation, decision making in the nation-state is through a cycle. Birch's modelling of the political system depicts a set of processes, which are sequential steps operating in a closed loop where inputs are articulated as demands that are aggregated and then authoritative decisions (depicted by a black box), are made to implement these demands as outputs which are policies. These policies are assessed by society, and then, through a feedback loop new demands may be re-articulated to start a new decision-making cycle. This basic systems theory model for Birch is central to a representative democracy where the institutions like elections, political parties and pressure groups are replaced by a process where interests and values are articulated, aggregated and then met by the political system through policies (Birch 1993: 220). However, Birch argues that Easton failed to take into account that in a modern democracy, inputs are from the bureaucrats and government executives not the people (Birch 1993: 222). What is needed is a people-centred approach and that is what the book argues SE is capable of fixing. These arguments are developed in more depth in Chapters 5 and 6. But, here it is necessary to provide a brief overview of what this book takes to be SE.

In the 1940s, SE originated from General System Theory as a military need. It was an engineering governance of multi-disciplines that was applied to develop technical solutions for the military (Blanchard and Fabrycky 1998). Eisner (2011: 1) defines SE as 'an interdisciplinary management process to evolve and verify an integrated, life cycle balanced set of system solutions that satisfy customer [end user] needs', where system is defined as 'an integrated composite of people, products, and processes that provide a capability to satisfy a stated need or objective'. This means that SE as a technique allows the development of effective solutions to meet a specified need or objective. SE also provides the tools to generate a set of solutions, and so there are alternatives to choose from for those who specify the need. The SE technique integrates people, product and processes such that the capability (system) is developed through a set of ten core processes (USAF 2010). Of these ten, two areas for the interface of relevance proposed for this book are (1) 'Configuration Management' and (2) 'Requirements'. The USAF (2010: 19) defines Configuration Management (CM) as the core overarching process to manage the baseline of the product under an end user's control. CM manages the 'information' that forms the system which is made up of people, processes and products that interact. CM is a technique where information is used to allow control to be retained by the end user, the military as the end user of the system, where the system is, for example, a fleet of aircraft. It allows the end user to have management oversight of an approved product baseline for a system and therefore approval of that baseline. The military use of the CM process is to manage and authorise complex changes to a system's baseline, and this stops scope creep, thus the end user authorises cost of changes to a system (USAF 2010: 6).

The 'Requirements' process converts users' needs into specifications that allow a product or service to be contracted for development. Requirements analysis is used to develop an understanding of what might be required, like for aircraft by using an iterative process to develop increasingly detailed derived requirements as the multiple layers of the complex product are defined from macro to micro or the other way round (Blanchard and Fabrycky 1998; USAF 2010; Department of Defense 1993). The requirements are transparent as they flow up and down. A requirement can be traced back to its source from the top down or from the bottom up. SE also enables replication if the system needs to be built again, since SE has the capacity to articulate traceability from specifications to an actual prototype during the life cycle of the system. The iterative nature of the Requirements process allows a negotiated outcome and incorporation of approved (legitimate) changes to a baseline through the CM process. The iterative process results in a 'set of solutions', which is provided to the end user for consideration and selection (USAF 2010: 45–52). The agreed solution emerges from a bottom-up approach driven by the end user based on transparent and repeatable processes. In this way, CM ensures that any changes to an agreed baseline (set of requirements) are authorised at the appropriate level (from the end user) during the life of the product (Lacy 2010; INCOSE 2007; Clifford 2010), thus legitimation and persistence of system is possible as the product's obsolescence is managed.

A product that could possibly be generated using SE techniques is a 'Social Contract' with citizens' (end user) involvement rather than elite's (representative) agenda or programme pushed from the top (Estlund 2002: 155). CM provides a decision-making framework to manage the evolution of the product baseline, the product being the social contract or people's agenda formulated through SE techniques and processes. A product baseline cannot be changed without seeking approval from a Configuration Control Board (CCB). Each proposal is justified on its merits to the board members to seek its ratification. Only approved ratified proposals are allowed for incorporation into an approved baseline. Proposals that are rejected cannot be implemented into the system through this management framework, as it is illegal. CM may allow a political system to have the capacity to persist over time due to an inbuilt system of processes that allow the evolution of the system to avoid obsolescence within an appropriate decision-making framework involving citizens (CCB) exercising their sovereign rights. These SE technique applications are explained in more detail in Chapters 5 and 6 of the book.

The SE argument of the book is that CM and Requirements have the potential to solve some of the operationalisation problems of system e-democracy. The argument explores how a system of processes from SE as used in engineering could be applied to the political science context in the spirit of a systems theory framework approach normally associated with Easton (1965a, b). E-democracy with SE processes augments democracy through a systems approach which is both reductionist and expansionist using Configuration Items (CIs) and Functional Configuration Items (FCIs), an atomistic and holist concept that is explained in Chapter 5 and applied in Chapters 6 and 7. To the problem of augmenting democratisation that Norris uncovered through her study of the democratic deficit, SE also brings together a dual thinking that is analytic (data) and synthetic (outcomes, concepts). A SE framework allows elements of the system like representatives and the people to work together towards a shared goal, the good life (in the Aristotelian sense) to be delivered by the political system. A communal outcome, such as pursuing the good life, allows the ruled and ruling to empower each other through the Requirements and CM techniques. SE processes create the shared empowerment within a political system like an e-democracy; technology allows SE to further empower as technology becomes more capable. The technology, as will be discussed in Chapter 5, is closely related to SE.

As noted earlier, SE is a cross-disciplinary technique which manages complexity both of a technical and managerial nature while using what is called a 'system of systems' (SoS) approach (Eisner 2011: 10). The analysis of human activities for systems that are made by humans is not an 'end in themselves but are a means of satisfying human wants' (Blanchard and Fabrycky 1998: 14). SE's ability to expand and adapt, in many diverse situations, allows human beings to create a complex world that reflects the contemporary values from the society where a SE-designed system is embedded. With the advent of interconnectivity from ICTs, even larger and more complex systems become possible. SE's reductionism on the other hand is an analytical way of thinking of this world in terms of how its disassembled parts can be decomposed and the behaviours of these parts can be explained partially to aggregate into an explanation of the whole system (Blanchard and Fabrycky 1998: 10).

To deliver a successful solution (system), SE's top-down and bottom-up design approach is dualistic, it uses a paradox of expansionist and reductionist views that coexist. While this paradox is expanded in Chapters 5 and 6, the key point to note here is that the design analysis uses data from the bottom up to create the function that the system delivers, and a synthesis mode drives the design for the functional level at the top to integrate its components in a specific manner that is top down. These two views using a purposive structured design process are made to converge in an effective design process with an evaluation step closing the design loop. Success is defined as meeting a user-agreed solution (Blanchard and Fabrycky 1998: 32). Synthesis starts from the top-level function, which is from a defined need, and then, analysis follows or detailed design at the lower levels or subsystems for creating the function, the evaluation step completes the design cycle as it is the step to ensure the system meets the need or the function.

To create a function each component at the bottom is synthesised from the specified operational role through an operational requirement which at the highest level, for example a system role, is an e-democracy that democratises. Now, this e-democracy or system function requires specific inputs at the lower levels that are integrated precisely in a manner such that each input is aggregated from the bottom up in a structured manner to deliver the complex function at the top, a people's democracy. The top-down view allows the function's requirements to be developed, as this goal allows the whole system to be fit together in a given way or configuration. This top-down or synthetic view drives how the components are made to fit together for a desired outcome (the top function). The function at the top level can be traced to each components' input at the bottom that contributes to that function, and this traceability is in both direction. These two approaches from bottom and top are combined in the SE systems approach to allow the analysis of collective phenomena for successful design explained later in the book (Blanchard and Fabrycky 1998).

This approach has been applied in the aerospace sector for generating a system (capability) (USAF 2010; Eisner 2011). A capability is created to meet an operational role (function or need) from end users' (people) requirements (Blanchard and Fabrycky 1998: 33). For example, the USAF's operational role is based on the changing environment where a designed aircraft is to operate, predicting this environment requires a degree of foresight that requires creation and integration of new technology. Now transitioning a system 'from the past, to the present and the future technological states, is not a one-step process' (Blanchard and Fabrycky 1998: 13). The SE process is iterative and needs skills to integrate the synthetic and analytic inputs, such that the technology delivers the desired function. To address this effort, the USAF has developed a SE framework to assess and assist the SE design process so that it is as transparent as possible, and repeatable and hence accountable (USAF 2010). A technical system designed by SE is complex, and to maintain its capability, it requires upgrades, for example, technological obsolescence or role changes due to the environment may be drivers for change. These capability upgrades can be met either through the development of new systems or through modifications of legacy systems to meet the new role (function). SE techniques are also used to manage the ongoing upgrades needed during the system's lifetime and thus allowing the system (capability) to persist. Upgrades are against agreed requirements, and they are in the form of approved modifications to an existing system baseline or configuration. Thus, a legacy system can be modified to improve a system's capability to meet new functions, and such changes are traceable through the CM process. The SE techniques are dynamic as they enable a system to adapt to changing environments after a baseline is struck.

Managing a set of complex requirements is usually undertaken through the use of relational databases, which provides upward and downward traceability. This traceability also ensures that changes required to renegotiate limitation from technology, like reduction of an agreed role, are approved by the end user. This is normally in the form of a user-approved modification in the form of new or changed requirements. Thus, SE is adaptive to a user's evolving needs as by allowing upgrades to an existing baseline it allows the system to persist, and resists obsolescence on the other hand. Obsolescence may also be due to a drop in performance of the system, like democratic deficit. The SE processes articulate and aggregate people's inputs to decision-making start from an atomist level to shape the functional or performance requirement of an e-democracy. These are explained in Chapter 5 and conceptually applied in Chapter 6 and constitute the core of the thought experiment.

Of course, this prompts two key objections. The first is why is the SE approach developed in this book not simply a variation on the theme of the long-discredited structural-functionalist approach? And second, the SE approach may be all very good in theory but how will it work in practice? Given that the book frames its overall discussion in terms of system theory, and 'political system' in particular, this first question seems quite appropriate. Once that has been dealt with, I will turn to the second question.

The idea of a 'political system' has been explored by sociological and political scientists using quantitative and qualitative approaches. Almond notes that during the twentieth century, the state as a concept gave ground to terms like political system or government (broadly interpreted) due to the need to account for the role of institutions that went beyond the strictly coercive institutions of the state like the military to include other extra-legal institutions like political parties, the media, pressure groups and the family. These various institutions also impacted on and shaped political processes (Almond 1988: 855–856). In a very real sense, the concept of political system thus included not just the idea of the state but also these newer entities that were understood to be outside the state's formal governing apparatuses. It is in this context that the idea of structural functionalism (SF) and system theory gained their explanatory appeal within political science in that they did not treat the 'state and governmental institutions' in a reductionist fashion (Almond 1988: 855). System theory posited that 'the political process was a set of interdependent subprocesses' (Almond 1988: 856) and a SF approach enabled political scientists to study how those processes actually operated.

It needs to be noted here that SF was popular when it was introduced in the 1940s as it argued that 'systems contain members who each held a function within the system; [and] the overall purpose of each member was to keep the system in balance to allow for its continuance' or equilibrium so it could persist (McMahon 2013). It was an approach that seemed to enable social scientists to explain the social reality of organisations. After World War II (1939–1945), it was popular for social scientists to 'ground sociological theory in an explanatory framework covering all forms of human behaviour seen in a social context' (Badcock 2014: 17). Talcott Parsons in sociology attempted to create an all-embracing framework for sociological theorising, and David Easton undertook a similar attempt at a grand theory within political science. The use of systems theory to frame social and political inquiry has had many strands spawning various other theories as alternative strategies for social research, (Easton 1966a). In political science, Easton's systems theory was a significant attempt and is discussed further in Chapter 4. For the moment it is necessary to highlight here some of the key differences between SE (as understood in this book), SF as a general theory and Easton's own system theory approach.

As noted, in the USA, SF had its roots in Parson's action theory and Merton's sociological theory (Easton 1972: 131–132). However, the functionalist part of SF 'became subordinate to a truncated form of [Easton's] systems analysis' (Easton 1972: 130). For Easton, this meant that Parsonian functionalism had limitations when applied to political science. One key limitation was that it could not explain system change such as a regime change (revolution for example). A democratic system some argued requires organic growth, and this organic growth instigates political change which could be a revolution or new Constitution agreed by the people. SF theory could not accommodate the political changes from such organic growth. Human interaction and behaviour is not structured like that in an 'ant colony' which according to McMahon (2013) provides a good example of SF's application. In practice, human social organisations have different degrees of hierarchical organisation with different degrees of agreement to social and political change over time. As explained by Easton, a systemic change may be prompted for personal gain by destabilising the political regime rather than SF theory's assumption that change is to maintain the order and balance of the system. Some people may want to disturb the system to a new regime so they could gain from the new systemic equilibrium, this dynamism could not be accounted through Parson's SF theorising (which posits a form of stable hierarchy that replicates itself for equilibrium) (Easton 1972).

In terms of voting behaviour, Parson's SF assumes that people in a country vote as part of their functional membership of a certain group. This has been found to be incorrect in practice as some people do not vote or if they do, it is because of their own volition rather than their function as a citizen with a civic duty to vote. In a country like the USA (where voting is not compulsory and where Parsons and Easton were working), voting outcomes may be estimated but not prophesied. The limitations of SF led to its demise in the 1970s. SF was criticised for being too abstract. It described those systemic functions in static terms, a worker in an ant colony remained a worker in that hierarchy (static), but this could not be said to be the same for a human being who could have various roles during their life cycle based on different motives of their own (dynamic).

In politics SF after a period of hegemony lost its appeal because its 'view of society as a system made up of interrelated parts, all interacting on the basis of a common value system or consensus about basic values and common goals' was inadequate to explain political phenomena in practice (McMahon 2013).

Now the theoretical perspective that SE takes is from modern system theory, it is one of functional differentiation where the political system is considered as an autonomous functional system of society (Luhmann 1995). Easton's political system approach is distinct as it treats anything which takes place outside the system as non-political (Easton 1971). In doing this, Easton seemingly differentiates between different types of social systems, namely functional systems and organisations. Systems theory has evolved considerably from its early days and has moved past the deficiencies of Parson's SF. One strand has led to Luhmann (1995) theory of social systems which explores autonomous system dynamics that shape and make up society as a system. Luhmann differentiates functional systems like politics at the macro-level on the basis that such systems involve certain tasks at the level of society and its organisations. In this context, organisations are considered as 'decision machines' (Nassehi 2005). The modern system reproduces itself, and as a functional system, it observes its environment made up of other functional systems, organisations and interactions for relevant (political) exchanges of interest to it. In such a functionally differentiated system, there is no politics (function) without political parties or government and, for example, no business without companies. Even though one tends to refer to the political system as a collective actor, the political system is not an actor as it cannot act or communicate on its own.

From a modern system theory perspective, a functional system allows the conceptualisation of a political system even though it cannot be seen, and secondly, it has decision support organisations (machines) that conduct their tasks through functional systems. Differentiation creates complex dependency such that organisations must refer to various functional systems as they do not exclusively belong to one functional system. Differentiation allows organised complexity to develop from the interaction of parts or units that form the system. This organised complexity allows order to emerge from the chaos that the system seemingly operates within. But significantly, during its operation the system is not equal to a mechanical device that controls everything within it, rather it is a system from the dynamics of what happens within it. This modern system theory is differentiated from that of the Parsonian approach as there is a 'shift of emphasis from the conditions of stable systems to the dynamics of an emerging order' (Nassehi 2005: 181).

In the book, the concept of 'system' is understood as something that continues to work itself out in detail and where modes of interpretation provide descriptions of the system itself or its forms. When interpreted as a totality of the political, the idea of 'system' includes everything that makes a political system a unity, and which can then serve as a framework for political observation and hence as the framing unit of political analysis. 'System' as the object of enquiry was to shift from inferences that were traditional and historicist to the kind of research where one poses open questions without being prejudiced in advance by possible consequences (Easton et al. 1995; Easton 1966a, b, 1971, 1981, 1985).

As discussed later in the book, Easton's allegedly scientific approach was not adopted as a guiding framework as it could not offer a methodology distinct from other positivist approaches like SF. Social inquiry involves assumptions and modes of interpretation which are not necessarily amenable to the positivist approaches of an Eastonian system theory. SE, as presented in this book, is imagining a purposive (goal-oriented) design method around people's (citizens') needs that is new to politics. For example, SE treats technology like software as a service to the people's needs. In this context, technology is subservient to meeting design goals towards creating a genuinely democratic *praxis*. SE's system e-democracy is conceptualised towards an imagined axiom of democracy (as people's rule) in an iterative manner that the book described. This differentiates SE from Easton's approach of trying to develop a scientific-oriented system theory.

As for the second question about the posited SE approach will work in practice, this is dealt with in some detail in Chapter 7. For present purposes, the following will capture what is at stake. The book looks at particular features of the political system of the Republic of Mauritius because they appear to be amenable to the implementation of the identified SE techniques that might enable a working e-democracy to be operationalised successfully. There are a number of reasons for this view.

First, the size of Mauritius lends itself to providing the means to answer some questions about SE's feasibility in practice. Mauritius is a nation-state consisting of two main islands, mainland Mauritius and a small autonomous dependency called Rodrigues. Rodrigues is a system within a larger system (Republic of Mauritius) that could potentially be used as a small-scale test case for validation of the SE developed approach. In effect, it provides an opportunity for SE with a pilot site. Second, Mauritius has a government body known as the National Development Unit (NDU), the role of which is to facilitate people's requests (inputs) to the government. The NDU's operational function makes it a good candidate for a SE-driven framework. Third, the Mauritian government, in its 2017 Budget Papers, has placed a priority on the use of developing the capacity of ICTs in a range of government services and activities, designated by the government as i-Mauritius (Government of Mauritius 2017). This policy imperative at the national level provides a rationale for testing the viability of this SE study's approach to digitising governance processes to an e-democracy. Thus for future digitisation, any NDU processes upgrade for an e-democracy is possible through the government's policy for i-Mauritius. Finally, a number of politicians (across party lines) and key government officials in Mauritius are keen to explore processes and institutional solutions that can deliver better democratic outcomes. In this context, the SE techniques discussed in this book would come into their own (given the discussions thus far held with key decision makers in Mauritius).

The core exploratory question framing the book is: *how might a Systems Engineering approach address currently perceived problems with e-democracy*? The answer to this question is provided in the book by first contextualising the problem in its first part (Chapters 2, 3 and 4) and developing a SE-informed design solution using changes from digital communication capability that opens potential for an e-democracy in the second part (Chapters 5, 6, 7, 8 and 9).

Chapter 2 explains how representative democracy as currently configured suffers from a democratic deficit. This suggests that a need exists for an improvement to democracy. Current scholarship suggests that e-democracy is a solution to democratic deficit. The rapid proliferation of ICTs has seen the emergence of the technological capacity for every citizen in the large nation-states to engage in their political systems. However, a framework to adapt the technology seems missing. Demonstrating this is the focus of Chapter 3 as it explores some of the contemporary problems with establishing various forms of e-democracy. Measurements of democracy content in digitised processes reveal that even established democracies have problems with digitisation that can mostly be attributed to the lack of an appropriate framework to adapt the available technology. This suggests a need for a different approach. The book proffers a new approach, namely the systems approach of SE, that it argues is capable of delivering technical solutions adapted to operationalise an e-democracy. Systems approach is not new to political theory. Therefore, Chapter 4 covers the concept of systems in political theory. This provides a theoretical and contextual framework for an understanding of SE's novel systems thinking in political theory. The first three chapters thus provide the context for an analysis of how (and what aspects of) SE can enable the system of representative democracy to be transformed into an effective e-democracy where people rule.

The second part of the book explains key features of SE that can be adapted to the needs of developing a workable e-democracy, one that creates a democracy that democratises. As noted above, SE is a design paradigm linked to technology that has a design methodology that can deliver large complex technological systems for the life cycle of the system. For SE, e-democracy is a solution in the form of a system construct or set of processes where the people are the end users or decision makers. So, Chapter 5 explains the SE framework and its key concepts. It also discusses how these are able to link a need to a system where the end user, in this case the *demos* or the people, is able to decide for the whole life cycle of the e-democracy. Chapter 6 then applies the abstract principles of SE to the idea of an e-democracy to show how they can deliver on e-democracy's promise and overcome some of the problems identified in Chapter 3. Chapters 5 and 6 thus provide the substance of the thought experiment. However, left at that level the discussion remains far too speculative. Hence in Chapter 7, the discussion turns to applying these abstract principles to an actual political system, the Republic of Mauritius. The chapter serves as a proof of concept for the SE methodology's adaptability to improve e-democracy. As noted above, Mauritius has a number of institutions in place, most notably the NDU, which could be adapted to an SE-focused approach. In addition, there is a strong commitment from the Mauritian government to take full advantage of the benefits of ICTs.

In sum, the book provides a thought experiment that argues that SE is a design paradigm which is novel to political theory, one that can upgrade an existing system (legacy system) in a purposive manner to deliver a desired artefact, an e-democracy that democratises. The SE design approach, as explained in the book, can provide a conceptual roadmap to augment democracy in practice, case of the Republic of Mauritius Chapters 7 and 8 and Australia in Chapter 9. This is in the spirit of

Hindess's interpretation that democratic deficit is an integral part of the design of representative democracy (Hindess 2002). The book, therefore, recommends further research, case of Australia Chapter 9, using the SE techniques to build on this work to design appropriate political systems as specified by the people in the context of their needs. This adapted SE methodology links the system e-democracy function to the need of the people. This book is viewed as part of a longer journey to create a democracy that democratises and importantly a system e-democracy that remains a democracy for its life cycle. A SE design methodology can provide the means to empower the people to be in control of their democratic institutions and hence their system of government. It is now time to turn to a brief discussion of how this book understands the idea of democracy. That is the task of Chapter 2.

References

Almond, G.A. 1988. The Return to the State. *The American Political Science Review* 82: 853–874.

Backus, M. 2001. EGovernance in Developing Countries. Research Report No. 3, April 3. International Institute for Communication and Development (IICD). https://scholar.google.com.au/scholar?hl=en&as_sdt=0%2C5&q=Backus+RESEARCH+REPORT+No.+3%2C+April+2001&btnG=.

Badcock, C.R. 2014. *Levi-Strauss (RLE Social Theory): Structuralism and Sociological Theory*. London, UK: Routledge.

Barber, B.R. 1997. The New Telecommunications Technology: Endless Frontier or the End of Democracy? *Constellations: An International Journal of Critical & Democratic Theory* 4 (2): 208–228.

Beetham, D. 2012. Defining and Identifying a Democratic Deficit. In *Imperfect Democracies the Democratic Deficit in Canada and the United States*, ed. R. Simeon and P.T. Lenard. Vancouver and Toronto: UBC Press.

Birch, A.H. 1993. *Concepts and Theories of Modern Democracy*, 2nd ed. London: Routledge.

Blanchard, B.S., and W.J. Fabrycky. 1998. *Systems Engineering and Analysis*, 3rd ed. Upper Saddle River, NJ: Prentice Hall.

Boyd, O.P. 2008. Differences in eDemocracy Parties' eParticipation Systems. *Information Polity* 13 (3): 167–188.

Chadwick, A. 2006. *Internet Politics: States, Citizens, and New Communication Technologies*. Oxford: Oxford University Press.

Chen, P., R. Gibson, and K. Geiselhart. 2006. *Electronic Democracy? The Impact of New Communications Technology on Australian Democracy*. Canberra: Australian National University.

Clifford, D. 2010. *ISO/IEC 20000 an Introduction to the Global Standard for Service Management*. London: Ebook Library.

Crespo, Rubén González, Oscar Sanjuán Martínez, José Manuel Saiz Alvarez, Juan Manuel Cueva Lovelle, B. Cristina Pelayo García-Bustelo, and Patricia Ordoñez de Pablos. 2013. Design of an Open Platform for Collective Voting through EDNI on the Internet. In *E-Procurement Management for Successful Electronic Government Systems*. Hershey: Information Science Reference (an imprint of IGI Global).

Dahl, R.A., and R.E. Tufte. 1973. *Size and Democracy*. The Politics of the Smaller European Democracies. Stanford, CA: Stanford University Press.

Dahlberg, L., and E. Siapera (eds.). 2007. *Radical Democracy and the Internet: Interrogating Theory and Practice*. Basingstoke: Palgrave Macmillan.

Easton, D. 1957. Traditional and Behavioral Research in American Political Science. *Administrative Science Quarterly* 2 (1): 110–115.

Easton, D. 1965a. *A Systems Analysis of Political Life*. New York: Wiley.

Easton, D. 1965b. *A Framework for Political Analysis*. Prentice-Hall Contemporary Political Theory Series. Englewood Cliffs, NJ: Prentice-Hall.

Easton, D. 1966a. *Varieties of Political Theory*. Prentice-Hall Contemporary Political Theory Series. Englewood Cliffs, NJ: Prentice-Hall.

Easton, D. 1966b. *A Systems Approach to Political Life*. Lafayette, IN: Purdue University. http://www.eric.ed.gov/contentdelivery/servlet/ERICServlet?accno=ED013997. Consulted 28 January 2018.

Easton, D. 1971. *The Political System: An Inquiry into the State of Political Science*, 2nd ed. New York: Alfred A. Knopf.

Easton, D. 1972. Some Limits of Exchange Theory in Politics. *Sociological Inquiry* 42 (3–4): 129–148.

Easton, D. 1975. A Re-assessment of the Concept of Political Support. *British Journal of Political Science* 5 (4): 435–457.

Easton, D. 1981. The Political System Besieged by the State. *Political Theory* 9 (3): 303–325.

Easton, D. 1985. Political Science in the United States: Past and Present. *International Political Science Review* 6 (1): 133–152.

Easton, D., J.G. Gunnell, and M.B. Stein. 1995. Democracy as a Regime Type and the Development of Political Science. In *Regime and Discipline: Democracy and the Development of Political Science*, ed. D. Easton, J.G. Gunnell, and M.B. Stein. Ann Arbor: University of Michigan Press.

Eisner, H. (2011). Systems Engineering: Building Successful Systems. *Synthesis Lectures on Engineering* 6 (2): 1–139.

Estlund, D.M. 2002. *Democracy*. Blackwell Readings in Philosophy. Malden, MA: Blackwell.

Government of Mauritius, M.o.F.a.E.D. 2017. Budget Speech 2016/2017. Government of Mauritius. http://budget.mof.govmu.org/budget2017/budgetspeech2016-17.pdf. Consulted 29 January 2017.

Grönlund, Å. 2001. Democracy in an It-Framed Society. *Communications of the ACM* 44 (1): 22–26.

Grönlund, Å. 2004. State of the Art in e-Gov Research—A Survey. *Electronic Government* 3183: 178–185.

Grönlund, Å. 2008. Lost in Competition? The State of the Art in e-Government Research. In *Digital Government*, ed. H. Chen et al. Berlin and Heidelberg: Springer.

Grönlund, Å. 2011. Connecting e-Government to Real Government—The Failure of the UN Eparticipation Index. *Electronic Government* 6846: 26–37.

Grönlund, Å., and T.A. Horan. 2004. Introducing e-Government: History, Definitions, and Issues. *Communications of the Association for Information Systems* 15: 713–729.

Hague, B.N., and B. Loader. 1999. *Digital Democracy: Discourse and Decision Making in the Information Age*. London and New York: Routledge.

Hilbert, M. 2009. The Maturing Concept of E-Democracy: From E-Voting and Online Consultations to Democratic Value Out of Jumbled Online Chatter. *Journal of Information Technology & Politics* 6 (2): 87–110.

Hindess, B. 2002. Deficit by Design. *Australian Journal of Public Administration* 61 (1): 30–38.

Hindman, M. 2009. *The Myth of Digital Democracy*. Princeton, NJ: Princeton University Press.

Hoff, J., I. Horrocks, and P. Tops. 2003. New Technology and the 'Crises' of Democracy. In *Democratic Governance and New Technology: Technologically Mediated Innovations in Political Practice in Western Europe*, ed. J. Hoff, I. Horrocks, and P. Tops. London and New York: Routledge.

INCOSE, I.C.O.S.E. 2007. *Systems Engineering Vision 2020*. San Diego, CA: International Council on Systems Engineering.

Kneuer, M. 2016. E-Democracy: A New Challenge for Measuring Democracy. *International Political Science Review* 37 (5): 666–678.

Lacy, S. 2010. *Configuration Management*. Swindon and Biggleswade: British Computer Society, The Turpin Distribution Services Limited.

Lambert, G. 2012. *In Search of a New Image of Thought: Gilles Deleuze and Philosophical Expressionism*. Minneapolis: University of Minnesota Press. http://ebookcentral.proquest.com/lib/newcastle/detail.action?docID=1047458. Consulted 26 February 2019.

Lee, C., K. Chang, and F.S. Berry. 2011. Testing the Development and Diffusion of E-Government and E-Democracy: A Global Perspective. *Public Administration Review* 71 (3): 444–454.

Levi, M., and L. Stoker. 2000. Political Trust and Trustworthiness. *Annual Review of Political Science* 3 (1): 475–507.

Luhmann, N. 1995. *Social Systems*. Stanford: Stanford University Press.

McCall, G., and K. Widerquist. 2017. *Prehistoric Myths in Modern Political Philosophy*. Edinburgh: Edinburgh University Press.

McMahon, M. 2013. *Structural Functionalism*. Salem Press. http://ezproxy.newcastle.edu.au/login?url=http://search.ebscohost.com/login.aspx?direct=true&db=ers&AN=89185764&site=eds-live. Consulted 26 February 2019.

Meier, A. 2012. *EDemocracy & EGovernment: Stages of a Democratic Knowledge Society*. Berlin: Springer.

Mezey, M.L. 2008. *Representative Democracy: Legislators and their Constituents*. Lanham, MD: Rowman & Littlefield.

Moss, G., and S. Coleman. 2014. Deliberative Manoeuvres in the Digital Darkness: E-Democracy Policy in the UK. *The British Journal of Politics & International Relations* 16 (3): 410–427.

Mulder, B., and M. Hartog. 2013. Applied E-Democracy. In *Proceedings of the International Conference of E-Democracy and Open Government*, ed. P. Parycek and N. Edelmann. Krems an der Donau, Austria: Edition Donau-Universitat Krems.

Nassehi, A. 2005. Organizations as Decision Machines: Niklas Luhmann's Theory of Organized Social Systems. *The Sociological Review* 53: 178–191.

Nielsen, Claus Ballegaard, Peter Gorm Larsen, John Fitzgerald, Jim Woodcock, and Jan Peleska. 2015. Systems of Systems Engineering: Basic Concepts, Model-Based Techniques, and Research Directions. *ACM Computing Surveys* 48: 11–18.

Norris, P. 1997. Representation and the Democratic Deficit. *European Journal of Political Research* 32: 273–282.

Norris, P. 2001. *Digital Divide: Civic Engagement, Information Poverty, and the Internet Worldwide*. Port Melbourne, VIC, Australia: Cambridge University Press.

Norris, P. 2004. *Electoral Engineering: Voting Rules and Political Behavior*. Cambridge, UK: Cambridge University Press.

Norris, P. 2011. *Democratic Deficit: Critical Citizens Revisited*. New York: Cambridge University Press.

Norris, P. 2012. The Democratic Deficit Canada and the United States in Comparative Perspective. In *Imperfect Democracies: The Democratic Deficit in Canada and the United States*, ed. R. Simeon and P.T. Lenard. Vancouver and Toronto: University of British Columbia Press.

Norris, D.F., and C.G. Reddick. 2013. E-Democracy at the American Grassroots: Not Now Not Likely? *Information Polity: The International Journal of Government & Democracy in the Information Age* 18: 201–216.

OECD. 2003. *Promise and Problems of E-Democracy-Challenges of Online Citizen Engagement*. Paris, France: OECD.

OECD. 2010a. *The Development Dimension—ICTs for Development—Improving Policy Coherence*. Washington, DC: Organization for Economic Cooperation & Development, OECD iLibrary. http://www.oecd-ilibrary.org.ezproxy.newcastle.edu.au/docserver/download/0309091e.pdf?expires=1517024328&id=id&accname=ocid194270&checksum=61AB79A0E079DDDBBFE3B6E4568A2B0F. Consulted 27 January 2018.

OECD. 2010b. *Good Governance for Digital Policies—How to Get the Most Out of ICT—The Case Of Spain's Plan Avanza*. OECD Publishing. www.oecd.org/gov/egov/isstrategies. Consulted 27 January 2018.

OECD. 2013. *Reaping the Benefits of ICTS in Spain Strategic Study on Communication Infrastructure and Paperless Administration*. Paris: Organisation for Economic Co-operation and Development, OECD iLibrary. http://www.oecd-ilibrary.org.ezproxy.newcastle.edu.au/docserver/download/4212081e.pdf?expires=1517023432&id=id&accname=ocid194270&checksum=28A93F4721E71AADB84F3539BCB992DE. Consulted 27 January 2018.

Päivärinta, T., and Ø. Sæbø. 2006. Models of E-Democracy. *Communications of the Association for Information Systems* 17 (37): 1–42.

Parsons, T. 1956. Suggestions for a Sociological Approach to the Theory of Organizations–I. *Administrative Science Quarterly* 1: 63–85.

Qvortrup, M. 2007. *The Politics of Participation: From Athens to E-Democracy*. Manchester: Manchester University Press.

Rios Insua, D., and S. French. 2010. *E-Democracy*, vol. 5. Dordrecht, Heidelberg, London, and New York: Springer.

Sainsbury, L. 2017. But the Soldier's Remains Were Gone: Thought Experiments in Children's Literature. *Children's Literature in Education* 48: 152–168.

Scholl, H.J. 2006. Is E-Government Research a Flash in the Pan or Here for the Long Shot? In *Electronic Government*, ed. M.A. Wimmer, H.J. Scholl, A. Grönlund, and K.V. Andersen. Berlin and Heidelberg: Springer.

Steele-Vivas, R.D. 1996. Creating a Smart Nation: Strategy, Policy, Intelligence, and Information. *Government Information Quarterly* 13 (2): 159–173.

Stokes, G., and A. Carter (eds.). 2001. *Democratic Theory Today: Challenges for the 21st Century*. Malden, MA: Blackwell Publishers.

Transparency Mauritius. 2013. *Corruption Perception Index 2013*. Port Louis, Mauritius: Transparency Mauritius. http://www.transparencymauritius.org/corruption-perception-index/corruption-perception-index-2013/. Consulted 25 December 2016.

Transparency Mauritius. 2015. *Corruption Perception Index 2015*. Port Louis, Mauritius: Transparency Mauritius Organisation. https://www.transparencymauritius.org/wp-content/uploads/2016/01/CPI-2015.pdf. Consulted 25 December 2016.

Transparency Mauritius. 2017. *Corruption Perception Index 2016.* Port Louis, Mauritius: Transparency Mauritius Organisation. https://www.transparencymauritius.org/wp-content/uploads/2017/06/CPI-2016-25-01-2017.pdf. Consulted 30 October 2017.

U. S. Department of Defense. 1993. *Military Standard Systems Engineering MIL-STD-499b.* EverySpec. http://everyspec.com/MIL-STD/MIL-STD-0300-0499/MIL-STD-499B_DRAFT_24AUG1993_21855/. Consulted 28 January 2018.

United Nations. 2005. *UN Global E-Government Readiness Report from E-Government to E-Inclusion, UNPAN/2005/14.* New York: United Nations.

USAF. 2010. *Air Force Systems Engineering Assessment Model.* Air Force Center for Systems Engineering, Air Force Institute of Technology: Secretary of Air Force, Acquisition. http://www.afit.edu/cse/. Consulted 8 September 2013.

Ward, D. 2002. *European Union Democratic Deficit and the Public Sphere.* Amsterdam, The Netherlands: IOS Press.

Weinberger, J. 1988. Liberal Democracy and the Problem of Technology. In *Democratic Theory and Technological Society*, ed. R.B. Day, R. Beiner, and J. Masciulli. Armonk, NY: M. E. Sharpe.

White, S., and N. Nevitte. 2012. Citizens Expectations and Democratic Performance the Sources and Consequences of Democratic Deficits from the Bottom Up. In *Imperfect Democracies the Democratic Deficit in Canada and the United States*, ed. R. Simeon and P.T. Lenard. Vancouver and Toronto: University of British Columbia Press.

Wilhelm, A.G. 2000. *Democracy in the Digital Age.* New York: Routledge.

Zweifel, T.D. 2002. *Democratic Deficit? Institutions and Regulation in the European Union, Switzerland, and the United States.* Lanham, MD: Lexington Books.

CHAPTER 2

The Idea of Democracy in Theory and Practice

1 Introduction

'Out of the dark and from very long ago has come a word' (Dunn 2005: 10), a disruptive word called democracy to describe a form of government that gave the *demos*, or the people, sovereign power in its originating city-state, Athens. Citizens could (indeed were expected to) represent themselves regardless of their wealth or class (Held 2006: 14). This was a radical departure from earlier forms of government in the ancient Greek world as democracy, as understood at the time, empowered the ordinary citizens to participate in their own rule, effectively citizens of the ancient democracies were both rulers and ruled. Democracy is complex and its many problems were highlighted by both Plato and Aristotle (Dunn and Harris 1997; Dunn 2005; Held 2006), two conservative critics of democracy amongst others who resisted the idea of a political system where the people ruled. While Plato wanted to abolish democracy, Aristotle wanted it restricted as much as possible.

Two thousand years later despite the seemingly widespread popularity of democracy, it remains an essentially contested concept (Gallie 1955). What Gallie meant by the idea of an 'essentially contested concept' was that there were some political terms whose meanings could never be settled. Gallie illustrated his argument about such concepts with an extended discussion of the concept of democracy. The point for the purposes of this book is not just that democracy as a term has contested meanings,

S. Bungsraz, *Operationalising e-Democracy through a System Engineering Approach in Mauritius and Australia*,
https://doi.org/10.1007/978-981-15-1777-8_2

but that such contestations can never be settled once and for all. This is as much a matter of political philosophy as it is of the actual practice of trying to implement democratic processes and institutions. Despite the seeming popularity of democracy (both as an ideal and preferred system of government), the idea of democracy in empowering the people is still resisted in contemporary political systems, democratic or otherwise.

In modern times, the development of democracy has taken a number of forms, the most common of which is the representative democratic system. In terms of empowering people, the system of 'representative government gives no institutional role to the assembled people' (Manin 1997: 8). Although presenting itself as based on political equality, representative democracy appears to accept and reproduce various forms of inequality, as citizens do not have a meaningful say in decision making except for voting who gets to rule. This minimalist form of participation in the democratic process is one of the chief flaws in modern democracies. In this chapter, the evolution of democracy is briefly reviewed. This is followed by a discussion of one of the key problems identified with contemporary democracy in the West, namely the democratic deficit (Zweifel 2002; White and Nevitte 2012; Ward 2002; Norris 2011, 2012; Beetham 2012), since this is what many proponents of e-democracy argue that e-democracy is able to fix.

2 Decision Making and Democracy

In the year 594 BCE Solon, an Athenian archon, facilitated the emergence of democracy by putting appropriate legislation in place that stopped rich Athenian farmers from stripping the poorer farmers of their citizenship and then due to their debt burden forcing them into enslavement (Dunn 2005: 32). This economic protection from the land-grabbing activities of the rich provided a conducive environment to set the scene for further reforms. Several decades later in 507 BCE, another legislator Kleisthenes, followed Solon's initiatives to build the framework to create '*demokratia*' (democracy or people's rule), what we today understand as the Athenian democracy (Dunn 2005: 33). Kleisthenes' new system introduced the right for rulers to be selected from the citizens' pool (*demos*), which was supported by the majority of the citizens, who were the poor farmers. Kleisthenes's system empowered every citizen who had served in the military, regardless of wealth or birth, to volunteer to participate in political rule. Henceforth, Athenian democracy was born.

Democracy became the 'single world-wide name for the legitimate basis of political authority' (Dunn 2005: 15). This shifted both the basis for and the balance of political power as one did not need to be from the nobility to rule in Athens. The Athenian city-state was a relatively small community within which traditional democracy operated as a form of direct citizens' rule for male citizens only, women were not included (Held 2006: 19). The Athenian *polis* operated, in principle at least, as 'a community of equals' where government decisions were openly debated amongst citizens, or the *demos* (Garnsey and Winton 1997: 1). In effect, Athenian citizens aggregated individual preferences to make group decisions for their city-state, a feature of self-rule that in my view is lost in contemporary representative democracy.

Democracy established itself in Athens because it somehow managed to organise power in the right way (Dunn 2005: 18). Now while democracy was operating in fifth to fourth century BCE Athens, the other states which were mainly monarchist disliked this concept of involving the citizens in governing. In the ancient world democracy, as revealed in the writings of Plato and his pupil Aristotle, reflected this dislike for this idea of a government where all citizens made decisions for the rest. Plato went to great lengths in his works such as the *Republic* to describe his 'loathing' of democracy (Dunn 2005: 47). The views of Aristotle also reflected his distaste for people's rule since he reserved 'the term *demokratia* for pathological' rule (Dunn 2005: 49), which he understood as a chaotic form of rule that did not have the common good as a goal. Medieval Europe continued the Aristotelian claims that traditional democracy did not work, and it persisted for generations of thinkers in the West where the wealthy were projected as better and virtuous, and bent on serving for the common good (Dunn 2005: 50). Perpetuating this belief both justified and allowed power to be concentrated with a few to the expense of the many. Scholars opposed to sharing power with the many kept the myth alive, like, for example, to support the US capitalist system against people's rule (Schumpeter 2010). The development of a capitalist economy almost hand in hand with that of democracy, whether in its nascent liberal or social democratic or neo-liberal forms of representative democracy, appears to be detrimental to creating further equality on a global scale (Morgan 2016).

As part of his survey of political constitutions and systems, Aristotle went on to develop an analysis of several forms of government apart from democracy (Held 2006: 11–28). He analysed the nature of various *poleis* to find what was common amongst them and produced a framework of six forms of government (Huxley 1997: 191). Aristotle's framework described what he thought was right and wrong forms of government (Huxley 1997: 189–199); he defined three good ones along with each of their opposites, three bad ones. An outcome of this analysis was the *Politics* as his legacy to define political philosophy (Dunn 2005: 49; Held 2006: 15). The *Politics* was developed from observing people's political interactions as a community and through analysing the Constitution (*nomos*) of 158 *poleis* (Garnsey and Winton 1997: 13). Aristotle's views were based on what he observed in practice from his enquiries of what was happening in the human world when governing was for the common good as in Athens (Dunn 2005: 46–47).

Aristotle used the term *demokratia* to define a bad form of government. He preferred another form of democracy called *politea* that he promoted as a good form of government (Dunn 2005: 47). *Politea* was a representative form of constitutional government and was regarded by Aristotle as the opposite of *demokratia*. For Aristotle, *demokratia* amounted to the tyranny of the poor, a term popularised later by other writers like Alexis de Tocqueville in the early nineteenth century (Mennell and Stone 1980). Aristotle opposed democracy because he thought it was unfeasible in practice. Aristotle's notions of government forms still shape scholarly debate. His insights inform European political theory about improvement to government capabilities (Barry 1991; Dryzek et al. 2008; Dunn and Harris 1997; Dunn 2005; Elliott and McDonald 1965; Held 2006).

The Athenians were citizens who got actively involved in the process of self-government taking turns to rule 'as governors were to be the governed' (Held 2006: 14). There was no distinction in the sixth century BCE Athens between the people and government, all citizens were involved in the process of decision making which was conducted openly. Citizens debated to decide for all and the outcome was based on the best arguments presented. Strong oratory skills were needed to convince fellow citizens about the merits of one's arguments. This direct participation in decision making was different to other forms of rules as every Athenian citizen had an equal right to speak in the *ecclesia* or popular Assembly.

Decisions were not from 'custom, habit or brute force' which was seemingly the norm of the time, and it appears that most Athenians were proud of their system of government to the extent they considered it to be the exemplar city-state for the rest of Greece (Held 2006: 15). Democracy as practised then meant that there needed to be political equality for all citizens. The poor as well as the rich could rule, one's economic circumstances did not preclude one from governing. A commitment to political equality as a principle in government allowed one to rule and then be ruled in turn. As a key democratic principle for equality to be maintained, citizen's participation in government was facilitated. It would seem that barriers like the financial cost of participating were removed. Those on official duties were compensated for their time away from their own affairs. The reforms introduced by Kleisthenes included 'payment for public office and payment for attendance in the Assembly' (Held 2006: 18). This economic compensation freed the citizens who wanted to participate to serve the community by having an equal chance to rule. It opened ruling to everyone.

Equality for all provided a moral basis to the notion that the people needed to be free to speak, to associate and to argue in order to have equal share of ruling. Both developments, political equality and freedom, provided the basis for support of the Athenian democratic system of government (Hall 2013: 58). Political equality and the freedom to engage in political activity and association reflected in both its democratic system and the system's processes. The Athenians were relatively free compared to, for example, the Spartans' obeisance to rigid rules. Sparta was a rival city-state 'with a rigidly controlled culture instilled in its citizens' to the point of making this rule-based culture a virtue (Hall 2013: 56). Thus, a representative system within a *polis* where all citizens were treated equally, enabled citizens to enjoy political equality (Held 2006: 17). This new mood from equality when introduced by Kleisthenes seemingly emboldened some people sufficiently for them to open their mouth for the first time to participate in the debates at the Greek assembly (Hall 2013: 58). This idea of political freedom for citizens who felt equal has and still inspires current political thinking. However in practice, some limits were necessary to ensure that the freedom of one person would not unjustly infringe on the freedom of another. But, if each citizen could rule and then be ruled, then it was assumed the risks associated with such freedoms could be minimised.

The ancient Athenians also understood that voting when used was not the best option for decision making. This was because a split decision meant controversy still remained as a lack of consensus invites disharmony or political cleavage. At the *ecclesia* or assembly, a consensus though was sought based on the understanding that the common interest would prevail to deliver the best solution for the *polis*. The *ecclesia* was the place where majority rule decisions were made that committed the Athenian state to particular policies and courses of action. However, when there were intractable issues to resolve or when differences of opinion arose and there were clashes from individual interest, voting was the mechanism that was used. It would seem that the Greeks invented formal voting procedures to legitimise decision making in what is now a parliamentary typesetting (Held 2006: 17).

These innovative decision-making processes for its time set the political scene for things to follow. The concept of self-government became embedded in the West but its forms were still morphing and there is still a long way to travel to create a government where the people rule and are ruled in turn like traditional ancient Athenian democracy. The context where this system was applied was an agriculture-based economy where slaves conducted most of the work, also there was mining which with slave labour supported the expenses of the state (Hall 2013: 59). Citizenship was restricted to adult polis born men, with women, children and slaves being excluded (Held 2006: 11–28).

Athenian democracy opened possibilities to improve systems of government that could include every citizen in decision making. It has inspired humans to explore new forms of governing to try to implement a better system of government. Changes in forms of government have followed over time as people sought ways and means to resist exploitation by an elitist few. However, these changes and the development of new systems of government with a democratic impetus to enable people to live in freedom were not always smooth or successful. In the next section, the trajectory of democracy is explored.

3 Evolution of Democratic Thought

The historical struggles leading to the changes that transformed governing from tyrannical systems in feudal times to modern forms of democracy were complex (Keane 2010). The uprisings that followed from people's demands for change reduced the grip of feudal traditions and customs. As these controls were loosened, there was apparently a renewed interest in the classical political ideas from the ancient Greeks (Held 2006: 1–122). To manage the affairs of society, a set of rules was necessary such that people did not have to be subjected to tyrannical rule mainly in the absolute form of monarchist control. In the early twelfth century BCE Italian city-state experiments resisted the feudal yoke (Held 2006: 32), with the economy booming the city-states aspired for autonomy. The conditions of sustained economic boom from improved trade created pressure for political change to reduce papal and imperial control. This resistance against the feudal system of authority in Europe was an extraordinary event as it brought into question the notion of ruling with absolute powers invested in the hands of a few like monarchs or similar individuals and the aristocracy (Held 2006: 32).

The Italian republican city-state innovations to oppose absolutist forms of control were not of the same order as the Athenian democracy but they were a step in the direction of people's rule. Historians and political scholars have depicted these struggles and these models as a chaotic form of government, partly the re-emergence of the writings of Aristotle allowed contemporary and later critics to associate these developments with Aristotle's scepticism and negative portrayal of democracy (Held 2006: 33). These city-republics helped shape political thinking in Europe and America towards new forms of government (Held 2006: 33–34). More struggles were to follow, like those in 1640s Britain, which challenged the arbitrary authority of a monarch to override the will of the Parliament. Although the English Parliament at the time was not representative of the population as a whole, this struggle and the debates it provoked underscored the growing influence of democratic principles.

The effect of the resistance by some scholars to the idea that the people need to have a say in how they are governed has tended to obscure the fact that 'there are questions we have ceased to ask, or interpretations of concepts we have ceased to entertain, which prove to be well worth excavating, dusting down, and reinserting into current debates' (Skinner 2012: 128). The peoples' struggles against excessive domination had to find outlets. Just to name a few, the revolutions in France in 1789 and USSR in 1917 against feudalism, along with the independence war of the USA in 1776 against colonial rule showed this human need to free themselves from the yoke of the powerful. These are by no means the whole list of human struggles for freedom. The people aspire to be free, the idea of democracy meant they could be free (Dunn 2005; Keane 2010).

In the sixteenth century, European political thought became preoccupied by the notion of the state as a coercive entity. At that time, there were around 500 independent political units which started to consolidate into larger units. The larger entities were becoming more powerful from the mergers. Also, as imposts from customs and feudalism waned, in this environment, the idea of the rights of the state and the duties of the subjects became a source of political debates. As a result, political thinkers began to debate the nature of sovereignty and rethink the justification for constituting and justifying political rule as well as that of the political obligation of both ruler and ruled (Held 2006: 56). A new space was created from the retreat of these traditional ways of governing. The political debates to fill this vacuum initiated changes that contributed to rules that defined the modern state and its system of government. One key question that arose concerned the limits of the state's authority and the extent to which it could interfere in the life of its citizens.

This question was prompted from a legacy of coercively imposed traditions from a feudal system where the majority, peasants in agriculture tied in various ways to landholders, were seeking reforms to improve their lives so they could eke out a livelihood. Due to the demands for change arising from the expanding colonial empires and the industrial transformation of agriculture, some key ideas to be discussed below that emerged, shaped political life centred around liberal ideas that were from peasants demanding reforms. Then, as the industrialisation process spread beyond agriculture to other areas of human activity and workers' rights became an issue, other ways of thinking about political rule, such as socialism, Marxism and anarchism amongst others, emerged to challenge liberalism and its limited views on democracy.

Liberalism challenged traditions and customs which did not promote the individual's freedom and what many liberals claimed were their natural rights (Held 2006: 56–95). Liberalism introduced the idea of freedom of choice about things that shaped everyday life, like religion, the economy and political matters. Central to liberal thought was the idea that there was a private sphere, civil society, separate from the state in which citizens were free to go about their business without interference from the state. Hence, there was a need to set some boundaries, to constrain the state's power.

This new thinking aimed at developing a set of rules to define the boundaries between the civil society and the state. For many liberal thinkers (and some not so liberal such as Jean-Jacques Rousseau), this was effected through the idea of a social contract where people agreed to form a system of government in return for which their life and property would be guaranteed along with a range of rights. People did not become completely free as they were still constrained within the framework that the state imposed, and the range of rights was never completely settled and always contested. This hypothesised social contract allowed liberal thinkers to provide an answer to the question of the legitimacy of particular political regimes and forms of government based on the premise that the authority for government rests on the consent of the governed as first articulated by proto-Liberal thinker John Locke (1960: 374–375, 381). In short, the liberal state consisted of a framework of a (supposedly) mutually agreed set of rules that at the same time aimed to ensure that the state did not get to exercise unfettered powers, like a despotic monarchism. These arrangements also had to ensure some degree of space whereby the governed could have some degree of participation and input into the mechanisms of governing. There had to be a reasonable opportunity for the possibility of self-government or self-regulation whereby people as citizens would have some opportunity to bring about changes in government decisions and policies. The solution, effected gradually over many decades, was a system of representative government in which some were elected to represent the many. This was a complex process and there was much debate about the form that the liberal state could take without infringing on the rights of free individuals. For some, 'liberalism was a revolt against that vast network of economic restrictions and privileges which comprised feudalism' (Pennock 1978: 11).

Institutions that were wielding absolute powers, for example, through religious tradition and from monarchies, were seen as constraining the notion of a liberal state. As a first step, these institutions had to be separated from the state and reconceptualised as belonging to the private sphere of civil society so that a liberal, as distinct from a feudal state, could emerge thereby allowing the idea of a state as 'a legally circumscribed structure of power separate from ruler and ruled with supreme jurisdiction over a territory' (Held 2006: 58). This demarcation was to distinguish between rulers and ruled more clearly. The idea was to move away from absolute powers like those from the sixteenth-century religious doctrines and the feudal system to consensual rules defining the rights of the state and individuals within that state. As noted, this process of developing a new set of common rules meant that the emerging liberal state had to manage the ongoing tensions arising from the need to ensure that maintaining individual liberty and state power could co-exist. The liberal rules that were developed to establish this balance saw the development of modern representative democracy. They defined the framework within which a state (government) and civil society could operate and made it possible for the people to be free to carry on their private business within rights defined in the Constitution of that state. Each state developed its liberal constitutional framework which varied in their degree of democracy.

It was the work of seventeenth-century liberal thinker John Locke that in the eighteenth century helped shaped the liberal constitutionalist tradition to influence the US Constitution and to a lesser degree European politics (Held 2006: 59). Locke's work challenged the thinking that those who governed or ruled could be trusted hence their power needed to be constrained and constitutional limits were necessary to its authority. If the people held sovereign power in the sense of being the source of the consent that authorised the state, then those who ruled had their power licensed by the people. That delegated power was conditional on the state abiding by the people's will. In an independent state, through collective consultation, the model was that a government was to emerge from the people to exercise the will of the people. The delegation of the people was voided if when using the state's apparatus the agents who were delegated power failed to pursue the needs of the governed or operated outside the Constitution or collectively agreed limits.

The idea was that ultimately the will of the people must prevail within the constitutional representative state. The notion of a temporary delegation paved the way for a parliamentary system of government where people renewed their trust in the existing government based on the government's performance at each election cycle. Through this constitutional arrangement for temporary delegation, liberal democracy emerged to deliver accountable government so as to address the trust issue unearthed in the liberalism idea. Trust was based from prior performance and the cyclic representation embedded in the Constitution acted as a control for its renewal. The liberal democracy concept extended the liberal concept to create an accountable government. This idea of trust from liberal democracy furthered the democratic intent to a certain degree. It allowed, for example, individuals to still pursue their own private interests but liberal democracy also demanded a 'right of citizens to participate in the determination of the collective will' through elected agents or representation from the people in a cyclical manner (Held 2006: 94). This conflation of democracy to the representative model of democracy as noted by Keane (2010) meant people decided only at elections and referendums.

Representative democracy became a practical means to deliver democratic outcomes but in doing so it shifted the meaning of democracy away from its ancient Athenian origins, and with different contexts where the idea was applied it gave different results. In larger representative democracies, different models were adopted to form what became known as nation-states. These nation-states adopted a form of the representative democracy that was to suit their unique context and the forms that emerged were shaped by the existing systemic constraints that were in place at the time in those nation-states. An argument, for example, that representation was a pragmatic means to rule rather than direct democracy was that the size of the nation-states constrained the forms that could operate in practice (Dahl and Tufte 1973). Many modern nation-states were significantly larger in size than those of the city-state that had Athenian-style democracy. In the new context of these large nation-states, this made representative democracy the feasible option (Dahl 2005; Dahl and Tufte 1973). We expand on this practical form of democracy below.

4 Democratic Deficit

Democratic deficit is a term used to define the gap between what people expect from democracy and what is actually delivered by the political system (Norris 2011: 5). This aspiration for democracy varies between countries, as the extent to which citizens value democratic ideals to reject autocratic alternatives. The perceived quality of democracy can be monitored by the public satisfaction with democracy's performance in a country. Democratic deficit indicates that while people 'aspire to democracy as their ideal form of government' they remain sceptical about how democracy works in their country (Norris 2011: 5). A demand exists for democracy. Democratic deficit is seemingly not confined only to the context of the European Union (EU) where the discussions about the deficit arose during debates about legitimacy of the decision-making process in the larger EU. It is reported that from the 1990s there was an encroachment by the EU onto the policy areas of its member states, like 'economic governance, foreign and defence policies, social and employment policies and justice and home affairs' (Carammia et al. 2016: 809). These policy areas were centralised to the EU entity by the nation-states giving them up, albeit reluctantly. This process was perceived as opaque when compared to the standards applied for decision making in the member states forming the EU. The EU decision-making process was seen to be short of the standards of transparency and accountability that were in place amongst the EU member states at national level. A democracy gap was perceived to exist between the institutions of the EU and the EU citizens' democratic expectations of those EU institutions (Ward 2002: 1).

This democracy gap was caused in larger part from the integration of countries with varying standards of democracy. This variance in standards meant EU citizen's expectations could not be met by a supranational like the EU. A lack of similarities from the different understandings of democracy from EU citizens from different states that makes the EU system was a contributing factor (Beetham 2012: 87). Furthermore, democratic deficit for some experts was about the democratic process and this process required EU's institutional reform, however, for the people it was about the EU policy outcomes. It was argued that the process producing the outcomes was flawed (Beetham 2012: 88). Flawed systemic processes will produce unsatisfactory outcomes and create democratic deficit as the system fails to meet public aspirations of democracy. This deficit required answers. An explanation offered links democratic deficit to a set of issues that is a combined effect from 'growing public expectations, negative news, and/or failing government performance' (Norris 2011: 5).

Research in the area of satisfaction with democracy and trust in government has been synthesised to develop a theory to explain the democratic deficit in any regime as a set of sequential steps that used the support framework created by David Easton (Norris 2011: 5–8; Easton 1965a, b). Easton's work seemingly catalysed the issue of political trust even in the 1960s through his two levels of support, namely diffuse support that was for the support of the regime or system and specific support which was for the incumbent authorities (Levi and Stoker 2000: 477). Using a market model, Norris explains the demand side is from the rising public aspirations while on the supply side there is the failure of democratic or policy performance from the state in matching the demand. Norris extends the model of support from Easton as she finds the role of the media to shape citizen's support is that of an adjudicator between the supply and demand sides. She argues that the negative coverage the media provides of the government through news media seemingly shapes citizens' perception. These perceptions then shape trust and satisfaction with government.

Now, Norris offers that those who focus on supply-side theories tend to argue public dissatisfaction result from the process or policy performance of democratic governments and also the government's democratic institutional arrangements (Norris 2011: 7). The quality of democratic government itself though is seemingly from citizens' satisfaction from their experiences of the processes in place. The policy outcomes, they define dissatisfaction, are based on delivery of public goods and services or policy outputs. She also argues that due to an inadequate perceived power share in representative government processes the constitutional arrangements can be linked to democratic deficits (Norris 1997). The failure of performance is an issue as to whether democratic deficit will lead to a lack of faith in the institutions of representative democracy.

The issue of representation was raised from the formation of the EU entity. A key question was how citizens' preferences could be linked to decision making in the EU (Norris 1997: 273). A channel of communication was suggested to strengthen consultation between the European Commission and the national Parliaments of the EU member states. It was anticipated that this channel would improve transparency and accountability in decision making by the EU. Voters would then throw out the nationally elected members for EU decisions that had an adverse effect at the local level. This election decision to throw out the elected members at national elections was complicated by the fact the decisions made at EU level were complex to interpret by the public for it was not possible to link EU policy to outcomes (Norris 1997: 275). The EU decision-making opacity reduced citizen's support for the EU and their political trust in their government as a part of the EU fell.

Extending Easton's conceptual framework for diffuse and specific support, Norris conducted a study for political support that was developed to assess democracy's support globally using existing public opinion polls data from the 1960s to the mid-1990s (Norris 2011: 10). The survey was extensive in its collaboration with 93 countries to provide insights about citizens' views of democracy (Norris 2011: 251–253). Interpretations of the data showed some trends. In many countries, citizens were sceptical about how well key institutions of democracy worked, listing amongst the prominent causes their political parties, parliaments and governments' operations (Norris 2011: 10). However, the survey also noted that the demand for democracy was universal. Democracy was a global phenomenon. People still aspired for the ideals that democracy represented and there was a rising tide of dissatisfied citizens becoming critical of the workings of democracy in their countries.

The survey was designed to analyse public opinion at a cross-national level from 1981 to 2007, amongst the 93 countries listed for the World Values Survey some African countries were included (Norris 2011: 251–253). The results from the study indicate that claims for a rising tidal wave of disillusionment with government from 1994 to 1999, listed as the third wave era, are unfounded as the data showed that confidence in the public sector tends to ebb and flow. It indicated that for the period surveyed interpretations of ever-growing disenchantment in established democracies were misleading. The findings in the USA showed that support for government institutions and that of the leaders had both fallen and risen in recent decades (Norris 2011: 12–13).

A study of the democratic deficit in Canada also showed trendless fluctuations with a similar conclusion that citizen's support ebbed and flowed. The study recommended that the over-simplified claim of long-term trends of a reducing confidence in democracy due to social changes should be rejected. Dissatisfaction with democracy it suggested is due to increasing aspirations of citizens who are comparatively more educated (Norris 2012: 47). A hypothesis was that a more critical appraisal of government performance and negative news tend to shape democratic deficits. A trendless fluctuation of support for democracy is also mentioned by another scholar in another comparative study where the EU, Switzerland and the USA were surveyed (Zweifel 2002). Further, studies have been carried out by others to assess support for democracy (White and Nevitte 2012).

Using a different conception of democracy, in a bottom-up survey analysis for the USA and Canada, the findings were that democracy was desirable but that in both Canada and the US citizens did not perceive their democratic system as exceptional (White and Nevitte 2012: 57). Living in a democracy was important, and these criteria scored the highest figure, 10 which stood for very important on the scale being used. However, as the idea of democracy varies from person to person it may be hard to find a pattern. The study advanced another hypothesis that it is people's understanding of democracy as a liberal democracy that may be the root cause of the democratic deficit in terms of perceived democratic performance. This bottom-up approach suggests that when democracy is understood as a liberal democracy, that understanding creates a sense of democratic crisis in industrialised democracy. It suggests that liberal democracy may also be contributing to the perceived lack of performance of democracy and the democratic deficit.

There is no consensus about what the crisis is except that most of them have a common element which is a lack of government ability to cope with an increasing number of citizen-led demands (White and Nevitte 2012: 64). The Neville and White investigation suggests that a growth of public demands makes government all things to all people, a situation which makes it difficult for governments to deliver and this may potentially undermine a government's legitimacy. It seems this legitimacy could be linked to the integration problems of the EU as a supranational entity where the larger model cannot link the EU citizens from a nation-state level demands to the decision-making process of the EU supranational in terms of policies. Thus, it seems that the EU formation is creating the 'biggest challenge to any perceived democratisation' (Ward 2002: 1), as at that higher level the EU entity decision making seems very complicated. The changing context of a more critical citizenry and the increased complexity of the decision-making processes that are being followed is causing disconnects of the clear links between policies and outcomes in modern political systems and as a result the limits of the representative democracy system are being reached.

5 Representative Democracy a Compromise

Many see the state and the representative system of government as adequate. Representative forms of democracy are projected as a pragmatic form that achieves the best outcome in the circumstances (Dahl 1995, 2005). But there are also those who point to how representative democracy could be improved (Rios Insua and French 2010; Päivärinta and Sæbø 2006; OECD 2003; Hoff et al. 2003; Mulder and Hartog 2013; Hague and Loader 1999; Dahlberg and Siapera 2007; Moss and Coleman 2014). The new form that is proposed is to fix some flaw in the representative democracy or its processes but the model's key weakness is that it is untested. Now representative democracy is itself an evolutionary design based on ideas that seem to work in practice. There has been no shortage of suggestions for improvements. Many models have been proposed over time and some even in the context of a possible future, but the momentum for positive change seems stalled (Held 2006). Assuming that democracy as described in Athens is the view that most people hold, or at least the idea that government is for, by and of the people (to paraphrase Abraham Lincoln), this raises issues of a lack of democracy in representative democracy. This is a democratic lack that goes to the core of representative democracy's design. This core design problem was inherited when the term democracy became applied to what was not classical democracy. Representative democracy as the commonly accepted form of democracy is not a democracy that democratises, even if the problem of trust is partly overcome through inbuilt cyclic elections to hold governments to account. But it is only partially overcome (if at all) because there is a perception of declining political and social trust in advanced industrial representative democracies (Levi and Stoker 2000). Despite this perceived decline, the status quo seems to continue even though a decline in citizen support ensues for the non-performing system.

Despite changes in context and flaws emerging in representative democracy, the evolution of democracy to its form for a people's rule continues to be undermined. After a surge in support for change post-World War II, a regression towards democracy is noted by some in the 1970s with the appearance of a top-down approach called citizen's juries (Qvortrup 2007: 70). This concept of politics using a small group of experts to gauge what mattered to the rest proved difficult to sustain as it became problematic to know what every citizen required. Polling was a mechanism to gauge the pulse of the electorate but it was inadequate so government brought a new form of participation citizen juries. But as the citizen's juries were often critical of government the funding for seeking inputs was reallocated to focus groups which were more controllable. As Qvortrup wryly noted, 'Democracy is always difficult when the guardians wish to protect themselves' (Qvortrup 2007: 72). This behaviour to resist reform from those in control is as old as democracy. Various excuses are offered to delay empowering the people but we single out some key areas for consideration here: size of nation-states, private property, democratic deficit and emerging party behaviours. There are many more issues that could be discussed with respect to this tacit resistance to improve democracy. However, to the ones just mentioned are sufficient to illustrate that the tacit status quo is a pattern of ongoing denial away from traditional democracy. It is a pattern to perpetuate the dominant mindset that democracy in any but a representative sense is unworkable. Basically, a contemporary representative democracy is designed for supporting a paradigm that people's rule does not work. We argue at this point that it is a self-imposed limitation, a design constraint meant to undermine better forms like e-democracy.

The issue of size of modern nation-states (Dahl and Tufte 1973), is used by the opponents of change to the current system to justify representative democracy for contemporary large nation-states (Hague and Loader 1999: 6; Hilbert 2009), when a gigantic online polis is feasible (Hilbert 2009: 93). Size is a key issue for resisting change to a different system of government like Athenian democracy. But more importantly it is argued that representative democracy in nation-states should never have been called a democracy in the first place. This problem of terminology where democracy is 'applied to systems as radically different as the city-state and the nation-state', which is a form of 'multi-unit polity', is according to Dhal and Tufte stretching the word democracy (Dahl and Tufte 1973: 25). They argue that ideals and institutions suitable for the city-state should not have transferred to ideals and institutions for the representative government of nation-states.

The issue of size is brought to the attention of the political community to justify representative democracy as different to democracy at the city-state scale (Dahl and Tufte 1973). This issue of size was not to dispute the merit of classical democracy but rather it was about more improvement being required to ensure democracy prevailed within the context of the state's constraints. The challenge was for people to select consensually the best possible options that were feasible, a design issue. However, when the best option, representative democracy, was chosen it was understood that its design was to be upgraded as things changed so as to reach the intent of traditional democracy. To show the unfeasibility of the Athenian direct decision-making model, an example used was the fact that Athens was sufficiently small, and therefore it could gather all its citizens in one place in a modern town hall type area for decision making. This depicts a lack of foresight, limitations by promoting constraints that may at times no longer exist, and is a tacit resistance that benefits a few opposing change. Using the obvious, they hide the real issue, which is an undermining of people's rule.

Fitting every citizen of a nation-state into a town hall is not feasible for the large numbers of citizens that make up a nation-state. Some sceptics, for example like Max Weber, apparently in questioning the 'applicability of ancient' traditions to the modern large states seemingly allowed such arguments to pave the way for Schumpeter's elitist view of democracy (Shaw 2008: 33). Such scholarly views are seen to assist the creation of a school of thought to support the current type of representative system. In a large nation-state, a core design aspect of the representative democratic system seems to be that whatever the form of delegation, this delegation must be accountable to the people (reflect the people's will). This accountability from the opacity of decision making in practice is questionable in many large modern representative democracies, which we noted above is reflected in the emergence of what is termed 'democratic deficit'.

Another argument about size was that citizenship was a restricted right in Athens, women were not considered citizens, '[o]nly Athenian men over the age of 20 were eligible' (Held 2006: 19). So the initial representative models derived from this traditional conception also had a restriction designed in it about who could vote. Initially based on the Athenian models all those who owned property were considered citizens, only they could chose representatives. This group's exclusivity was contested later to make voting applicable to every adult man citizen of the country. Over time with intense struggles, women and minority groups like blacks and people of colour who had been excluded were given the right to vote in western democracies. There was concern amongst the rulers that such a move to open up voting would create a tyranny of the poor, thinking which seems to have a root in Aristotle's elitist views for *demokratia*. Strong voices emerged for women, for example, Mary Wollstonecraft (Held 2006: 49–54), and later in England John Stuart Mill, in the USA Victoria Woodhull, Elizabeth Cady Stanton and Susan B Anthony, and in Australia Henrietta Dugdale and Catherine Helen Spence. Exclusions were slowly removed to allow better participation in voting, a key process in representative democracy.

The issue of size is a constraint in a system for a face-to-face interactions but that representative democracy system with unfettered power to delegates are ideal is suggestive of a myopic view in promoting the Aristotelian justification against traditional democracy. The constraints of size should not continually distract from the fundamental need to reform an inadequately designed system of delegation and that keeps persisting. Now, a system in the form of representative democracy idealised from size arguments also allows only a few to decide about reforms. Hilbert argues the debate should not be about a dichotomic view of direct or representative democracy but rather a view of how democracy can be improved to overcome the challenges of equal access, free participation and technological design (Hilbert 2009: 87). Arguments about size when technology and its capability still allow control to the few are a system design obstructive to a goal of people's rule. With technology, to democratise, size is no longer a constraint, digitisation of a representative system where a few rule is meaningless for democracy.

To deliver representative democracy, the state was linked to democratic values (like freedom and equality). With the ensuing political reforms that detached citizens' obeisance to monarchist rule, these democratic values when prosecuted found new questions being raised about further reform to representative democracy and the state as an entity embodying these values. As education was universalised, a culture change in thinking was in progress amongst citizens for a range of freedoms and rights. There was a need to ensure that the state powers could be curbed further to transform liberal democracy. There was a demand emerging for opening access to every citizen in traditional democratic fashion. However, the idea of liberal democracy as a representative form of government with its acceptability and ongoing support as a pragmatic form became more powerful in the West. Issues of reform to this form were resisted until it was challenged by Karl Marx.

A focus on the state as a form of public power that was supreme within a defined geographical boundary had in the process of building forms of delegation sidestepped the further empowering of the majority of the citizens. The poor who were delegating their power to the state were never freed except to vote for the delegates who usurped these powers without delivering to the people's will. The flaw in the representative system remained. Private property rights were never extinguished as in the liberal representative system it perpetuated through capitalism ways and means for wealth to shape decision making. Protection of private property rights removed certain kinds of choices from the democratic arena. Private property reduced the 'democraticness (Wright) from democracy' as political power were less equally distributed in the population (Wright 1994: 536).

The ideal of liberalism had emerged initially to maximise the individual's freedom with the idea of limiting government's power in the large nation-states (Held 2006; Gutmann 1980; Galston 1991). This liberal idea was strengthened later with the addition of the protection of the concept of private property. The private property concept seemingly provided the incentive (economic motive) for individuals to incorporate the liberal system in various *polities* (Dunn 2005; Held 2006). Establishing liberalism was a project to provide resistance to domination (for equality or *isonomia*) and it evolved creating contemporary representative forms of liberal democratic government. The poor who were the voting majority were unequal from the start. In the liberal democracy state with the 'institution of inheritance having been retained, there was from the start a vast source of inequalities which prevented' that the equality being established was for a few, as for many it remained unequal due to economic discrepancy, for example (Pennock 1978: 11). The representative democracy in the form of liberal democracy never extinguished the inequality that already existed in the form of private property.

It was an opportunity for designing true reform that empowers the people. Instead, Aristotle's work is rolled out to promote the concept of *politea* or representative form of government as the good form of democracy (Dunn 2005: 48–49). This interpretation helps perpetuate the alternative of representative democracy with its mainly liberal version in the West. This is despite the working class still being subjected 'to laissez-faire government policy and ruthless free enterprise' that promoted a private property concept that profited a few (Corcoran 1983: xi).

Now it seems that 'The Greek champions of democracy praised and fought for rule by the multitude (*to plethos*), by a broad array of political arrangements' but there were no recordings of these arrangements or if there were then they have been lost (Dunn 2005: 49). This vacuum of written evidence allows the thoughts that have followed to promote traditional democracy as a disliked concept using the Latin word *demokratia* that was first created in the 1260s from Aristotle's translation of *Politics* (Dunn 2005: 58). Democracy took three centuries from the time of its creation to become the more acceptable representative system called *politea* as democracy's acceptable twin (Dunn 2005: 58–59). The struggles to establish liberal democracy mentioned above have been complex and tortuous. In the West, *politea* as a concept shaped the hegemonic liberal representative democracy form. A real challenge to liberal democracies was brought about from technology when Marxism utilised a deepening class divide arising from the rapid industrialisation of human work.

Technology, namely mechanisation, through the industrialisation era saw a deepening class divide that erupted as a class warfare to challenge the claims of liberal democracy (Held 2006: 96–122). The industrial technology firstly generated an economic surplus (improved productivity), and secondly, it influenced the call for freedom and equality (printing to communicate to the many). New means of production provided a conducive environment for change to the class disparity emerging from the unequal wealth distribution in liberal democracies. Industrialisation, through mechanisation, was providing those not involved in production to enjoy the surpluses made under conditions of excessive exploitation of the labour class (Held 2006: 97). To stimulate economic activity through these new means of mechanised production, the market was freed even further, liberalised in these Western industrialised representative democracies. The mass exploitations upon which it was based created significant pressure from the workers seeking reforms.

The production of a surplus and its appropriation by a few created the means for Marx to question this form of labour exploitation that exacerbated an already existing inequality. Marx used traditional democratic ideals to challenge the liberal model's neutrality in the face of the excessive dominance of the masses by an economic elite class. Humans aspired to autonomy and their conditions of excessive exploitation of labour by a capitalist class brought a created pressure for political change towards reducing the economic elites' control.

In the Western liberal system, the elite's economic domination that was exercised by controlling the means of production was challenged by a new contender, Marxism (Held 2006; Birch 1993). The economic elites' excesses had prepared the ground for a political change in the Western representative democracies. Karl Marx and Friedrich Engels' works provided a basis for workers to dispute the idea that the free market economy and a neutral liberal state were conducive to genuine human freedom (Held 2006: 96). This paradigm shift was based on targeting the exclusivity of the elitist systems and challenging its class structure (Held 2006: 96–122). The lives of a large segment of citizens were being limited by the economic might of a few and these significant majority seemed to have no say in the matter.

Marx and Engels believed that the liberal state perpetuated a system that was lacking in equal justice to its citizens. Liberal democracy according to them was perpetuating a capitalist system where a few historically controlled the many due to their private property rights over the means of production and that in turn gave them excessive political power. The economic gap between workers (selling their labour) and owners (controlling capital) was enormous, it led to social unrest. Marxism, in principle at least, promoted a form of direct democracy to become a new contender to the elitist models that existed before. Marxism provided hope for the masses for a power share arrangement, but the changes from workers' demands that ensued did not lead to a direct democracy like Athens to deliver the desired power share.

To garner support, we note there was a strategic use of the poor to drive change. This Marxist attack on hierarchical society was about control over the means of production and its carrot was the promise to distribute to the many the created surplus (Dunn 2005). To win support from the masses and emerge as an alternative to liberal democracy, Marxism seemed to provide the direct democracy concept in the form of communism to the West. A key contribution is that it shifted the focus from liberalism's hegemony towards an alternative (Birch 1993). Marxism's failure seemingly reinforced views that representative democracy was the only appropriate way democracy could work.

Kant suggests that 'Man wishes to live comfortably and pleasantly, but nature intends that he should abandon idleness and inactive self-sufficiency', thus natural impulses that encourage humans 'towards further development of his natural capacities' should be fostered (Kant 1991: 45). As individuals, humans aspire to be free, but they also want to have adequate means to support a good life through their own endeavour. This was Aristotle's teleological assumption from his observations, and his departure from Plato's harmonisation to Supreme Being clones who are levelled (equalised) by the state. Kant points to a need for balance between freedom and equality, and he proposes it is to be guided by reason (Kant 1991). Seemingly representative democracy was a realistic means for individuals to achieve a good life they desired through their personal endeavours.

On the other hand, Plato assumes a hierarchical distribution of 'appetitive elements of man's nature throughout society' and that 'general altruism cannot be relied upon to check man's self-seeking proclivities', and some more recent democratic elites like 'John Stuart Mill, Walter Lippman and Joseph A. Schumpeter' views on the matter seemingly align (Pennock and Chapman 1977: 4). They are reflective of views that seem to justify the state's control rather than people's control of the state. The natural human desire for self-autonomy to develop their unique individual capacities to meet their needs, however, creates a significant tension between the communist and liberal systems, while its class attack generates deep divides between the poor and the rich. As 'good government requires the consent of the governed' (Wallas 1962: 21–22), and if the consent is motivated by some preconceived end (good life), the human needs and wants for a good life make them competitive. Apparently, according to Pennock and Chapman (1977: 5), Marx's optimism was his assumption about human nature being 'potentially cooperative and communal to the exclusion of any but the most benign competitiveness'. However, a communal system as a concept was not new.

Plato proposed a communal system through harmonising every individual in ancient Greek Gods' image while a pragmatic Aristotle proposed an individualistic system to counter Plato's argument as unfeasible in practice (Dunn 2005; Dunn and Harris 1997). Even though the tension between these two views still exists, a common thread is that both philosophers had treated *demokratia* or the rule of the poor as unfeasible. They had articulated their reservations about the chaos from such systems in practice (Dunn and Harris 1997; Dunn 2005). Marxism by projecting to empower the poor went against this thousands of years old thinking, it seemed similar to the strategy applied by Kleisthenes to garner support from the many that had brought about *demokratia* against his fellow elites in Athens (Dunn 2005), but the context had changed. In the West, Marxism failed to succeed against its stronger US version of liberal democracy, this failed attempt at reform was used to continue the negative perception of people's rule.

A perception remains that representative democracy could be improved further to deliver a better form of democracy as the neo-liberal form of representative democracy in practice is yet to meet the intent of democracy, which is people's rule. An effort to improve seems to have faltered the post-1970s with inequalities re-emerging from unchecked market forces (Morgan 2016). Neo-liberal's system of excessive freedom of the market creates massive inequalities and exacerbates existing ones, and today's representatives still manage to suppress the voice of the many who are unequally subjected to this free market forces. For the vested interest of the few, the representatives keep perpetuating through a status quo a political system that protects the interests of the free market against those of the many. The market failures highlight the systemic problems that representative democracy contributes to in its inability to curb the market forces reducing the core value of democracy like equality for all citizens.

6 Legacy and Promise of E-Democracy

Despite the promises of e-democracy to fix democratic deficits (Dahlberg and Siapera 2007; Rios Insua and French 2010), there is ongoing crisis in 'liberal democracy in the early twenty-first century' (Coleman and Blumler 2009: 1). We expand on this crisis considering political parties and their behaviour. In the neo-liberal era, with a market-oriented approach the representative democratic system has produced a certain way of doing things in a state that has become entrepreneurial (Dahl 1995: 391). An entrepreneurial paradigm may suit a market's needs but its processes may seem lacking for a democracy meant to reflect people's needs or people's rule that e-democracy promises. A market-oriented approach perpetuates a certain thinking amongst those elected that were meant to protect the interests of the public, representatives are elected in a political system that now seems to lack the capacity to deliver. Furthermore, if the system that needs reform accepts that capitalism has become too powerful then even if the political party in place wants to change the system they are unable to change it. The way the party and individuals in the party think about the problem becomes the limiting factor. This thinking generates a crisis of a disengaged public which then allows the system to persist as the public are unable to change it due to being trapped within the representative system.

In the representative democracy system, where the powerful market is free there is a need to repair the relationship between the electorate and politicians by improving trust. There is an eroding trust, which some politicians perhaps believe they can overlook and thereby defer change to a more people inclusive approach in government decision making. More to the point is how one empowers the people in this new environment so they engage using the capability from digital technology to improve trust. The Internet capability offers the potential to allow new thinking to improve participation in representative democracy.

The powerful market on the other hand constrains as the political party needs to operate within the framework of the existing market oriented system. A party therefore becomes and promotes the capitalistic dependency existing in the representative democratic system. In Western representative democracies for example, there is no room for reform unless the actors making decisions move away from their inherent capitalist thinking (Wright 1994). This means that 'unless a group [political party] has the capacity to overthrow the system completely, then even' political parties that are opposed to the system have to support it due to the material benefit they draw from it (Wright 1994: 543). Political parties thinking within a paradigm may have become wittingly or unwittingly captives of wealth and its benefits that they assume are necessary. Thinking within the design of a system is shaped by the system's power, thus a representative model that projects democracy as a neo-liberal free market type cannot change to a traditional democracy unless the system is uprooted (Wright 1994). New thinking requires a new paradigm else it tends to restrict the potential that new and emerging technology provides, like with an e-democracy.

In a free market-oriented system, the market imposed changes of the representation system has shaped the party behaviours. The political parties when operating within the contemporary representative democracy require politicians to be supportive of the free market ideology. Due to their inherent need to be elected sometimes using money donated for electioneering activities raised from the market, politicians are bent on serving market politics rather than the people. These elected representatives as agents of the people have become entrepreneurs in a party organisation that is required to win power so it can rule to its own agenda seemingly dictated by a powerful free market rather that an agenda that reflects the people's will.

To keep the citizens under control, over the years an idealised speculative notion about what people want seems to have been promoted by mainstream political parties as is seen in their 'daily failures of communication and unfulfilled promises' (Coleman and Blumler 2009: 2). The promises are made at electoral events but they are rarely delivered in practice. The ease of making such promises, when pragmatism dictates it is otherwise, seems troubling in a democracy. To get elected, this behaviour occurs even in the well-established democracies where representative democracy processes allow such exaggerations (freedom of expression). This can be perceived as an issue of communication failure as there are no means for the people to ensure that the promises are delivered by these parties. Accountability is only at the next election cycle, with parties relying on the short memory of the electorate. Over time this communication failure leads to the people distrusting parties and the representative system they operate within as the people's 'experiences and expertise often seem to be diminished and marginalised' (Coleman and Blumler 2009: 3). Keman even reflects if the office-seeking party behaviours are not undermining legitimacy in these established modern democracies.

Keman (2014: 312) identifies two democratic performance dimensions that political parties are undermining. These dimensions are unlawful behaviour within the limits of the system's rule of the law, that is the party exceeds the authority vested by the people without breaking the rules. Secondly, the policies from parties that are irrelevant to the people. These undermine the authority which people entrusted on representatives to run the affairs of government. A reduction in trust with parties increases the volatility in the modern democratic system given the system in operation. To understand these behaviours, we look deeper into the party as organisations in the business of seeking power.

These parties as systems emerged from factions, at times ideologically driven, so democracy evolved to its modern form of representative government to become a system that learnt about becoming enterprises to win power. These new enterprising representatives created a system where the modern democratic system decision making seems broken (Dahl 2005). However, as these systems of government become more complex the citizen's voice is becoming lost even though improved communication technology is being created.

The Internet has 'inverted the few-to-many architecture of the broadcast age, in which a small number of people were able to influence and shape the perceptions and beliefs of entire nations' (Coleman and Blumler 2009: 8). The existing constitutions which are largely reflective of an industrial era where the parties have learnt and adapted to optimise their operations may now be assisting the political parties to defer changes if it is not to their benefit like through maintaining the current electoral system of representation. The electoral system as part of the Constitution to ensure cyclical elections are held is to bring representatives to account. The entrepreneurial parties seem to be able to circumvent accountability to the people.

To address the changing behaviours of political parties, Norris calls for an electoral engineering as an option for she suggests that with new electoral rules reform is possible to shape the behaviours of 'politicians, parties and citizens' (Norris 2004: 8). Electoral systems she argues are the most basic democratic structures but elections 'are not sufficient by themselves for representative democracy' (Norris 2004: 3). Electoral engineering disrupts existing behaviours as, for example, in New Zealand where the switch from a First-Past-the-Post system to a mixed member proportional system in 1993 caused sudden fragmentation to the two-party system. This need to reengineer according to Pippa Norris is from latent 'issues of effective democratic design' as her engineered reforms to an electoral system were meant to shape actors' behaviours (Norris 2004: 5). Electoral reforms have been attempted to improve the First-Past-The-Post system to reducing the power of the two main political blocs in some countries but with some limited success.

Electoral engineering attempts to achieve reforms that have a balance between seeking an increased accountability in majoritarian system or more pluralism through proportional systems (Norris 2004: 5). She argues that electoral rules have mechanical effects that shape which candidates are elected to Parliament and which parties enter government. Altering the rules has a psychological effect that shapes the behaviours of the citizens and political actors for vote-maximisers in the electoral marketplace. These behaviours shape voting choices a core activity in the representative democracy as she argues that in the representative system politicians who are not vote-maximisers are less successful in getting elected. Politicians adapt to mass political behaviour rather than try to shape them, but in an apathetic electorate there is no incentive for the politician to change the system to an e-democracy that democratises, for example (Norris 2004: 21). Alternatively, Dahl points to a focus on the minimal set of institutions that must work in the nation-states' system to deliver democracy (Dahl 2005). Now in these representative democracies, political parties that tend to behave as enterprises with their own agendas to prosecute rather than those reflective of the people's will on behalf of which they pitch their electoral appeal, through their parties' behaviours, they seem to add to the problem of democratic deficit.

A degree of disconnect has emerged between people's needs and politicians performance to deliver to those needs. Concerns are also from a technology that is capable of change yet incapable to change the representative democracy to a form that democratises, an e-democracy. According to Coleman and Blumler, some are making some 'excessive and deterministic reactions' claiming the direct Internet democracy cannot be stopped while others argue the Internet will be made to mould into traditional politics (Coleman and Blumler 2009: 9). The next chapter explores in depth some of the issues surrounding e-democracy's failure to operationalise.

7 Conclusion

Classical democracy seems undermined through a systemic resistance from ongoing negative justifications of why people's rule could not work rather than how it could be made to work. This inward thinking is despite the shortcomings of the current form of representative democracy which is not reflective of the people's will as the people have no real say. The cyclical voting system is undermined without people's input in decision making, accountability lags as people decision making can only be exercised at the next electoral event. The underlying trust issue implied for such a system requires a means for channelling citizens wanting more say like for the UK where they are demanding inputs to setting health, education, elderly care, policing and local economy policies (Coleman and Blumler 2009: 7). At the core of the need for reform is the issue of the gap between what the people believe democracy should deliver and what contemporary democratic forms actually delivers. The democratic performance problem seems to be in the translation of the ideal of democracy into a desirable system like a democracy that democratises, one whose function is designed to maximise people's contribution to their own rule. In the evolution to a representative democracy, there is what might best be described as a design issue concerning what the people feel they need from a democratic system and what they actually get, or a designed democratic deficit (Hindess 2002).

With the development of ICTs, new ways of involving citizens in the democratic process appear possible. The possibility now exists for every citizen to become their own representative. It has led to a number of scholars to consider e-democracy as an improved form of representative democracy system that enables much greater participation by citizens in their own government (Meier 2012; Lee et al. 2011; Moss and Coleman 2014; Coleman and Moss 2012; Coleman and Blumler 2009). They suggest that the concept of e-democracy holds promise and that this form of government has the potential to overcome the democratic deficit (Grönlund 2001). The capacity of the ICTs to allow citizens to communicate with each other through the Internet also overcomes the issue of the size of the polity as it enables decision makers to hold virtual town hall meetings (i.e. meetings in cyberspace) (Moss and Coleman 2014; Coleman and Moss 2012; Coleman and Blumler 2009). ICTs thus have the potential to change behaviours by creating new possibilities. ICTs are the key to the development and implementation of e-democracies. The next chapter explores the issue of e-democracy and reviews debates about the success or otherwise of attempts to implement actual forms of e-democracy.

References

Barry, B.M. 1991. *Essays in Political Theory*. Clarendon Paperbacks. New York: Clarendon Press.

Beetham, D. 2012. Defining and Identifying a Democratic Deficit. In *Imperfect Democracies the Democratic Deficit in Canada and the United States*, ed. R. Simeon and P.T. Lenard. Vancouver and Toronto: UBC Press.

Birch, A.H. 1993. *Concepts and Theories of Modern Democracy*, 2nd ed. London: Routledge.

Carammia, M., S. Princen, and A. Timmermans. 2016. From Summitry to EU Government: An Agenda Formation Perspective on the European Council. *Journal of Common Market Studies* 54 (4): 809–825.

Coleman, S., and J.G. Blumler. 2009. *The Internet and Democratic Citizenship: Theory, Practice and Policy.* Communication, Society and Politics. Cambridge: Cambridge University Press.

Coleman, S., and G. Moss. 2012. Under Construction: The Field of Online Deliberation Research. *Journal of Information Technology & Politics* 9 (1): 1–15.

Corcoran, P.E. 1983. *Before Marx: Socialism and Communism in France, 1830–48.* London: Macmillan.

Dahl, R.A. 1995. Justifying Democracy. *Society* 32 (3): 386–392.

Dahl, R.A. 2005. What Political Institutions Does Large-Scale Democracy Require? *Political Science Quarterly* 120 (2): 187–197.

Dahl, R.A., and R.E. Tufte. 1973. *Size and Democracy.* The Politics of the Smaller European Democracies. Stanford, CA: Stanford University Press.

Dahlberg, L., and E. Siapera (eds.). 2007. *Radical Democracy and the Internet: Interrogating Theory and Practice.* Basingstoke: Palgrave Macmillan.

Dryzek, J.S., B. Honig, and A. Phillips (eds.). 2008. *The Oxford Handbook of Political Theory.* Oxford: Oxford University Press.

Dunn, J. 2005. *Setting the People Free: The Story of Democracy.* London: Atlantic.

Dunn, J., and I. Harris (eds.). 1997. *Aristotle Volume II.* Great Political Thinkers, 2. Cheltenham, UK and Lyme, NH: Edward Elgar.

Easton, D. 1965a. *A Systems Analysis of Political Life.* New York: Wiley.

Easton, D. 1965b. *A Framework for Political Analysis.* Prentice-Hall Contemporary Political Theory Series. Englewood Cliffs, NJ: Prentice-Hall.

Elliott, W.Y., and N.A. McDonald. 1965. *Western Political Heritage.* Englewood Cliffs, NJ: Prentice-Hall.

Gallie, W.B. 1955. Essentially Contested Concepts. *Proceedings of the Aristotelian Society* 56 (1955–1956): 167–198.

Galston, W.A. 1991. *Liberal Purposes: Goods, Virtues, and Diversity in the Liberal State.* Cambridge Studies in Philosophy and Public Policy. Cambridge and New York: Cambridge University Press.

Garnsey, P., and R.I. Winton. 1997. *Political Theory*, vol. II, ed. J. Dunn and I. Harris. Cheltenham, UK and Lyme, NH: Edward Elgar.

Grönlund, Å. 2001. Democracy in an It-Framed Society. *Communications of the ACM* 44 (1): 22–26.

Gutmann, A. 1980. *Liberal Equality.* Cambridge and New York: Cambridge University Press.

Hague, B.N., and B. Loader. 1999. *Digital Democracy: Discourse and Decision Making in the Information Age.* London and New York: Routledge.

Hall, J.M. 2013. The Rise of State Action in the Archaic Age. In *A Companion to Ancient Greek Government*, ed. H. Beck. Chichester, West Sussex: Wiley.

Held, D. 2006. *Models of Democracy*, 3rd ed. Stanford, CA: Stanford University Press.

Hilbert, M. 2009. The Maturing Concept of E-Democracy: From E-Voting and Online Consultations to Democratic Value Out of Jumbled Online Chatter. *Journal of Information Technology & Politics* 6 (2): 87–110.

Hindess, B. 2002. Deficit by Design. *Australian Journal of Public Administration* 61 (1): 30–38.

Hoff, J., I. Horrocks, and P. Tops. 2003. New Technology and the 'Crises' of Democracy. In *Democratic Governance and New Technology: Technologically Mediated Innovations in Political Practice in Western Europe*, ed. J. Hoff, I. Horrocks, and P. Tops. London and New York: Routledge.

Huxley, G. 1997. On Aristotle's Best State. In *Aristotle Volume II*, ed. J. Dunn and I. Harris. Cheltenham, UK and Lyme, NH: Edward Elgar.

Kant, I. 1991. *Kant: Political Writings*, 2nd ed. Cambridge Texts in the History of Political Thought. Cambridge, UK and New York: Cambridge University Press.

Keane, J. 2010. *The Life and Death of Democracy*. New York: Simon & Schuster.

Keman, H. 2014. Democratic Performance of Parties and Legitimacy in Europe. *West European Politics* 37 (2): 309–330.

Lee, C., K. Chang, and F.S. Berry. 2011. Testing the Development and Diffusion of E-Government and E-Democracy: A Global Perspective. *Public Administration Review* 71 (3): 444–454.

Levi, M., and L. Stoker. 2000. Political Trust and Trustworthiness. *Annual Review of Political Science* 3 (1): 475–507.

Locke, J. 1960. *Locke's Two Treatises of Government/A Critical Edition with an Introduction and Apparatus Criticus by Peter Laslett*, Rev ed. New York, NY: Cambridge University Press.

Manin, B. 1997. *The Principles of Representative Government*. Themes in the Social Sciences. Cambridge and New York: Cambridge University Press.

Meier, A. 2012. *EDemocracy & EGovernment: Stages of a Democratic Knowledge Society*. Berlin: Springer.

Mennell, S., and J. Stone (eds.). 1980. *Alexis de Tocqueville on Democracy, Revolution, and Society: Selected Writings*. Heritage of Sociology. Chicago: University of Chicago Press.

Morgan, G. 2016. New Actors and Old Solidarities: Institutional Change and Inequality Under a Neo-Liberal International Order. *Socio-Economic Review* 14: 201–225.

Moss, G., and S. Coleman. 2014. Deliberative Manoeuvres in the Digital Darkness: E-Democracy Policy in the UK. *The British Journal of Politics & International Relations* 16 (3): 410–427.

Mulder, B., and M. Hartog. 2013. Applied E-Democracy. In *Proceedings of the International Conference of E-Democracy and Open Government*, ed. P. Parycek and N. Edelmann. Krems an der Donau, Austria: Edition Donau-Universitat Krems.

Norris, P. 1997. Representation and the Democratic Deficit. *European Journal of Political Research* 32: 273–282.

Norris, P. 2004. *Electoral Engineering: Voting Rules and Political Behavior*. Cambridge, UK: Cambridge University Press.

Norris, P. 2011. *Democratic Deficit: Critical Citizens Revisited*. New York: Cambridge University Press.

Norris, P. 2012. The Democratic Deficit Canada and the United States in Comparative Perspective. In *Imperfect Democracies: The Democratic Deficit in Canada and the United States*, ed. R. Simeon and P.T. Lenard. Vancouver and Toronto: University of British Columbia Press.

OECD. 2003. *Promise and Problems of E-Democracy-Challenges of Online Citizen Engagement*. Paris, France: OECD.

Päivärinta, T., and Ø. Sæbø. 2006. Models of E-Democracy. *Communications of the Association for Information Systems* 17 (37): 1–42.

Pennock, J.R. 1978. *Liberal Democracy: Its Merits and Prospects*. Westport, CT: Greenwood Press.

Pennock, J.R., and J.W. Chapman. 1977. *Human Nature in Politics*. Nomos. New York: New York University Press.

Qvortrup, M. 2007. *The Politics of Participation: From Athens to E-Democracy*. Manchester: Manchester University Press.

Rios Insua, D., and S. French. 2010. *E-Democracy*, vol. 5. Dordrecht, Heidelberg, London, and New York: Springer.

Schumpeter, J.A. 2010. *Capitalism, Socialism and Democracy*, 1st ed. Florence: Taylor & Francis.

Shaw, T. 2008. Max Weber on Democracy: Can the People Have Political Power in Modern States? *Constellations: An International Journal of Critical & Democratic Theory* 15: 33–45.

Skinner, Q. 2012. On the Liberty of the Ancients and the Moderns: A Reply to My Critics. *Journal of the History of Ideas* 73 (1): 127–146.

Wallas, G. 1962. *Human Nature in Politics*. Lincoln: University Nebraska Press.

Ward, D. 2002. *European Union Democratic Deficit and the Public Sphere*. Amsterdam, The Netherlands: IOS Press.

White, S., and N. Nevitte. 2012. Citizens Expectations and Democratic Performance the Sources and Consequences of Democratic Deficits from the Bottom Up. In *Imperfect Democracies the Democratic Deficit in Canada and the United States*, ed. R. Simeon and P.T. Lenard. Vancouver and Toronto: University of British Columbia Press.

Wright, E.O. 1994. Political Power, Democracy, and Coupon Socialism. *Politics & Society* 22: 535.

Zweifel, T.D. 2002. *Democratic Deficit? Institutions and Regulation in the European Union, Switzerland, and the United States*. Lanham, MD: Lexington Books.

CHAPTER 3

E-Democracy

1 Introduction

Proponents of e-democracy suggest that it will empower citizens politically and allow them to participate directly in government decision making (Rios Insua and French 2010; Mulder and Hartog 2013; Grönlund 2001; Dahlberg and Siapera 2007). Despite this optimism, it remains the case that e-democracy experiments have yet to produce an operational digital democracy that empowers citizens in decision making (Päivärinta and Sæbø 2006). To understand democracy, Dunn (2005: 39) mentions 'we need to think our way past a mass of history and block our ears to many pressing opportunities'. Others explain that democracy is not a singularity event that will happen, it is a promise that 'there is an engagement with regards to democracy' and that democracy will be realised in the future (Mouffe 1996: 83). Digitising this promise presents some challenges to create the democratic engagement processes, as for even the most proficient programmers this is complex as the political norms can change significantly between (and within) societies. The digitisation of existing democratic processes like voting in representative systems may not deliver democratic outcomes. The OECD reports that significant future work remains, as barriers continue to exist to engage citizens in policy making (OECD 2003: 24). The barriers are also contextual as each government and its citizens have different cultures, organisational histories and constitutional developments (OECD 2003: 9).

S. Bungsraz, *Operationalising e-Democracy through a System Engineering Approach in Mauritius and Australia*,
https://doi.org/10.1007/978-981-15-1777-8_3

This chapter provides an overview of e-democracy and attempts made to implement it. The chapter begins with a review of literatures about how scholars have conceptualised the term 'e-democracy'. The discussion then turns to an examination of where e-democracy (or variations on the theme of e-democracy) has been attempted. The aim is twofold. In the first place, the discussion aims to identify what was tried, what outcomes occurred and what these instances might have had in common. The second aim is to synthesise key issues to provide an insight into what might be the practical barriers resisting implementation of e-democracy.

The ICTs change behaviours by creating new possibilities, amongst them a form of democracy given the name e-democracy. Increasingly, citizens expect the political system to adopt the digital technologies to understand and meet the peoples' needs. Government systems may have to respond to this expectation arising from the people. E-democracy is a term that addresses this promise through the use of ICTs. A significant argument has emerged since the mid-1990s around various ideas to develop e-democracy. Supra-nationals such as the World Bank, the OECD and the EU have been active in supporting the very idea suggesting that better government can result from digitisation of political system (Grönlund and Horan 2004: 719). For scholars, ICTs offer new design opportunities (Hoff et al. 2003), to enable opportunities towards developing a radical form of democracy (Dahlberg and Siapera 2007). Others see with the Information Age, the advent of a new era, as new forms of government are required through the capabilities that are being created (Hague and Loader 1999). However, some remain unconvinced; they treat the idea as mythical arguing that the current system's gatekeepers have adapted ICTs' resources in new ways that do not democratise (Hindman 2009). The access to Internet, broadband availability and online security according to OECD (2010a), constrain an e-democracy to be practised (OECD 2003). Given e-democracy's potential to improve representative democracy, the chapter explores the ongoing issues of e-democracy in terms of its implementation and challenges. E-democracy is a political system where ICT technology is itself designed to allow control to shift to citizens or the people for policy decision making.

2 The Overview of E-Democracy

The advent of ICTs introduced a new informational capability that creates opportunities for innovation and the potential for the improvement in government (Wilhelm 2000; Scholl 2013; Rios Insua and French 2010; Parycek and Edelmann 2013; OECD 2003; Mulder and Hartog 2013; Hague and Loader 1999; Hoff et al. 2003; Hilbert 2009; Grönlund 2004; Dahlberg and Siapera 2007; Boyd 2008). Internet strengthened the concept of global village by connecting peoples over the nation-state boundaries. This communicative capability and its reach made scholars realise that there was a new potential to improve the current form of representative democracy. Some even saw from the use of the Internet by every citizen 'a paradigm shift' in the process of democracy (Hague and Loader 1999: 3). The mid-1990s saw the emergence of various concepts of digital democracy or e-democracy as an attempt to replace the representative model with direct citizens' participation (input) (Grönlund 2001: 1).

Discussions on how digitally (ICTs) enabled governments might be made possible abound; the scholarly literature grew from a few in 2000 to about 6000 by mid-2013 (Scholl 2013: 2). Despite this prolific literature, e-democracy has not created a practical government system where the people rule, leading some to question if e-democracy is a flash in the pan (Scholl 2006). In addition, government efforts to implement an e-democracy seemingly lag the technological capacity (Moss and Coleman 2014). E-democracy is commonly defined as referring to the use of 'ICT in political debates and decision … making processes, complementing or contrasting traditional means of communication, such as face-to-face interaction or one-way mass media' (Päivärinta and Sæbø 2006: 3). The idea is that ICTs create the potential for the engagement of every citizen. The assumption is that in this way, ICTs allow for a better system of government because it involves more of the voting public in the communication of political ideas and debates. For Päivärinta and Sæbø, it is about the communicative potential of ICTs. On the other hand, Meier (2012: 3) defines e-democracy a little more narrowly as 'the support and enhancement of civil rights and duties in the information and knowledge society'. This partially agrees with the view of Päivärinta and Sæbø but here the emphasis is placed on the idea that enabling citizens to make informed choices through greater communication is part of upholding their civil rights and their civic duties as citizens in electing representatives. For Meier, this is what is central to and for an information—and knowledge-based society, and it centres on making it possible for citizens to elect better representatives. These are two different approaches. The first definition explores the communicative potential of the ICTs capabilities towards a participative and deliberative democracy involving the people in an Athenian-type democracy. The second assumes information is transformed into knowledge to empower citizens for participating in decision making by selecting appropriate representatives to form government. The range of concepts for potential use of the ICTs in an e-democracy varies between these two forms, direct democracy versus representative democracy, which we explore further below.

The first (an Athenian type) approach looks at direct inputs from citizens using ICTs to engage in governing. The second (an Aristotelian) approach is an outsourcing of the governing to the best candidate who is selected using ICTs at an election event. Both have merits. And these are by no means the only proposals nor the only ways to apply digital technology to create an e-democracy. They do however flag two different ways one could conceptualise and apply ICTs to develop an e-democracy, though these examples are by no means exhaustive. Other definitions include digital democracy, electronic democracy, and teledemocracy, which seems synonymous with the use of the ICTs to improve the engagement of politicians with citizens through information for democratic processes like polling, voting and discussion (Grönlund 2001: 1). The ICTs communicative capability through these or similar terms generated a prolific scholarship that explores ways to improve the representative democratic system experience for citizens' and politicians' engagement in governing.

The term 'e-democracy' is not the only term that could be used to describe the adoption of ICTs to improve democratic practices. There are many contenders. For example, some prefer the term 'digital democracy' (Hague and Loader 1999: 3). They argue digital democracy is a term to explore designs for a strong democracy to emerge from the relationship between technology (ICTs) and society. In their view, the design of digital democracy requires a paradigm shift because the Information Age creates ever new technological relationships in society. These relationships can be used to re-invigorate democracy and its processes (Hague and Loader 1999: 4). The existing technology drives change to improve existing political processes.

Others such as Hoff et al. (2003: 7) prefer a broader term, electronic democracy, to allow them to explore beyond the technology to discuss the direct democracy potential that the ICTs offer for what they term as a 'cyberdemocratic' model for an Information Age. The terminological stance is to differentiate based on application of the technology, like ICTs that allows innovations to other electronic forms. ICTs, they argue, can reinforce in various ways different democratic values; for example, it can reinforce the existing 'electoral chain of command or the constitutional model' (Hoff et al. 2003: 5). They suggest four models to describe how an electronic democracy might be envisaged. These are: (1) a consumer model, (2) a demo-elitist model, (3) a neo-republican model, and (4) a radical version or cyberdemocratic model (Hoff et al. 2003: 186). These may seem speculative as they assume that political parties and politicians are embracing the technology in new ways to engage with the public. They argue that the particular form of ICT engagement defines the model that will prevail. They note that in practice a mix between these models tends to occur as the models tend to overlap with each other. Nonetheless, in each model there is a specific relationship between ICTs and the political system, a role which they claim it is crucial to investigate (Hoff et al. 2003: 189).

An e-democracy (or even its derivatives) has not succeeded in replacing the hegemonic liberal representative democracy in Western countries or in other parts of the world (Dahlberg and Siapera 2007; Hague and Loader 1999; Hoff et al. 2003). The working marriage between liberal democratic forms and the market (as noted in the previous chapter) remains as solid as ever, leaving some scholars to suggest that in contemporary times under neo-liberalism the market is moving globally, unfettered and out of control (Morgan 2016). As already mentioned, supra-nationals, like the UN and OECD, also recommend the use of digital technology capability as they suggest it may lead to good government (Grönlund and Horan 2004), but for better government they do not offer detailed guidance. There have been changes suggested to some representative processes. For example, digital voting has been initiated by some countries in the form of e-voting. However, the OECD reports that the e-engagement or the effective use of the ICTs to engage the citizens in policy making in practice requires improvement (OECD 2003: 32). E-engagement, according to the OECD, requires governments to adopt ICTs to provide information, consult with citizens, and allow citizens' participation in decision making. These e-engagements apparently are inadequate in most OECD countries (OECD 2003). In a political context, governments' application of ICTs to democratise is seemingly lacking in the public sphere.

In a society that is informed through the public sphere, some suggest that ICTs could facilitate citizens' activities towards developing consensus (Hoff et al. 2003: 7). With ICTs consensual ideas generated through the public sphere to inform government decision making are culturally diverse and potentially more open to pluralism. An implication is that involving the people in this public sphere is important to democratisation. Scholars claim ICTs enable a more democratic process. Apparently, the technology can make 'quantum leaps' to improve the political system with novel ways of engaging so the public has a say (Hague and Loader 1999: 3). There appears to be a case for change as embracing these new ideas potentially addresses the shortcoming in representative democracy's performance. Democratic deficit, highlighted earlier in Chapter 2, is a problem with representative democracy. Democratic performance when it is addressed by e-democracy re-engages the people after 2000 years to operationalise traditional democracy in a modern context.

The OECD mentions that leadership and political will are important to pursuing improvements to democracy using ICTs (OECD 2003). It suggests that exploiting this particular capability of ICTs can be done for both economic and political reasons, but progress seems lacking in the political direction. Despite the OECD recommendation for a commitment from political leaders and senior public servants to develop an e-democracy, that commitment is seemingly missing. Several cases of experimentations to digitise processes like voting, policy formulation and discussion were studied by the OECD, and it found that no OECD country has a systematic way to conduct public online engagement in policy making (OECD 2003: 90). To apply the capability of the ICTs, it suggests a necessary precondition is leadership so that a framework emerges. In fact, the OECD notes that significant work remains to be done to build an effective partnership between politicians and the people. To improve the partnership building, it recommends the use of traditional offline methods, meaning face-to-face interaction. The OECD report suggests that the online engagement is to be used to enhance the face-to-face interactions (OECD 2003). This would mean that for an e-democracy engagement a totally digital process is not yet an option as some elements of the face-to-face interactions ostensibly remain necessary.

ICTs are opening the potential for creative and innovative forms of government, and it is generating strong scholarly interest for digital government (Scholl 2013; Grönlund 2008). E-democracy is of such significance that both supranational bodies and scholars recommend that states explore digitisation capabilities for better government (van der Hof and Groothuis 2011; Ubaldi 2013; Scholl 2013; Rodríguez Bolívar et al. 2010; Rios Insua and French 2010; Reddick 2005; Parycek et al. 2013, 2012; OECD 2003, 2010b; Lanvin et al. 2010; Hilbert 2009; Grönlund and Horan 2004; Grönlund 2004, 2008; Goos et al. 2002; Finger and Sultana 2012; Coleman and Moss 2012; Burke 2012; Boyd 2008). To study e-democracy initiatives, the OECD formed an Expert Group recruited from twelve OECD countries (OECD 2003). They collaborated to study the government attempts that used digital technology for an e-democracy (OECD 2003). The report found that the online engagement of citizens, for example, must ensure that people's inputs will be used for policy making. The OECD report argued that there is a legitimate expectation from the public for improvement to the old practices of policy making else it will result in 'widespread' public disillusionment (OECD 2003: 90). We look further into the ICTs to provide some further insights to contextualise these claims for decision-making improvement through changes to democratic performance.

Even as early as 1997 Barber noted that technology brings a new imperative, a level of speed to which representative democracy may not be able to adapt. He argues that '[p]olitical reasoning is complex and nuanced, dialectical rather than digitally oppositional' with a simple yes and no (Barber 1997: 209). The technology has potential but in a society bent on privatisation, the commercialisation of the Internet becomes a major influence on how the new technologies will be used. Thus, the introduction of ICTs into politics in those circumstances will reproduce the same defects, especially if the political will to reform is absent (Barber 1997: 212). While things stagnate in politics despite a potential for reform, the technology on the other hand is evolving and integrating itself further into society, but as a commercial tool that may be reflective of the market's requirements rather than those of a democratic political system.

The Internet of Things (IoT) is a new term used by Greengard (2015) to describe new digital capabilities or applications that can be hosted online as Internet technology becomes part of mainstream society. There is a significant body of work being conducted in the field of communication engineering to deliver this IoT platform (OECD and Gaël 2016).

Policy makers from various countries explore this new IoT capability to ensure that the policies are in place to enable this virtual platform to be used to improve both the private and public sector processes. For example, it is predicted that a 1% improvement using IoT would result in a USD 2 billion savings in the commercial airline sector (OECD and Gaël 2016: 11). IoT is predicted to provide significant improvement in performance for both the economic and political sectors of society. New forms of political participation hitherto impossible without ICTs can provide the momentum for the e-democracy concept to innovate using the IoT platform. However, the IoT is driven by commercial imperatives like e-commerce and e-business and although it may also host democratic processes these democratic processes must be created first. A further concern is how the digitisation might shape systems of government and initial digital form given democracy is understood in many ways and it can therefore be digitised in various ways. Digitisation may not democratise as intended if the design is aimed at furthering state or market control rather than empowering the people.

Some scholars attribute the lack of progress in pursuing the introduction of ICTs as being due to an absence of government will to provide a coherent e-democracy policy (Moss and Coleman 2014). Developing that government will may take time. However, the fact that the Internet is providing ever-changing possibilities in the public and commercial spheres may provide extra pressure on governments to at least to be seen to be catching up. In addition, this pressure may well create the public will to demand that their governments do more to catch up. Having said that, it is as well to acknowledge that some scholars also warn that the local and national government initiatives currently hold positions that seem to suggest that their views have been captured by the commercial mindset of the private sector (Dahlberg and Siapera 2007: 5). They argue that without radical change to a deliberative type regime where the twin values of 'liberty and equality' for ruling and being ruled for all are pursued the potential of the Internet for creating a public will may be wasted (Dahlberg and Siapera 2007: 7–8). There seems to be a need for ensuring that the initiatives from the state in digitisation are directed to a traditional form of democracy.

Some authors though are wary about the promises of technology alluding that the political aspects of the technology must be understood (Oates et al. 2006), others though depict pessimism (Norris and Reddick 2013). The way politicians respond to the technology matters as do their decision making about it. Some technologies tend to define an era like the Stone Age, today the Information Age, and in future the knowledge era. Technology brings diachronic and synchronic change as the term describing it evolves. Technology relationships from the social and technical interface are in some disciplines reductionist, archaeology and anthropology, for example, tend to impose terms that reduces society in a given era to its use of technology, for example, Bronze Age where the material was used for tools and weapons. The shaping of society from technology gave rise to technology determinism (Wyatt 2007). As e-democracy is linked to technology like the ICTs, one would assume that e-democracy from its use of the ICTs technology would shape society, this aspect requires some clarification.

Two schools of thought emerge about technology; one assumes technology is developed to meet a need in society while another adheres to the view that technology determines the type of society created around technology. The opponents to technological determinism posit that society's needs (problem) are met by creating the appropriate technology (solution), for example, to meet a need identified by society SE develops a design to deliver a solution to the specified problem from society, an e-democracy in this case. Now consider, for example, drugs or Medical Resonance Imaging (MRI) technology that are developed to treat a medical condition, it would be ludicrous to profess that the medical condition that the drug treats is created to meet the requirement for the use of the drug. A view that technology controls society is disputed as with SE, technology is developed and applied to make the life of people better, a service to meet society's demand. As discussed in the book, a SE improvement to people's lives from technology meets a need defined through the specifications from the people, in this sense technology created by SE is designed to meet a need in society, an e-democracy system in the book. Over time, the e-democracy (solution) changes through SE to fit the changing need of society which defines the e-democracy specifications. This differentiates SE from the technological determinism assumption of a society that becomes captive to changes beyond its control from the technology. SE in the book shifts control to society as it empowers it. SE processes lead towards a *demos* construct where technology is designed for use by the *demos* to provide it with e-democracy, a self-rule service to society in the form of a type of direct democracy using technology.

We explore below if past e-democracy experiments exhibit a tendency to reproduce past practices (thereby reproducing the problem they claim to be solving) or if they have succeeded in moving towards creating the democratic community or demos. The following explores some of the e-democracy attempts mentioned in the literature.

3 E-Democracy Moments

The rationale for reconfiguring an existing representative democracy into an e-democracy through the use of ICTs is to enable citizens to participate directly in decision making and hence is intended to empower the people in decision making. However, this is yet to occur though some experiments with technology (towards an e-democracy) have taken place, as reported by the OECD for Australia (OECD 2003). Some authors have explored technology for deliberation and participation in the form of Internet politics for an e-democracy (Chadwick 2006: 83). The key aspect to creating an e-democracy is that the ICTs should facilitate better deliberation and participation for all citizens (Rios Insua and French 2010: 1). For some, these ICTs capabilities are not being used to create an e-democracy (Moss and Coleman 2014). Decision making by citizens shifts the power to citizens and seemingly there was an attempt to use the technology's capability in the case of the experiment 'Today I Decide' that occurred in Estonia (Coleman and Blumler 2009: 90). Such initiatives seem rare though, as e-deliberation and e-participation as key aspects of an e-democracy is not a common norm in the OECD countries (OECD 2003).

To join the EU, the Estonian government had to improve multiparty participation following the dismantling of its communist regime. It had to show that liberal democracy could work in Estonia and thus there was an attempt to use ICTs to show its democratic credentials for a representative democracy (Pruulmann-Vengerfeldt 2007: 171). With the advent of the Internet in 1992, Estonian academics seemingly put computerisation on the political agenda. ICTs were seen as one of the tools that would enable the country to move from a one-party system to the dominant model of liberal democracy in which multiparty competition is the norm (Pruulmann-Vengerfeldt 2007: 171–172). The application of ICTs seeking citizen's inputs was a means to increase the plurality of the party system within a competitive representative democracy. However, it did not increase the power of the individual to represent themselves like in an ancient Athenian democracy. In effect, it was only a transition to a more plural system where more parties could participate and compete in the electoral process.

The OECD reports that only a very few OECD countries have started to experiment with online tools for core democratic processes of deliberation and participation and much of the interaction in such cases has been left to the participating citizens (OECD 2003: 16). The use of electronic discussion boards to raise electronic petitions or e-petitions is noted in the case of the Scottish Parliament (OECD 2003: 56). The people could raise issues for tabling in the Parliament, but the issues were filtered by the Public Petition Committee, which also provided feedback to the petitioners. This petitioning system already existed in many countries, but it was digitised by the Scots (OECD 2003). Petitions, as a means to influence the Parliamentary Agenda, were to provide active participation or e-Participation through electronic means. There has also been the use of electronic voting or e-voting experimentation by the UK government (OECD 2003: 32). However, the OECD noted that policy inputs were not yet structured in a decision-making process. There was no evidence mentioned or any suggestion that any form of a direct democracy was being proposed in the e-democracy experiments reviewed by the OECD (2003). Rather, ICTs were used to render existing processes to become more efficient in collecting citizen's inputs, but these were not used to enable those citizens to become involved in actual decision making.

Now, e-petition and e-voting are existing processes used in the representative system. E-voting processes for elections are where digital technology has made some inroads into improving their operation, but as mentioned earlier voting (or petitions for that matter), is a blunt, indirect decision-making instrument. Voting at every electoral cycle only delegates power to a representative based on their perceived performance, and even then, in an entrenched two- or three-party system, it is not clear that actual performance necessarily features in the calculations of voters. Citizens' delegation for decision making without accountability and transparency still creates a trust issue with decisions to select someone at elections. The Smartvote case study is a significant approach for use of technology in the direction for improving accountability, and it is covered in more detail below. Now, it seems that with digitisation there remains a tacit resistance to moving towards implementing a more direct form of democracy in favour of promoting or improving the liberal version of representative democracy. In Australia, according to Australian Electoral Surveys (AES), the people (citizens) are averse to using technology aimed at improving the voting process by using digital forms (e-voting) to make voting easier (Smith 2016). The paradigm shift from a paper-based system requires trust which is seen as problematic for electronic voting technology delivered through the use of Smartphones. However, countries like India have successfully transitioned to e-voting. Smith's study of the AES e-voting surveys is discussed further below. It is sufficient to note here, for now, that Australia's version of representative democracy is accountable only at election cycles when the representative can be re-elected or removed depending on how their performance has been perceived by voters. If the system in place (voting for representatives) is inadequate, then digitisation tends to amplify the existing inadequacies, such as a lack of trust.

The crucial point here is that voting is the only time when citizens decide, but they only decide who will represent them. That in itself is problematic as for Birch representation in politics raises more questions than it answers. Birch contends that, with the possible exception of the US political system, representative electoral systems do not require that people's specific needs should drive the representative's mandate (Birch 1972). A representative democracy allows the representative to follow their own interests once elected. An e-democracy promises to enable a people's mandate to be developed through a degree of direct democracy that has the prospect of improving the accountability of the elected to those who elected them.

As Chapter 2 highlighted, some of the challenges democracy has faced in practice in the past continue in the digital era. As was also noted there, democracy is contextual as it is based on values allocation (norms) within the society (*polity*) it operates in. These democratic values allocation in each *polity* is problematic for actualising an e-democracy that democratises as democracy varies within each nation-state due to being constrained through the ways they are pursued by the elected representatives and their parties. This variance in the quality aspect of democracy was flagged during studies of the democratic deficit (Beetham 2012: 76), and Beetham links the degree of heterogeneity in democratic regimes to democratic deficit. He suggests that the democratic performance be linked to standards, as he adds that if the original EU members were to seek admission against the EU's democratic standards these original members would fail, 'it would be refused entry' (Beetham 2012: 77). Standards would provide a means to tackle what democracy is or must be, a point that is taken up later in the book.

The seemingly stringent EU standards were applied during EU membership application. It is interesting to note that when these standards were applied to existing EU member countries it revealed that they would themselves fall short of those standards of democratic good practice. Now these good practices may seem appropriate and well-intended as it is what new members had to comply with when they sought membership. These standards could be used to improve the institutional processes to become efficient for those who fall short. However, there is seemingly no need to fix these deficiencies for existing EU members. The existing paradigm assumes that the existing system of representative democracy is not part of the problem. Yet given the opinion surveys on the current standing of democracy and the OECD's upholding of particular democratic standards, there seems to be a clear disconnection (Beetham 2012). According to Beetham, the opinion surveys measurement fundamentally fails to seek improvements to empower the public to have an input in shaping outcomes like state's decision making. Despite the democratic standards not being met, the democracy opinion surveys show there are no issues with democracy (Beetham 2012).

A similar discrepancy between policy guidance and standards implementation will be flagged later below in the discussion of the Smartvote case study. Perhaps, what they do show is how different measures yield different results when assessing democracy. However, what stands out is that the public still has a lack of input into, let alone control over, the state's decision making. In this context, the digitisation for the state's legacy processes continues the variance of what the politician assumes the public needs and what the public requires. The identification of a democratic deficit highlights the variance but cannot get it fixed without the political will (Beetham 2012). As with the EU standards for example, democracy depends on those in power to put in place the steps for reform, but this is something that these representatives are not obliged to act upon given they are using what the system allows and has allowed for decades. Representatives can select the opinion polls survey to claim that nothing is wrong with the system, rather than explain the standards which show the system's failings.

Apparently, there is a tacit understanding by those like the opposition to keep quiet about such issues. So even if voting becomes digitised the fact that the public perceive that their vote does not count will not incentivise people to participate until their voices matter (Beetham 2012). For the OECD through its study of twelve countries, it identified two core issues recommended from digitisation, e-engagement and e-voting, which we discuss further below. Another comprehensive approach to the use of ICTs is Meier's e-democracy concept from the University of Fribourg, Switzerland, which will be treated separately below (Meier 2012), but first we discuss the ICTs use in the twelve countries reviewed by the OECD.

The OECD gathered a national team of experts and studied e-democracy in twelve member countries namely Australia, Canada, Czech Republic, Finland, Germany, Italy, Mexico, Netherlands, New Zealand, Slovak Republic, Sweden, and the UK (OECD 2003). The goal in these twelve OECD countries was to include people in decision making or policy formulation through ICTs. These experts concluded that two core aspects of the e-democracy experiments that emerged were about e-engagement and e-voting (OECD 2003: 32). These findings are discussed below. The team of experts explained that e-engagement was about creating the mechanisms to allow public participation in policy making when using ICTs. E-engagement was understood to provide the means to explore a relationship between government and citizens in processes using digital technology that is seemingly created for informing, consulting and actively participating. In these virtual relations, it is assumed that engagement using ICTs or e-engagement is to improve policy formulation; however, the final decision making still rests with the government. The aim of the e-engagement process is a shared outcome involving the creation of better quality policy, an improvement in trust, and greater acceptability of the policy and the policy making (OECD 2003: 33). It is a form of digital cooperation between citizens, politicians and senior bureaucrats to make policy. A further aim was that of eliminating corruption as citizens can supervise government's policy implementation through e-engagement and thereby strengthen representative democracy. In the digital engagement process, the OECD claims a relationship emerges where government delivery of policies can be monitored through the people's supervision of the government policy-making process. The assumption is through e-engagement, the ICTs makes feasible both closer engagement and also a continuous improvement in government and citizens' relationship.

Now the policy-making process for better policy making has to be built first, and furthermore when built, it must both be understood and used by all stakeholders. The OECD suggests that communities be built that breaks up the large number of citizens into discrete groups (OECD 2003). These groups are then aggregated through online tools forming online communities that can engage with government to be discussed below. It explains how an eCommunity is built around ICTs. In the context of the Smartvote software eCommunity and assumptions about an information era and a future knowledge era allows an eDemocracy to emerge (Meier 2012). For now, the issue is that these digital relationships are within a paradigm of a government leading the debate or setting the agenda. The OECD anticipated that there will be a learning period for both government and the citizens to create the tools for policy making. For a process that sets the agenda, the report suggests a life cycle approach to policy making (OECD 2003: 88).

A life cycle for policy-making process is where the agenda is set, items on the agenda are analysed to formulate policy which are then implemented and monitored. The monitoring process provides the feedback that informs the reformulation of the policy until the outcome is the one that citizens are satisfied with, and the feedback from the last step forms a closed loop for a policy life cycle process (OECD 2003: 34). The OECD discussion gives the example of the Interactive Policy Making (IPM) experiment applied on 3 April 2001 by the European Commission to gauge the marketplace reactions for the EU policy-making process through an IPM project (OECD 2003: 37). This project spawned nine experiments that were intended to support the e-Commission initiative which was to modernise the institutions that created EU regulations with the inputs sought from citizens, consumers and businesses. The initiative was through a web page called 'Your Voice in Europe' (OECD 2003: 123). The intent was transparent policy making for the EU market.

The initiative was created because there was a need to know what people thought about new EU policy ideas. The EU Commission needed reactions to the policy proposals which they had already put into place and the Internet was seen as an appropriate means to help them to gather these views from EU citizens and businesses to make the policy more effective. The IPM was a digital front end for people to provide their views to analysts about policy making that would shape the Green Papers informing evaluations of existing policy. This web design supplemented the existing forms of communication like fax, emails and translation from one language into another given the EU was a multi-national entity. This initiative was the precursor of the current web page that is reported as work in progress and it was also seeking feedback for improvement, the web interface provides information and communicates the Commission's political priorities to EU members (European Commission 2018b).

To create the e-engagement systems, the design took into consideration two core requirements which were technology and democracy. While the democracy requirements seem to have explored existing democratic practices like voting, the use of the technology was to explore the participatory aspects. The role of the technology was for disseminating information with consultation and participation as guiding principles. To develop the e-engagement process, the OECD studies showed that e-engagement required the three sub-processes informing, consulting and participation. Now it was apparent that to reach and engage was complex and with the diverse audience it was a design challenge. The public e-engagement systems created some issues (OECD 2003: 38). For example, critical issues that emerged included accessibility, usability and security, each of which could clash with the democracy's need to be open and transparent. Another issue was the ease of use of the system to ensure that no person would be left behind, or that a digital divide would not increase or create inequality.

The democratic needs for equality, open and transparent design were complicated for the designers. The ease of use issue also extended to citizens' competency to use the system requiring that it be able to cater for citizens with the lowest technological proficiency. This makes some of the direct access claim for ICTs that is without assistance from a facilitator for inputs or use of the system to become complex to design. This problem of access for those with specific disabilities is discussed further below with one of the case studies from Meier (2012). With digital technology designs, there are trade-offs like anonymity to protect privacy and authentication to validate legitimate users as the digital system must know who the person is prior to allowing access. Privacy creates complexity to digitise for an e-democracy. Assisting people so they can understand complex policies with adequate information that is sufficient to provide clarity for the lowest requirements level is a challenge in the current digital context with web page interface.

To design the engagement systems, some countries like Canada and the Netherlands developed guidelines for online consultation. However, it was not obvious how the citizens' inputs were to be integrated into the government policy-making process (OECD 2003: 36). The problem was to find a mechanism whereby these inputs can be effectively shaped into policy outcomes, but such a mechanism was missing. Though the e-democracy literature seems exuberant at quoting pilot cases at the local and national level, the emphasis is placed on the input aspects of policy making. This front-end e-engagement towards policy making focuses on seeking comments and disseminating information. There was no process that described how the inputs from citizens 'integrated into policy and feedback provided to those who contributed' (OECD 2003: 36). These back-end issues seemed to be assumed as a black box. They are not treated in the literature to explain this key aspect of people making decisions. The focus remained on people providing inputs which are optional for the representative to consider in policy making. This is a failure of an e-democracy to democratise. There have been many pilot experiments conducted to capture information with the ICTs, but then they reflect a black box concept approach where somehow the technology converts these inputs into policy. The task is to explain how these inputs can be translated into requirements that are to drive the policy development. This issue is taken up in relation to the European Union's IPM case study that aimed to make policy making less complex through the use of ICTs.

For the OECD, the IPM was an experiment aimed at improving government using the Internet for collecting and analysing inputs from the marketplace to improve policy making (OECD 2003: 37). In much policy making in a pre-digital age, the role of the public has often been marginalised in favour of the role of the political representative. The IPM process aimed to allow the twenty-six Commission services to use the inputs from public to evaluate existing policies and also through the open online consultation seek inputs for new policies. This project was called an e-Commission initiative towards better policies for the EU. The goal was to produce better regulations for the member states and thereby improve how the EU market was run.

The intent of the Commission was based on the idea that better regulations could be created by participation from stakeholders as it was thought that the inputs would assist the EU central body to create better decisions. With this objective in mind, the EU Commission required a user-friendly system for inputs from the market place, which consisted of both business and citizens. Reactions to a policy proposal were invited from the market. The online inputs were intended for various steps of the policy-making process like agenda-setting, analysis, formulation, implementation and monitoring (OECD 2003: 127). It was noted that infrastructure such as broadband capacity became a constraint for online engagement as the OECD noted that the technology was 'still in the early stages of deployment' (OECD 2010a: 121). Another concern was that for those who had no access like the poor, this digital connectivity made them even less equal than others. Also, the system was meant to be transparent so everyone could understand the EU policy-making process. Though the intent of the project was laudable, in practice a comprehensive and effective system for the EU to develop policy making understood by the public failed to be created as anticipated.

When the process was put in place, it was assumed that people would be willing and able to join in the debate. The attempt with the IPM was within the paradigm of a representative system decision-making process. The technology seemed to have potential but as the OECD noted, 'specific technologies chosen for online engagement vary in their degree of sophistication – most examples feature a dedicated website with email options' and this raised new questions for governments' ability to innovate to engage using digital technology (OECD 2003: 13). The vague policies could not deliver in practice. A problem of synthesising the intent of the policy may also have been a contributing factor as the framework for moving from the need to solution seemed to be missing, making the OECD's observation about a framework mentioned earlier a key issue.

E-democracy experiments using ICT technology promised to improve representative democracy. However, the digital experiments failed to deliver traditional democracy in practice. Democracy is more than voting. Digital proposals from some of the experiments to digitise liberal processes can create efficient functions like e-voting and informed candidate selection, but an electronic liberal system or e-liberal democracy may not necessarily democratise (empower the people). In practice, democracy is a complex concept to digitise in a way that reflects modern needs of the current era. An upgrade to an e-democracy that democratises increases the complexity for what needs to be designed.

Significant progress was reported in the case of Switzerland through a customised digital application Smartvote. We now review the Switzerland polity case where representative democracy existed and there was no pressure to change from a mono-party system to a pluralistic party system as in the case of Estonia's digitisation for entry in 1 May 2004 (European Commission 2018a). The EU digitisation case creates complicated issues as it has to apply the ICTs to policy making in the context of a multicultural system smashed together to form a single market entity. The Smartvote software application is a case study that provides some more insights about ICT use and electronic participation in an existing representative democracy which has a digital policy for democratisation in place (Meier 2012). Exploring some of the ICTs attempts reported by the OECD and Meier (OECD 2003; Meier 2012), there is still significant work required to reach a stage whereby a system e-democracy for people's direct input in decision making emerges. However, the Smartvote application is forward-thinking compared to the OECD experiments noted above as some indirect citizen control is anticipated through provision of information to make better choices for decision making. We discuss this technological customisation to improve the representative democracy towards an e-democracy next.

3.1 Steps Towards a Swiss E-Democracy

The EU, through the Lisbon declaration, committed European heads of state to agree to set up an information society that would innovate to create a more competitive internal market with associated structural reform for an e-government (Meier 2012: 1). This project was to include every citizen and no EU citizen was to be left behind. According to Meier in Switzerland, this was anticipated to develop a change in participation models whereby the relationship between community and government would allow political control to shift to 'the citizens in the long term' (Meier 2012: 3). As a step towards this goal, web access was to be available to everyone. People with disabilities were to be included and there was to be no digital divide, and to ensure its effectiveness yearly measurements were to take place. A study of 50 government portals to assess the progress for barrier-free access in the Kantone area was conducted and it showed that though access improved from 2004, deficiencies remained for the 'Internet portals of federal states, municipalities and semi-public companies' (Meier 2012: 29). These portals apparently created unequal access for Swiss citizens. Though the Internet opened up opportunities for both the elderly or senior citizens and the handicapped citizens, they were reportedly 'confronted over and over again with insuperable obstacles' (Meier 2012: 29). According to the 'Access for all' foundation experts, none of the five biggest cities of Switzerland had portals 'that was sufficiently barrier-free' (Meier 2012: 30). For these experts, the websites surveyed did not meet the standard required to be considered barrier-free. People who were blind, for example, had problems. None of the websites measured by the 'Access for all' met the required standards to allow access for vision-impaired citizens.

This access issue resulted in a lack of participation by this citizen group in political life. The 'Access for all' tests found that the developed standard that was to be put in place was not met by the web designers. Instead, despite official policy, the design of the web even fell short of the international web standard. This is a standard which is the 'starting point for creating an accessible Web site' as laid out in 'the Web Content Accessibility Guidelines (WCAG) 1.0 of the World Wide Web Consortium' (Meier 2012: 30). Now, in Switzerland these WCAG guidelines are legislated, but they are seemingly too broad to implement. In a digital context even though the intent was to design free access in Kantone, Switzerland, a lack of public infrastructure created inequality for accessibility. For the countries making up the EU (European Commission 2018b), barrier-free web access for the EU is reported to be around 3% (Meier 2012: 28). The figure means that for access to the World Wide Web there is a growing disparity for some groups that is emerging, in Switzerland about 15% of the population are impaired (Meier 2012: 28). This access inequality becomes problematic for hosting applications on such platforms for every citizen. This is despite a government policy that is mandated to ensure access.

In an information era, a barrier-free web is crucial to make it possible for intellectually and physically impaired people to be autonomous. In the Kantone case study, although there was a large improvement yet for around 15% of the population in Switzerland, significant work is still required to achieve barrier-free access. To compound the problem, data privacy creates issues for people with disabilities on these platforms and these remain unaddressed. For example, the identification system becomes problematic for visually impaired people. Also, security for digital signatures is a problem (Meier 2012: 153). Significant progress was made in 2008 but more work is required to provide accessibility of the Internet to all citizens. Despite a government and European policy that no citizen will be left behind, the Swiss situation is yet to fully implement the same. With the precondition of access still progressing there is no case of an e-democracy, it is a work in progress. However, it can be considered here how the intent of democracy is being conceptualised.

The Swiss initiative aligns with the EU policy to digitise and to create virtual communities. Smartvote is an election assistance system for voting and its elections process is a long-term goal to shift control to citizens through an e-democracy (Meier 2012). The Smartvote software is an application that is currently used for elections in Switzerland to drive web portals' delivery of information about candidates and to assist voters to select candidates (Meier 2012: 163). It was used for the first time in 2003 by Swiss Federal Election. It is operated by a Non-Government Organisation (NGO) Politools, and the portal was accessed 1.3 million times in 2015, it also had a participation of 85% or about 3300 Swiss candidates that used the software (Smartvote 2018). This application is designed to work in what the author, Prof. Dr. Bruno Jeitziner, describes as a digital market where candidates bid for votes through their electoral programme. The software works on the assumption that a voting market operates and candidates put themselves up to bid for a mandate from the voters. In the context of Switzerland, a highly regionally fragmented electorate, and where people's votes are apparently based on local issues and the Swiss political system, the software claims it does not need to factor in the question of factional teams for decision making or governing (Meier 2012: 163). The software assumes the representative's political affiliation to a party is not a good indicator. The voter's market must use the political profile of the individual candidate as this profile provides transparency about the political position of the candidate on issues.

Smartvote is based on dating software that has been transformed to match citizens to their best candidate. Its goal is to support a democratic process to allow participation with e-voting, which is electronic voting, and e-election which covers the selection process of candidates. The Smartvote software simplifies information by displaying it for citizens' consumption in various ways to inform them about the candidate. Smartvote therefore substitutes itself in terms of informing citizens about who their best choice should be, like a dating match based on the requirements for identifying an ideal partner. There is no guarantee about the match, it is the best guess. Smartvote also claims that representatives themselves do not fully understand the voters' preferences and hence its information can also assist candidates to engage more effectively (Meier 2012). Using questionnaires and date matching algorithms, Smartvote processes the information using similar principles to dating agencies' software to create a match between candidates and voters. Candidates enter their answers to political topics that are saved in a database. Then it is the voters' turn to answer the same questions. Smartvote then calculates the scores to provide a match. Matching recommendations are based on the scores against each of the criteria from questionnaires. These scores can be displayed visually using spiderweb profiles which depict the answers of the voter against those of the candidates. The software can create a visual display where a high value depicts high agreement and a low value depicts low agreement.

The Smartvote software also creates a political profile of candidates that voters can access to consider their choice of candidate to represent them. The information required for this is from discussions (eDiscussion), elections (eElection), and posts (ePost) which can all be accessed by the voters (Meier 2012). By capturing and presenting the data in a form that informs voters about the candidates, these profiles help to inform voters during an election event as they can then compare the candidates to find a match to their requirements. Voters can also generate their ideal candidates and run each candidate through their ideal candidate profile to see the degree of match between candidates. The spiderweb displays the degree of match between candidates to the voter's ideal. This comparison allows the voter to distinguish between candidates to identify the ones who are closer to their ideal. Candidates' lists can also be arranged using charts to display which candidate align most closely with a voter's preferences. This is generated from the answers that have been provided by each candidate to the questionnaire. This software seemingly captures a candidate's political attitude to key issues creating a more transparent system. For example, a topic like gay and lesbian marriage can be entered into the database as a question and each candidate's position on the issue can be displayed on the web page for all voters to see. Smartvote seemingly allows a voter to choose a candidate who is closest to their views of their ideal representative (Meier 2012: 163).

The Smartvote software design is predicated on the assumption that the government would establish systems and infrastructure for an eCommunity. An eCommunity is a community that uses electronic processes to interact through the web for a future e-democracy. The project assumes eCommunities will exist and in that context the use of the Smartvote application is ultimately to lead to a customised or mouse-click representative selection (Meier 2012). It is the formation of an eCommunity that allows an e-democracy to operate. This development raises some issues about the various risks that might emerge. In the first place, in the digital environment vulnerable citizens like those who are mentally ill, physically impaired, or senior citizens are at risk, for example, of identity theft. Second, there is always the possibility for data manipulation and security breaches when using e-voting and e-elections as these problems cannot be fully eliminated (Meier 2012: 161). But there are also other problems for an eCommunity beyond (but related to) these issues.

A key problem is that of secure, electronic identification. The eCommunity project suggests that the issue of electronic identification requires a shift in paradigm because some people are yet to trust a digital system for security. Digital and online systems require new regulations for privacy and protection against data theft. It must also anticipate that other risks emerge as electronic processes become mainstream and better understood. The availability of the Internet itself is an issue as the infrastructure in place (broadband) must be able to support the simultaneous use of the technology by millions. An inadequate infrastructure can lead to new problems like electronic congestion. Granted, e-government projects are works in progress. Despite being reported that each year Europe is making good progress with the Lisbon agreement, the problems of Internet availability, accessibility and security remain to be overcome (Meier 2012: 70).

Digital technology does make the voting and information gathering and its dissemination efficient; however, this is not the same as empowering citizens to rule. The digital technology described above is effectively a means to rule within the existing arrangements of a representative system that itself is based on the privileging of market needs. Such an approach assumes that the market is an appropriate model for developing the digital processes needed to improve representative democracy. The persistence of a democratic deficit in Europe tells us otherwise. There is significant work yet to be done to improve democracy itself in those EU countries given that they would fail their own democratic membership standards as mentioned above (Beetham 2012: 77).

The Smartvote project is yet to deliver an e-democracy in practice. However, it provides a means to think through the various processes that could be digitised to create an efficient government that a customised software like Smartvote enables. E-democracy is a case of ensuring processes that exist are made to be more efficient so that the representative democracy also becomes more efficient. It would seem the key imperative is that the market will shape the democracy that emerges. The system is embedded within the EU framework and that constrains Swiss democratic aspirations. The Lisbon agreement is perhaps not what the people want as can be seen from the UK electorate's decision to exit from the EU despite many representatives' and the government's views to the contrary. Meier points out that it is only through the sub-processes of eVoting and eElection that an effective eCommunity can be formed so that it will have enough information and the digital infrastructures available to strengthen participation in the existing context of a vote market (Meier 2012: 154).

The Smartvote software seemingly solves a problem of information in an electoral market where political programmes are exchanged for votes. It assumes that voters have limited capacity to process information and they are unwilling to do so (Meier 2012: 163). It does not anticipate that voters want to represent themselves like in a direct democracy. The software meets to some degree the Aristotelian view of an ideal representative of the people at an individualised level. However, there are a couple of important problems. One is that the voting assistance system could be gamed to project an image to maximise votes through well-crafted answers to the questionnaires. The objective for Smartvote is to seek votes. A person seeking votes would normally present an ideal profile to convince the largest number of voters. In this sense, it is not necessarily about improving democracy or people's self-rule. The paradigm is still delegation through more information to facilitate the choice of a representative.

Another problem is that it does not allow people to be directly involved in policy making. Elections are a key feature of representative democracy when people decide what is to be delivered in the next cycle. Electoral programmes can make promises, but they may not be delivered post the election event and the elected representative cannot be brought to account until the next election cycle, there is an accountability lag. Also, if electoral programmes are vague, its delivery would be hard to measure during the election cycle. Programmes are proposed by the representative who may have inadequate knowledge of what the people want. The policies from these programmes may fail to address real issues, for example, policies are vague and blunt instruments to implement democracy and its inadequacy was highlighted by both Meier and the fifty portals that failed to meet government policy for 15% of its population.

Voting is a key process in current forms of representative democracy and these are being digitised. Indeed, it is a key marker of the democratic process in general. Some scholars exploring digital technology reflect the idea that voting potentially enables various features of an e-democracy, like electronic voting or e-voting and candidate profiling (Meier 2012). We discuss this aspect for the Smarvote support software for voting operating in a paradigm of a voting market, before moving to the informational aspects and informational risks that digitisation poses without reform.

3.2 *Digitising the Electoral System of Representative Democracy*

The electoral system and democracy as we know depend on the voting process. While digitisation of the process has been to replicate the system making it more efficient with technology through e-voting during an election event, others claim that this is a start of reform to a key feature of democracy where technology enables flexible voting (Alexander 2002). This flexible voting would improve participation they claim, eVoting then allowed others to claim according to Hilbert that ICTs allowed every citizen to vote on every issue (Hilbert 2009). While others like the Smartvote technology claims to generate, according to Meier (2012), the information to assist electorates in their selection process, but as has already been shown, this thinking is within the existing paradigm of a representative democracy. ICTs applied in this form of voting and election support limit the potential capability of the technology. It is reflective of a lack of political innovation as ICTs-enabled process repeats the existing representative democracy where the public inputs are limited to be digitally constrained through a defective system that suffers from a lack of democratic performance. So even though it may appear to embrace the ICTs it does so without improving the already existing democracy to transform into a democracy that democratises or towards people's rule.

For others, ICTs enable deliberation for decision making using a virtual public sphere (Coleman and Blumler 2009: 9), which uses the communicative capability of ICTs towards a *demos* that is self-articulating. A self-articulating *demos* however creates more demands that a sporadic voting system of existing representative democracy can barely comprehend and respond to from the rapid changes that society is undergoing using ICTs. The ICTs make the voting process of representative democracy seem inadequate to meet the needs of the people as before it can identify the need and develop an appropriate policy this need may already be obsolete. A new way to investigate how to respond to this new environment where voting to delegate becomes inadequate is to allow the people to be involved in decision making. An option is to explore what has worked in other models and create a new model from those things that work.

For modelling, a novel concept proposed by Päivärinta and Sæbø (2006) to eliminate voting is to create from the combined strengths of existing ideal models a new model, they call the Partisan model, which they suggest be digitised to deliver the ideal e-democracy. The Partisan model they assume will improve on existing models as the ICTs will be used to digitise the model's processes for decision making by citizens. They argue that the literature surveyed revealed that with digital means e-voting has focused on the technology and legal aspects of voting rather than the actual contributions of e-voting cases to democratise (Päivärinta and Sæbø 2006: 5).

With the Partisan model, citizens are the decision makers in a spontaneous non-government facilitated virtual unrestricted discourses setting the agenda through virtual communities. This model from citizen sponsored unrestricted communication they suggest would remove the voting for representatives as it is implicit that citizens are the decision makers. They argue that 'E-Democracy is communication uncontrolled by government and without clear connection' to the decision-making process (Päivärinta and Sæbø 2006: 12). Seemingly a form of this model was used in South Korea (Päivärinta and Sæbø 2006: 14), it was to promote the oppositions' views against a dominant government but the reports are that it was only partially successful. Details of the South Korean model had not been provided nonetheless it highlights a different manner of thinking that is greenfield as it does not assume that digital technology application must comply with a pre-set, hegemonic paradigm. This model creates a virtual public sphere for decision making initiated by active citizens that is independent of government.

An information issue arises which simplistic approaches to e-voting do not address. For example, when a system that is 'too well adjusted to lives organized around the struggle to maximize personal income' adopts digital technology for reasons of efficiency, the technology is likely to perpetuate the disenchantment and demoralisation felt towards the political parties we vote for and the political class we allow to rule (Dunn 2005: 184). E-voting will not restore the trust issue that has been created by a system which was based on the notion of competition which is representative democracy. The continuous and strident need to make news makes Schumpeter's electoral entrepreneurs lure the gullible to promote their promises to trade for votes, potentially using software like Smartvote. Negative news may in fact undermine the informing software if the people are not involved in decision making. In an era of the Internet with its reach, manipulating the public becomes easier due to the speed of access to the many. This may come at a cost though.

Negative news feeds the democratic deficit as shown by Norris and is mentioned earlier in Chapter 2. In such circumstances, trust is anticipated to become more elusive if the system is not overhauled. A different approach is suggested by the book that would improve the transparency in the decision-making process that seems the basis of negative news to feed the distrust of politicians and the political class. E-voting does not in itself remove the quality of information that feeds trust for those who vote, the people. E-voting is yet to be accepted by some people used to a different paradigm (Meier 2012: 153–154). People used to the current rituals of voting may require some time to adapt to a new system like an eElection before they will trust it.

As mentioned earlier, Smith's study of e-voting showed that people who are positively inclined to the political system do not trust technology like smartphone for voting despite its convenience when compared to the existing paper-based system with which they are already familiar. Using a 2013 Australian Electoral Study (AES), Smith (2016) explored citizen's trust for new and in existing channels using the AES survey data from 2004, 2007, 2010 and 2013. The old voting channels for Australia forming part of the survey were three channels that are well established, namely a paper ballot on election day, a pre-poll paper ballot and a posted paper ballot. The new voting channels in the AES surveys were not well known in Australia and these were Electronic Voting Machine (EVM), smartphone voting using text, and Internet voting. The paper-based systems were preferred due to their familiarity.

Smith found that younger voters were more inclined than older voters towards new channels of voting. He noted that older voters' experiences of the Australian political system positively inclined them to favour older channels. Using a Factiva search, Smith analysed Australians exposure to changes in voting channels from media reports and his findings suggest that Australians' assessments of the new electronic voting channels were most likely based on their level of familiarity with information technology. Amongst the new voting channels, the one that was least trusted was Smartphone. Smith used a 0–9 scale, where 0 indicated no trust and 9 full trust, to measure patterns of trust in paper-based and electronic voting channels. Using regression analysis, Smith's findings indicated that in Australia as 'a benign political context, there is widespread voter confidence in both existing voting channels and in possible alternatives', that is new channels were not excluded (Smith 2016: 80). The assumption was that Australian preferred a paper-based system as voters perceived the existing channel to be more transparent than an electronic one for counting and voting. However, if this was broadened with more options then new channels would become acceptable to the Australian public. Information about the new channels would improve trust in the new channels.

To improve trustworthiness, a government even with e-voting must innovate in how it uses information. It must explore ways and means so that information is managed to improve transparency as restricting information reduces transparency. Meier suggests that information generation and sharing is a means to improve representative democracy (Meier 2012). In Switzerland, Smartvote is an application as mentioned above which is for the provision and sharing of information to voters. A problem with digitising the electoral system is reflexivity (Webster 2007). This is a significant issue to emerge from digital communication. Reflexivity is the 'premium placed upon speed of action' from information, for example, in the context of warfare where it is an issue that provides a key advantage to those who can exploit it for their benefit according to Webster (2007: 216). In an information society or e-society, one should be concerned about information. Information and its use in various ways become the issue for an e-society (Webster 2007: 2). Decisions are made by pre-programmed computers such that speed of information can give an edge, for example, it can misinform using social media, using bots, for example (Freitas and Bhintade 2017). Communication technology is evolving, bots are an automated means to respond to a text message using a 'software that leverages artificial narrow intelligence to perform specific tasks in place of a human' (Freitas and Bhintade 2017: 8).

The use of social media informs when people are online and they are 'an addictive and highly engaging medium to communicate with' (Freitas and Bhintade 2017: 7). It may also give rise to a feeling of a heightened degree of state surveillance for welfare recipients that generates risks for a democracy (Webster 2007: 222). The state according to Webster in an information society collates massive amount of information that it can use to model its citizens' behaviours. Also, information when enabled by ICT makes the generation of demands within a political system instantaneous. The data rate stresses the political system which is not designed for such type of information processing. Actors delegated decision-making powers get overwhelmed by the rate of data being generated by the ICTs. These data from the ICTs may not be structured to support decision making by the actors giving an advantage to those who are resourced to exploit this competitive edge. Representative system like an election is based on free and fair competition. However, there is more to ICTs than providing a competitive advantage to win an election.

ICT applications can create new ways for governing like e-voting, but new problems like reflexivity become an issue due to the rate of demand creation. An instantaneous rate of creation of demands from too many inputs may cause strain to the political system due to the unmet demands. Unmet demands create stress to a system that can result in regime change (Easton 1971). In an information era, these stresses from unmet demand could be purposively manipulated for gain by an unscrupulous representative seeking political advantage. The e-democracy literature assumes that digitisation would automatically improve governing and democracy by empowering citizens when it is plugged into the black box called the Internet. However, issues like rate of demand creation from the speed, volume and also the varieties of demands remain unaddressed. ICTs data generation, for example, from social media potentially overwhelms a political system, it cannot respond. The system, for example, if a paper-based one, may not be designed for reflexivity inherent in the modern technologies. Modernisation may require more information for each citizen to make informed decisions (Webster 2007: 203–227). We explore some policy-making suggestions below that goes beyond e-Voting to explain this information aspect further.

Exploring the policy life cycle, an OECD report suggests that e-democracy can be a form of participatory democracy where if citizens are engaged in policy making they need to be informed and become active citizens (OECD 2003: 28). It assumes that the ICTs can both inform and motivate citizens to participate if there is no digital divide of information rich and information poor. The design for the technology must be right and if so the technology can support direct democracy (Hoff et al. 2003). While some look at e-democracy supporting virtual communities or e-community like Meier's case study (Meier 2012), others explore e-democracy as a model using online polling to gauge public opinion (Coleman and Gotze 2001). Coleman and Gotze report that these may have a place in good government but they fail to look deeper into the process of public opinion formation that the Internet technology is capable of. Most of the e-democracy experiments they mention are along the processes for online surveys, polls to referendum and e-petitions from citizens. Yet another suggestion is about using ICTs for policy deliberation (Coleman and Moss 2012). There are many ways that the technology could be applied to allow participation, the below is some of them.

Technology becomes an enabler to support the above participatory initiatives. The UK government surveyed senior policy makers and identified nine elements that needed to be addressed namely: 'be forward looking, be outward looking, be innovative, flexible and creative', in the context of twenty-first century be evidenced-based and furthermore: 'be inclusive, be joined up, ensure policy is constantly reviewed, enable systematic evaluation, and learn from experience' (OECD 2003: 31). This is a long policy list and it failed to enable citizens' participation in decision making in the UK. It led to a large number of experiments about information dissemination, consultation and participation. These experiments led to a review of the UK government's policy for electronic democracy—a policy which has been deemed a failure by Moss and Coleman (2014). Despite acknowledging some initial successes of policy experimentation, Coleman and Moss reported that in the UK a people-inclusive policy-making process remains more a matter of policy rhetoric than reality.

Actualising the e-democracy idea in practice using the Internet also has the potential to disrupt contemporary representative democracy as power shifts from representatives to citizens. When it is made feasible, e-democracy enables people's rule and a modern form of traditional democracy for large nation-states. This novel form of traditional democracy is to be on a scale that could not have been imagined by proponents of the currently existing representative system (Schumpeter 2013). In the light of e-democracy's potential, those who seem to justify support for representative democracy as being constrained by the size of nation-states advocate an argument that is reflective of a pre-Internet era (Dahl and Tufte 1973). Experiments in e-democracy have been conducted, though according to Päivärinta and Sæbø only few were aiming at direct democracy (Päivärinta and Sæbø 2006: 28). The OECD mentions that a policy for e-democracy is necessary but recommends rejecting this notion if it is a 'tokenistic policy' that results in meaningless e-friendly outcomes (OECD 2003: 159–160). The OECD argues that e-democracy is not a panacea through hyperbole to flaws in the representative democracy system. E-democracy is an opportunity to relate democracy to the Internet capability in new ways. Thus, only e-voting may not suffice to actualise this potentiality from digital technology. A framework for digitisation is recommended by the OECD to avert the issues encountered by the other countries attempting to implement e-democracy.

4 Digital Technology and Some Lessons for Democracy

There are common themes in the e-democracy literature and experiments that are grouped below, which include terminologies from a fast-moving technology to issues on access that creates a digital divide. In the new digital environment, a new form of poverty emerges, information equality. Another aspect is technological determinism where the assumption that ICTs will inevitably lead to an e-democracy is flawed without defining how an e-democracy is to emerge in that context. There is also the issue of the information itself as to why the state is to be trusted by citizens if the democratic aspect from the digital processes does not augment democracy. We expand on these elements below. The technology aspect permeates all and then there is a human agency aspect where the issues like political will and the state's intent for trusting the human agents in charge to embrace technology to democratise. We argue that both a technological framework for digitisation and the political will to empower the people in decision making are the core pre-requisites to e-democracy. Designing the system that enables access as well as reduces digital divide from information from vague policies is challenging. The below discussion is to support our shared view that political will and the design method are the core elements as representative democracy processes were never meant to democratise (Hindess 2002).

4.1 Diverse Terminologies at an Infancy Stage

The proliferation of e-terms is an issue in the digital era (Crespo et al. (2013: 2). This proliferation tends to fragment the literature about e-democracy and the experiments that describe the attempts (Päivärinta and Sæbø 2006). The pace of technology development tends to make the literature obsolete and as new terms emerge from the new technology this in itself creates complexity. The experiments described by the OECD (2003) and Meier (2012) have a common theme that despite the variations of terminologies the cases indicate there remain possibilities of new forms of participation to engage citizens in political life. While for the OECD, it is the economic imperatives that the technology allows driving new digital infrastructure which has resulted in some innovative common communication networks like the Internet or now IoT where autonomous communication between machine to machine is becoming feasible (OECD and

Gaël 2016), the pace of development of the technology is creating many new innovative constructs. For example, an IoT platform for commercial needs has the potential to create better engagement for citizens and politicians in policy making to improve transparency in decision making using technological capability to automate data transfer. However, the associated proliferation of terminology makes ideas about Internet technology described in the literature seems obsolete. An era of digital technology innovation creates its own issues of terminological obsolescence that must be tackled.

For Meier, the e-democracy is not about creating 'new rights and duties for the citizens, but an extended information policy, activation of citizens, community formation, and creation of transparency' (Meier 2012: 3–4). This e-democracy initiative aligns with the European Parliament's electronic government or e-government policy (Meier 2012: 1). Information interchange is facilitated through the transactions that the market services required from both e-citizens and e-businesses. Indirectly, the technology created for e-trade also allows some degree of e-participation for citizens using it. The technology created via this e-market is then aligned to reflect some informational flow improvements to representative democracy's processes of eVoting. Bringing together the imperatives of the market with those of the e-democracy project may not prove workable, if these innovations are driven by market principles. It is likely that the market will undermine initiatives that are against its interest to ensure that the digital terms linked to democracy reflect the market's need for efficient but not necessarily democratic processes. Emerging e-terms will reflect their roots from market-driven thinking.

4.2 The Application of ICTs: Context and Framework

Whenever new media technology has emerged, its proponents have often claimed for its significant capabilities that would shape democracy in positive ways. But liberal democratic theory has had little to say about technology in this sense (Weinberger 1988: 125). The potential application for new media technology raises many claims that tend to be speculative, bordering on the futuristic like the 'cyber-optimists' (Norris 2001: 11). The literature seems to follow a pattern with the two schools of thought emerging: those supportive of the technology and those disputing that the technology makes any difference. However, for liberals they are not forced 'to rethink their views of liberty, freedom, politics and society' with new technology (Weinberger 1988: 125). Technology it would seem is just another problem to avoid, or a neutral part of the tools of democracy. In a hegemonic liberal democracy, this attitude towards technology is akin to black box thinking. It would appear that liberal democracy sees technology as a promise for growth of the economy to which liberal democracy is wedded (Weinberger 1988: 126). This is despite the possibility that technology like ICTs could be a danger to liberal democracy's existence. A black box approach also misses the potential of improvement to liberal values. The effect of the technology is that it can create systems that individualise as suggested by Hoff et al. (2003) for the cyberdemocracy or digital direct democracy model mentioned earlier, and individualism is core to liberal democracy. Yet, it would seem that this thinking is yet to permeate existing liberal democratic thought.

ICTs do make individualism more prominent and therefore it is a core issue that liberal democracy may be missing by neglecting its transformative potential to improve the current liberal forms. For left-leaning liberal democrats, some claim that if society catches up to the full technological bounty then society can be run by discourse rather than force and exploitation (Weinberger 1988: 127). An active exploration of the potential of digital technology should therefore be a prominent liberal project given that it allows the needs of the individual citizen to be communicated to the state. A moral imperative to equalise seems to exist, yet the political will in most Western liberal states depict an apathy to e-engage through promoting a 'silence' that acts as a 'most effective agent for subliminal persuasion in mass communication' (OECD 2003: 156). Ignoring ICTs as not part of political theory and assuming the current issues will somehow work out to deliver better government is flawed. Contemporary technology requires a new way of thinking. It is an opportunity for change which democracy has been embarking on from its 2500 years of evolution to its representative mainly liberal democratic form. Politicians and scholars may need to understand this and embrace the improvement technology brings to improve democracy from a model that may be reaching its end of life as an obsolete artefact for a modern era.

4.3 *Political Will and Digital Leadership*

The OECD discussion of e-democracy identified that there is a key role for states' leaders to provide appropriate policy to enable the technology to be applied in decision making through online citizens' inputs. The

UK, one of the states, prominent in its use of ICTs, seems to be delaying such a move (Moss and Coleman 2014). Others seemingly are in a wait and see attitude for copying any successful e-democracy as the OECD reported that only a few countries even consider running such experiments. As noted above, only twelve countries have experimented with engaging its citizens in policy making through some form of e-democracy system (OECD 2003). Delays from leaders to address the informative power of the technology may strain the representative democracy. In a system that already has a performance issue, such delays to implement the technology exacerbates the disappointment amongst citizens increasing democratic deficit. Thus, a lack of political leadership towards change may lead to a lack of citizens' support and that further exacerbates the lack of trust people feel towards politicians.

To democratise the informative power of the technology is a challenge that the state must address. Individual citizens in a society are becoming better informed as newer and improved applications become available through their smartphones. Potentially, these devices allow every citizen's voice to be raised so it can be heard. This is something that has been lacking in the political environment within which liberal democratic representative systems have been operating. In many ways, this has served to date as a self-justifying rationale that liberal democracy's representative form is the only alternative in the West and other countries with similar regimes. However, the political context has changed and the technology is changing it further.

The OECD noted that when using ICTs to implement change, leadership is necessary. This means that there needs to be the political will to undertake and oversee such change. For the OECD, political leaders need to ensure they demonstrate commitment to improve the political system for transparency and accountability in decision making (OECD 2003: 10). Importantly, the OECD (2003: 13) also noted that states using ICTs should make 'an important start in developing a methodological framework that addresses how ICT can be designed and used to effectively and efficiently support information provision, consultation and participation in policy-making'. This advice for a framework is based on the lessons learnt by the OECD in its analysis of how various countries implemented ICTs for better government (OECD 2003). In addition, poor or improper use of ICTs may deliver a project that fails to address what is needed despite a government policy to digitise. However, the e-democracy lag is something that political parties may not be averse to as failure defers change to improve the system and is something that major parties may be profiting from as they are the established power brokers.

4.4 ICTs and Digital Divide

Use of digital technology seems ubiquitous yet there appears to be an emerging digital divide rendering a segment of the population as information poor (Norris 2001). Implementation of policy by government may not yield the expected results as the technology comes at costs that might exacerbate existing inequalities. Many people may not be able to afford the various devices that provide access to this technology, and even if they are able to access the devices they may not have the means to sustain the cost of upgrades. For an e-democracy, this cost issue will affect people's ability to participate in its public life leading to what Norris calls a 'democratic divide' (Norris 2001: 4) that is both social and political. Though the technology can connect, problems of access can disconnect.

The OECD has also warned that an inequality from the digital divide could become entrenched if governments do not put in place the appropriate policies and the right infrastructure to ensure access to all its citizens (OECD 2003: 60). In Europe, this social inclusion issue was identified and through the Lisbon Agreement in 1999 a Europe Action Plan was drafted to tackle it. Historically, when new communication technologies emerged similar issues affected their adoption. As a generalisation, the cost of new technology is a barrier which tends to get eliminated over time as the technology gets diffused and penetrates further. This led Pippa Norris to observe that even the digital divide may disappear in time if governments put in place the appropriate policy this digital inequality may then be 'a short time phenomenon' (Norris 2001: 11). However, this argument is valid only if the economic disparity is overcome. The social disparity may grow as Norris also notes that the gap between the rich and poor tends to increase if the economic issues are not addressed. Seemingly, the Internet is creating a class structure between those who are educated and have the income to access it and those who cannot have access due to economic barriers. However, if the government has the right processes and the right policies for breaking down existing barriers of access to the new technologies then there is an opportunity that the Internet can support an e-democracy. A point also raised by Norris is that a digital divide exacerbates the levels of trust in society between the governed and those who govern. This particularly affects those most vulnerable and lacking a voice in government. A digital divide left unaddressed exacerbates the trust issue where it exists. The divide recreates the old pattern of an informational class between those who have and the have-nots. Some scholars have presented the evidence that levels of trust tend to improve if the digital divide is removed. In the study conducted by Norris, online users exhibited more trust in democratic political institutions compared to non-online users (Norris 2001: 227).

Another related danger is that the same technology creates flash actions like protests and even revolutionary uprisings like the 'Arab Spring'. Restricting access to information will allow manipulation of the have-nots by the haves with the outcomes from such flash actions being unpredictable. Virtual revolutions from excessive negative news from information oligarchs with a vested interest in containing the democratic impulse are a potent danger for an e-democracy that democratises. Governments which are oblivious to this risk to stability may be allowing a volatile system to become the norm in representative democracies.

Barriers to participation exist in all democracies. For example, Dahlberg and Siapera (2007) argue that a performance issue that has constrained democracy to its contemporary form is access for every individual to the Internet. They suggest that in order for every citizen to participate two things must be in place: first, the values of liberty, equality and solidarity must be re-articulated within the Internet's connectivity and second, the digital divide must be removed. In their view, the problem is democratic access; they assume it is solvable by the state. They propose that states provide a channel for citizens' inputs to make their voices heard above those of a mass media captured by private interests. It is assumed citizens raising their voices through the Internet channel allows them to counter the mass media which has been captured by neo-liberal interests to marginalise the less-resourced voices (Dahlberg and Siapera 2007: 2). The government has an important role to provide access to the web for all citizens and eliminate the digital divide else change to improve is not realised. Secondly, the government must be inclined to use this technology to listen to its citizens. A liberal democracy implementation of the technology to democratise is problematic. We expand on this liberal mindset below.

4.5 Risks for Trusting the State's Digital Democracy

There are risks to trusting the state with the power that ICTs provides to inform. As we noted earlier, liberal democracy was an evolution from a lack of trusting the state's coercive power. However, the ICTs may not empower. According to Stahl (2008: 89), it is quite the opposite because 'the disempowering faculties of ICT lie at the heart of the design, plan and use of the technology'. ICTs can be designed to manipulate the public as its informational capability gives an advantage to the state which holds enormous amount of data on citizens. So, the way the technology is designed to inform can allow those in power to maintain their power. It can be also designed to coerce through surveillance especially for the welfare recipients over whom the state holds the maximum amount of data (Webster 2007: 203–227). A similar claim is made that the ICTs are not value-neutral (Stahl 2008), their impact depends on their application. It is possible that the ICTs will be used for empowering the state, rather than the people. This is claimed by Stahl (2008) from observation of his Egyptian case study. The Egyptian ICT gatekeepers were trained to assist the people with the usage of the system. Instead, they apparently became the power multipliers deployed to propagate the ruling hierarchy's version of the truth (Stahl 2008: 97). The OECD reports that ICTs policy in Egypt 'was a relative success' due to reduced costs which increased access to the web (OECD 2010a: 19). So, information technology by improving access to the technology could be used to manipulate and to control by those collecting the information. This concern about a big brother state using electronic surveillance is not new.

Digital collection of information is from digital tools, they are used in marketing to understand and influence consumer behaviour. The Egyptian case also creates a conundrum about why the state should be trusted by citizens to use the ICTs to provide true information. This is especially the case when the coercive powers of the state are still significant even in the neo-liberal democracies of the West. It may be that the benefits of the technology are being over promoted by a select few (Oates et al. 2006: 1–15). The technology companies may be perceived to be overselling the merits of digitisation with inadequate safeguards from its use that may currently exist for a representative democracy operational framework.

Another dimension of risk emerged from the World Trade Organisation (WTO) as questions about who profits from the ICTs 'as policies forced onto poor nations also benefit rich nations' (OECD 2010a: 30).

The technology has a risk. Technology does shape society in both negative and positive ways (Day et al. 1988). Another risk, noted by Oates et al. (2006: 15), is that ICTs debate matters for 'a body of literature to evolve and develop' that may become a self-perpetuating myth leading to an 'ultimately empty debate'. Furthermore, Lidén (2015: 1–2) mentions that some scholars are 'merely accepting the shortcomings of using ICTs in democracy, despite the risk that they will leave democracy an empty shell and assist in eroding social capital'. This may be cause for concern about a digital political system where democratic values are eliminated in an e-democracy or digitisation process that recreates an efficient neoliberal system of government.

The party system, for example, is now gaming the representative model to win elections (Norris 2004, 1997). This party behaviour may persist with digital tools because the Swiss e-voting example was effective only at improving the efficiency of an already existing representative system process. The software merely provided new digital tools that were adapted to contemporary liberal democratic processes.

The various experiments with e-democracy seemingly digitise legacy processes that become more efficient but there remains a challenge to find a means to measure when and if the virtual processes actually lead to a democracy that democratises. The EU countries digital government project is being driven by an efficient market paradigm. Though digitisation improves the existing processes, technology application can bring new problems like where 'certain interests may be furthered at the expense of others' based on how the policy is implemented (Tobias et al. 2003: 348).

5 Measurement of E-Democracy: A Virtual Process

Measuring progress towards digital democracy is problematic in the virtual world; the technology may promise something yet actually deliver a very different outcome when the policy is implemented. There are many claims of the technology's communicative ability that range from utopic ideas to dystopian views for an e-democracy (Kneuer 2016: 667). Some consider participation and engagement as key aspects of an e-democracy (OECD 2003). The UN and OECD have proposed a digitisation index, to measure democratic participation in a digital environment which is e-participation. The e-participation is used to quantify the extent that ICTs democratise in a State.

Furthermore, these eParticipation measures proposed by UN and OECD have been critiqued for their lack of measuring the democracy content in a state (Grönlund 2011). Digital indexes can be manipulated, and as the data can be faked by the state's mechanism it is flawed for measuring democracy (Grönlund 2011). The challenge is to create a democratic process (OECD 2003). The EU democracies have digitised many of their government processes and services but the goal seems to be for downsizing the public service for the 'use of ICTs facilitates reductions in government expenditures' (OECD 2013: 7). The use of ICTs is an economic one or e-government oriented. The states are yet to involve their citizens in decision making (OECD 2003).

In his analysis of the UN eParticipation index, Grönlund found that the relationship 'between the index and democracy and participation is non-existent' (Grönlund 2011: 26). He explained that an authoritarian regime could window-dress its web page in order to get a high eParticipation score. Rather, he argues that in order to gauge democratic participation it is crucial to investigate the underlying processes in the political system rather than just digital web pages Grönlund (2011: 35). His finding that even authoritarian regimes scored highly on the UN digital measures gives cause for concern about the validity of such measurements because of the potential to mislead about democratisation. He recommended a reality check was needed when using such digital indexes for benchmarking or citing.

Refinements are suggested by Kneuer, to build on the eParticipation measures proposed by the UN and OECD supra-nationals. Kneuer suggests a conceptual framework where e-democracy is based on equal and free access, digital technology, and e-participation where social media allows bottom-up and top-down communication between citizens and politicians, and e-government for e-services (Kneuer 2016: 672). A democratic process for people's decision making remains elusive though, despite a technology that can allow improvements.

The improvement suggested is a new set of measures that relates to eParticipation (Kneuer 2016). The set of new variables that is offered to refine the eParticipation measure are e-information, e-consultation and e-decision making. These new variables are applied by Kneuer to find some patterns in five established democracies, namely Canada, France, Germany, the UK and the USA; these countries use digital technology for e-Services. Through the new variables e-information, e-consultation and e-decision making, Kneuer explores 'whether governments are reducing or increasing their activities in specific types of engagement or whether citizens are becoming more active or not' (Kneuer 2016: 676). Kneuer identifies that online tools are not working well, and the atomisation of the public sphere is hard to assess.

Kneuer (2016) proposed set of measures to improve on eParticipation measures is on the basis of the assumption that digital media has a democratising performance. That is digital media is a key democratic need in the sense that it is a fourth dimension that must exist within the current digitised systems of representative democracy to support democratising. In this context, digital media like traditional media has a key surveillance and feedback role for democracy, and with online information such surveillance and feedback has the capacity to improve the system as it develops. Kneuer suggests two pillars to build the appropriate measures: eGovernment (eServices by government) and eParticipation (the UN and OECD measure). However, it was interesting to note that the scores of five countries (Canada, France, Germany, the UK and the USA) with similar information provision through ICTs revealed different scores for public consultation for decision making (Kneuer 2016: 676). Kneuer's measurements show that all five democracies are in the 96% for e-information. However, e-consultation varied from 45 to 77% with the lowest being Germany at 45%, for citizen e-decision making it mentioned that amongst the five the high range is about 89% for the USA and UK to a low range of 11% and 0% in the cases of Germany and Canada, respectively (Kneuer 2016: 676).

The eParticipation measurements showed even in well-established modern exemplar Western democratic systems that have digitised their government processes to e-services using ICTs, the democratic capacity building proved inadequate for decision making. From these measures, the quality of democracy would at the most report digital failings. An explanation is that the motives behind the use of the ICTs are important as the technology itself depends on how it is applied (Kneuer 2016: 667). Digitisation to an e-democracy with e-services may not fix the performance issue of democracy or democratic deficit. The e-services are a push-down strategy which Kneuer terms e-government (Kneuer 2016: 672). Kneuer argues that the technology is not democratising in itself for 'its effect on political structures, processes, actors, behaviour and norms depends on the motives of use, the content that is transmitted, the way that the technology is used' and 'on the political context in which the digital media are used' (Kneuer 2016: 667).

It would also be misleading to assume that even if the processes exist that they are being actually used by all citizens. The degree of citizens' engagement matters. For example, even if the digital processes exist, they may not be acceptable to citizens. These processes may be perceived with suspicion as part of a state's coercive suite of tools in those societies with such cultures. When digital participation is sought by the state, these barriers must be overcome to improve engagement, so people actually participate. Post massive investments in digitisation, this digitisation failure may not bode well in a resource-constrained and politically volatile environment. The state may have a digital system, but measures of digital democratic process must be treated with care ensuring the digital technology equalises access to every citizen.

6 Conclusion

The scholarship suggested e-democracy is the solution to democratic deficit. However, as explained by the book an e-democracy has failed to be sustained in practice. An assumption in the literature is that the ICTs democratises and this claim when investigated showed that the framework for implementing ICTs mattered, in fact the OECD recommended a digitisation framework (OECD 2003). An e-democracy is feasible based on the improved participation and deliberation capabilities that exist through the communicative capability of the ICTs (Grönlund 2001; Moss and Coleman 2014; Coleman and Gotze 2001; Coleman and Blumler 2009; OECD 2003). The literature suggests and also explained throughout this chapter that the concerns over e-democracy have been growing, notably through the development of various technologies and under various name; nonetheless, there are both challenges and risks in implementing e-democracy. The chapter also highlighted, inter alia, how a powerful actor like the state and political parties might misuse ICTs in the name of institutionalising e-democracy, e-voting or e-participation. These analyses strongly indicate that there is a need to revisit the very concept of emerging e-democracy.

Information technology is not being used in an innovative way to allow e-democracy to operationalise as the capability of the technology allows new means to govern with people's input in policy making (OECD

2003). The OECD anticipates that technology will empower individuals and is supportive of its increased use (OECD and Gaël 2016; OECD 2003, 2005, 2010a, 2010b, 2013, 2015, 2016). This book does not treat this OECD view of technological empowerment as a given but it agrees that technology does have a role to play. The role for technology is one where its capacity is applied to subordinate representative system thinking to the ideal of people's rule by design. As noted, the book's proposal to use SE to develop an e-democracy design is proposed such that the traditional idea of a participatory democracy is given primacy. In the book, the representative democracy system is a democratisation project awaiting completion through an e-democracy, a case for a systemic upgrade.

Given past attempts at implementing an e-democracy have not succeeded, there is a need to develop an alternative approach. Chapter 4 explains systems theory to set the stage for an alternative approach, which while drawing on systems theory does not pursue a structuralist-functionalist approach that is most commonly associated with the earlier manifestation of systems theory in political science in the 1950s and 1960s. The approach developed in the book aims at providing a bridge between the discussion of e-democracy explored in this chapter and the proposed alternative which is the 'thought experiment' developed through Systems Engineering (SE) to create an 'e-democracy that democratises'. SE is a systems approach that is novel to political theory, though the idea of systems theory is not, since it was once championed by David Easton in the 1960s. SE is explained in Chapter 5 as an alternative framework to develop an operational e-democracy, but first the next chapter discusses David Easton's systems theory in political science so that it can be seen that the proposed SE approach acknowledges but is not indebted to that of Easton.

References

Alexander, D. 2002. Democracy in the Information Age? *Representation* 39 (3): 209–214.

Barber, B.R. 1997. The New Telecommunications Technology: Endless Frontier or the End of Democracy? *Constellations: An International Journal of Critical & Democratic Theory* 4 (2): 208–228.

Beetham, D. 2012. Defining and Identifying a Democratic Deficit. In *Imperfect Democracies the Democratic Deficit in Canada and the United States*, ed. R. Simeon and P.T. Lenard. Vancouver and Toronto: UBC Press.

Birch, A.H. 1972. *Representation*. Key Concepts in Political Science. London: Macmillan Press.

Boyd, O.P. 2008. Differences in eDemocracy Parties' eParticipation Systems. *Information Polity* 13 (3): 167–188.

Burke, M. 2012. A Decade of e-Government Research in Africa. *The African Journal of Information and Communication* 12: 1–25.

Chadwick, A. 2006. *Internet Politics: States, Citizens, and New Communication Technologies*. Oxford: Oxford University Press.

Coleman, S., and J.G. Blumler. 2009. *The Internet and Democratic Citizenship: Theory, Practice and Policy*. Communication, Society and Politics. Cambridge: Cambridge University Press.

Coleman, S., and J. Gotze. 2001. *Bowling Together: Online Public Engagement in Policy Deliberation*. London: Hansard Society.

Coleman, S., and G. Moss. 2012. Under Construction: The Field of Online Deliberation Research. *Journal of Information Technology & Politics* 9 (1): 1–15.

Crespo, Rubén González, Oscar Sanjuán Martínez, José Manuel Saiz Alvarez, Juan Manuel Cueva Lovelle, B. Cristina Pelayo García-Bustelo, and Patricia Ordoñez de Pablos. 2013. Design of an Open Platform for Collective Voting through EDNI on the Internet. In *E-Procurement Management for Successful Electronic Government Systems*. Hershey: Information Science Reference (an imprint of IGI Global).

Dahl, R.A., and R.E. Tufte. 1973. *Size and Democracy*. The Politics of the Smaller European Democracies. Stanford, CA: Stanford University Press.

Dahlberg, L., and E. Siapera (eds.). 2007. *Radical Democracy and the Internet: Interrogating Theory and Practice*. Basingstoke: Palgrave Macmillan.

Day, R.B., R. Beiner, and J. Masciulli (eds.). 1988. *Democratic Theory and Technological Society*. Armonk, NY: M. E. Sharpe.

Dunn, J. 2005. *Setting the People Free: The Story of Democracy*. London: Atlantic.

Easton, D. 1971. *The Political System: An Inquiry into the State of Political Science*, 2nd ed. New York: Alfred A. Knopf.

European Commission. 2018a. EU Membership. Communication Department of the European Commission. https://europa.eu/european-union/about-eu/countries_en#tab-0-1. Consulted 4 February 2018.

European Commission. 2018b. Policies, Information and Services. European Commission. https://ec.europa.eu/info/about-commissions-new-web-presence_en. Consulted 28 January 2018.

Finger, M. and F.N. Sultana. 2012. *E-Governance: A Global Journey*, vol. 4. Amsterdam and Fairfax: IOS Press.

Freitas, E., and M. Bhintade. 2017. *Building Bots with Node.js*. Birmingham: Packt Publishing.

Goos, G. et al. (eds.). 2002. Electronic Government. First International Conference, EGOV 2002. Aix-en-Provence, France: Springer.

Greengard, S. 2015. *The Internet of Things*. The MIT Press Essential Knowledge Series. Cambridge: MIT Press.

Grönlund, Å. 2001. Democracy in an It-Framed Society. *Communications of the ACM* 44 (1): 22–26.

Grönlund, Å. 2004. State of the Art in e-Gov Research—A Survey. *Electronic Government* 3183: 178–185.

Grönlund, Å. 2008. Lost in Competition? The State of the Art in e-Government Research. In *Digital Government*, ed. H. Chen et al. Berlin and Heidelberg: Springer.

Grönlund, Å. 2011. Connecting e-Government to Real Government—The Failure of the UN Eparticipation Index. *Electronic Government* 6846: 26–37.

Grönlund, Å., and T.A. Horan. 2004. Introducing e-Government: History, Definitions, and Issues. *Communications of the Association for Information Systems* 15: 713–729.

Hague, B.N., and B. Loader. 1999. *Digital Democracy: Discourse and Decision Making in the Information Age*. London and New York: Routledge.

Hilbert, M. 2009. The Maturing Concept of E-Democracy: From E-Voting and Online Consultations to Democratic Value Out of Jumbled Online Chatter. *Journal of Information Technology & Politics* 6 (2): 87–110.

Hindess, B. 2002. Deficit by Design. *Australian Journal of Public Administration* 61 (1): 30–38.

Hindman, M. 2009. *The Myth of Digital Democracy*. Princeton, NJ: Princeton University Press.

Hoff, J., I. Horrocks, and P. Tops. 2003. New Technology and the 'Crises' of Democracy. In *Democratic Governance and New Technology: Technologically Mediated Innovations in Political Practice in Western Europe*, ed. J. Hoff, I. Horrocks, and P. Tops. London and New York: Routledge.

Kneuer, M. 2016. E-Democracy: A New Challenge for Measuring Democracy. *International Political Science Review* 37 (5): 666–678.

Lanvin, B. et al. 2010. Promoting Information Societies in Complex Environments: An In-Depth Look at Spain's Plan Avanza. *The Global Information Technology Report 2009–2010*, 127–140. World Economic Forum.

Lidén, G. 2015. Technology and Democracy: Validity in Measurements of E-Democracy. *Democratization* 22 (4): 1–16.

Meier, A. 2012. *EDemocracy & EGovernment: Stages of a Democratic Knowledge Society*. Berlin: Springer.

Morgan, G. 2016. New Actors and Old Solidarities: Institutional Change and Inequality Under a Neo-Liberal International Order. *Socio-Economic Review* 14: 201–225.

Moss, G., and S. Coleman. 2014. Deliberative Manoeuvres in the Digital Darkness: E-Democracy Policy in the UK. *The British Journal of Politics & International Relations* 16 (3): 410–427.

Mouffe, C. (ed.). 1996. *Deconstruction and Pragmatism/Simon Critchley, Jacques Derrida, Ernesto Laclau, and Richard Rorty*. London and New York: Routledge.

Mulder, B., and M. Hartog. 2013. Applied E-Democracy. In *Proceedings of the International Conference of E-Democracy and Open Government*, ed. P. Parycek and N. Edelmann. Krems an der Donau, Austria: Edition Donau-Universitat Krems.

Norris, P. 1997. Representation and the Democratic Deficit. *European Journal of Political Research* 32: 273–282.

Norris, P. 2001. *Digital Divide: Civic Engagement, Information Poverty, and the Internet Worldwide*. Port Melbourne, VIC, Australia: Cambridge University Press.

Norris, P. 2004. *Electoral Engineering: Voting Rules and Political Behavior*. Cambridge, UK: Cambridge University Press.

Norris, D.F., and C.G. Reddick. 2013. E-Democracy at the American Grassroots: Not Now Not Likely? *Information Polity: The International Journal of Government & Democracy in the Information Age* 18: 201–216.

Oates, S., D. Owen, and R.K. Gibson. 2006. *The Internet and Politics: Citizens, Voters and Activists*. London and New York: Routledge and Taylor & Francis Group.

OECD. 2003. *Promise and Problems of E-Democracy-Challenges of Online Citizen Engagement*. Paris, France: OECD.

OECD. 2005. *Report from OECD Forum 2005 to the OECD Ministerial Council Meeting*. Paris: OECD Publishing. http://www.oecd-ilibrary.org/docserver/download/0105101e.pdf?expires=1517024994&id=id&accname=guest&checksum=6E744C2E44301A5B640ADBF534EFC209. Consulted 27 January 2018.

OECD. 2010a. *The Development Dimension—ICTs for Development—Improving Policy Coherence*. Washington, DC: Organization for Economic Cooperation & Development, OECD iLibrary. http://www.oecd-ilibrary.org.ezproxy.newcastle.edu.au/docserver/download/0309091e.pdf?expires=1517024328&id=id&accname=ocid194270&checksum=61AB79A0E079DDDBBFE3B6E4568A2B0F. Consulted 27 January 2018.

OECD. 2010b. *Good Governance for Digital Policies—How to Get the Most Out of ICT—The Case Of Spain's Plan Avanza*. OECD Publishing. www.oecd.org/gov/egov/isstrategies. Consulted 27 January 2018.

OECD. 2013. *Reaping the Benefits of ICTS in Spain Strategic Study on Communication Infrastructure and Paperless Administration*. Paris: Organisation for Economic Co-operation and Development, OECD iLibrary. http://www.oecd-ilibrary.org.ezproxy.newcastle.edu.au/docserver/download/4212081e.pdf?expires=1517023432&id=id&accname=ocid194270&checksum=28A93F4721E71AADB84F3539BCB992DE. Consulted 27 January 2018.

OECD. 2015. *States of Fragility 2015 Meeting Post-2015 Ambitions*. Paris: OECD Publishing. http://dx.doi.org/10.1787/9789264227699-en. Consulted 27 January 2018.

OECD. 2016. *Development Aid at a Glance Statistic at a Glance 2.0 Africa*. OECD Publishing. http://www.oecd.org/dac/stats/documentupload/2%20Africa%20-%20Development%20Aid%20at%20a%20Glance%202016.pdf. Consulted 27 January 2018.

OECD, and H. Gaël. 2016. *The Internet of Things-Seizing the Benefits and Addressing the Challenges—2016 Ministerial Meeting on the Digital Economy*. OECD Publishing. http://dx.doi.org/10.1787/5jlwvzz8td0n-en. Consulted 27 January 2018.

Päivärinta, T., and Ø. Sæbø. 2006. Models of E-Democracy. *Communications of the Association for Information Systems* 17 (37): 1–42.

Parycek, P., and N. Edelmann. 2013. CeDEM13 Proceedings of the International Conference for E-Democracy and Open Government. Paper Presented to Conference for E-Democracy and Open Government, Danube University Krems, Austria.

Parycek, P., N. Edelmann, and M. Sachs. 2012. CeDEM12 Proceedings of the International Conference for E-Democracy and Open Government. Paper Presented to Conference for E-Democracy and Open Government, Danube University Krems, Austria.

Parycek, P. et al. 2013. CeDEM13 Proceedings of the International Conference for E-Democracy and Open Government. Conference for E-Democracy and Open Governement, Danube University Krems, Austria, Danube University Krems.

Pruulmann-Vengerfeldt, P. 2007. Participating in a Representative Democracy: Three Case Studies of Estonian Participatory Online Initiatives. In *Media Technologies for Democracy in an Enlarged Europe: The Intellectual Work of the 2007 European Media and Communication Doctoral Summer School*, ed. Nico Carpentier, Pille Pruulmann-Vengerfeldt, Kaarle Nordenstreng, Maren Hartmann, Peeter Vihalemm, Bart Cammaerts, and Hannu Nieminen. Tartu: Tartu University Press.

Reddick, C.G. 2005. Citizen Interaction with E-Government: From the Streets to Servers? *Government Information Quarterly* 22 (1): 38–57.

Rios Insua, D., and S. French. 2010. *E-Democracy*, vol. 5. Dordrecht, Heidelberg, London, and New York: Springer.

Rodríguez Bolívar, M.P. et al. 2010. Trends of E-Government Research: Contextualization and Research Opportunities. *The International Journal of Digital Accounting Research* 10 (16): 87–111.
Scholl, H.J. 2006. Is E-Government Research a Flash in the Pan or Here for the Long Shot? In *Electronic Government*, ed. M.A. Wimmer, H.J. Scholl, A. Grönlund, and K.V. Andersen. Berlin and Heidelberg: Springer.
Scholl, H.J. 2013. Electronic Government Research: Topical Directions and Preferences. In *Electronic Government*, ed. M. Wimmer, M. Janssen, and H. Scholl. Berlin and Heidelberg: Springer.
Schumpeter, J.A. 2013. *Capitalism, Socialism and Democracy.* Abingdon: Routledge.
Smartvote. 2018. Smartvote Web Page. Smartvote. https://www.smartvote.ch/about/ideaEGOV. Consulted 5 February 2018.
Smith, R. 2016. Confidence in Paper-Based and Electronic Voting Channels: Evidence from Australia. *Australian Journal of Political Science* 51: 68–85.
Stahl, B.C. 2008. *Information Systems Critical Perspectives*, vol. 2. Routledge Studies in Organization and Systems. Hoboken: Taylor & Francis.
Tobias, O., S. Håkan, and D. Peter. 2003. An Information Society for Everyone? *International Communication Gazette (Formerly Gazette)* 65 (4): 347.
Ubaldi, B. 2013. *Open Government Data: Towards Empirical Analysis of Open Government Data Initiatives.* OECD Working Papers on Public Governance, 1–61. OECD Publishing.
van der Hof, S., and M.M. Groothuis. 2011. *Innovating Government Normative, Policy and Technological Dimensions of Modern Government.* The Hague and The Netherlands: T.M.C. Asser Press.
Webster, F. 2007. *Theories of the Information Society.* International Library of Sociology. Hoboken: Taylor & Francis.
Weinberger, J. 1988. Liberal Democracy and the Problem of Technology. In *Democratic Theory and Technological Society*, ed. R.B. Day, R. Beiner, and J. Masciulli. Armonk, NY: M. E. Sharpe.
Wilhelm, A.G. 2000. *Democracy in the Digital Age.* New York: Routledge.
Wyatt, S. 2007. Technological Determinism Is Dead: Long Live Technological Determinism. In *The Handbook of Science and Technology Studies*, ed. E.J. Hackett, O. Amsterdamska, M. Lynch, J. Wajcman. Cambridge: MIT Press.

CHAPTER 4

Systems Theory in Politics

1 Introduction

A systems approach as a grand theory has been attempted in political science prior to the advent of the ICTs technology. This chapter explores David Easton's approach that sought to replace institutional analysis in political science with what he thought was an approach more firmly grounded in the then prevailing understanding of scientific method (Birch 1993: 211). Easton embraced a positivist systems framework encouraging scientists to study political systems (including democracy) as a whole in terms of outputs, inputs, and processes with feedback to understand regime types. Easton's main focus was not democracy as such but political systems. The Eastonian political system translated information from inputs into demands. In that sense, it resonates with Systems Engineering (SE) and its techniques. Thus, this chapter discusses the concept of systems theory, then explores Easton's adaption of it to the study of politics, and its contribution to the field of political science, and finally argues how SE as an interdisciplinary system thinking can be situated in political theory.

Systems theory in the 1940s branched out into Easton's positivist theory applied to politics and Parsons action theory resulting in structural functionalism (SF). Easton found SF was inadequate to system theorising for politics as it promoted a status quo and it was subsumed within Easton's systems analysis which for Easton reduced its analytical value for

S. Bungsraz, *Operationalising e-Democracy through a System Engineering Approach in Mauritius and Australia*,
https://doi.org/10.1007/978-981-15-1777-8_4

politics (Easton 1972). Chapter 4 provides an overview of systems theory to contextualise systems theory in politics. A key aspect flagged earlier is that SE is a modern systems theory which is not positivist as expounded in the rest of the book. SE's constructivism is a modern system construct and differentiates from SF's tendency towards a static equilibrium that replicates an existing system towards a status quo. SE's modern systems theory is one where dynamic equilibrium allows a system to morph into new systems with a new equilibrium that remains dynamic (evolutionary) (Luhmann 1995). SE's modern systems approach allows the system design to evolve, to adapt in a dynamic manner, to change into something new which SF theorising tends to resist from its tendency to replicate the existing system itself. That is SF clones a system, i.e. SF duplicates versus a SE that creates systems with new capabilities—evolutionary design. The SE's changes to the political system are specified from the dynamic interactions of the subsystems (organisations and societies) making up that system, SE allows autopoiesis. As we noted in the Introduction chapter to the book where it was made clear that the book was not adopting a structural functionalist approach, the discussion of Easton is not an endorsement of Easton's methodological position. Rather the discussion is this chapter is intended to do two things: one is to show that systems theory is no stranger to political science, and two, to lay out a context for the discussion of SE in Chapter 5 so that it will be clear that SE is quite different from SF in terms of both its focus and methodological implications. The following discusses systems theory's journey in politics before moving to SE as a novel approach.

2 An Overview of Systems Theory

Prof. von Bertalanffly in his classic article, *An Outline of General System Theory,* claimed system theory 'is a logico-mathematical discipline, which is in itself purely formal, but can be applied to all sciences' and that it 'can be applied to very different fields, such as thermodynamics, biological and medical experimentation, genetics, life insurance statistics' including in the field of political science to study the case of democracy as sought in this book (von Bertalanffy 1950: 138). This new basic scientific discipline, as Prof. von Bertalanffly claimed then, aimed to develop a universal law (isomorphic laws in science) to understand different fields like biology, social sciences, and natural sciences (von Bertalanffy 1950: 136). Science he argued studied phenomena by reducing them to the interplay of its elements which could be studied independently or as a whole (von Bertalanffy 1950: 132). It was assumed that there were common features in all the sciences. This systems theory, also known as systemology in military contexts (Drack and Pouvreau 2015; Blanchard and Fabrycky 1998), has been explored through SE to solve operational problems.

'Systems theory in the broad sense has the character of a basic science' and 'it has its correlate in applied science' and it is closely associated with automation (Buckley 1968: 13). Automation is from cybernetics, and to regulate it requires a feedback and control mechanism. The system construct, in cybernetics for example, is used to provide insight about system behaviour and solve complex interdisciplinary problems like large transportation problems where humans and machines interface (Dewan 1969). Cybernetics 'as a form of communication' system provides analytical possibilities for 'well defined elements of large enough systems' to allow predictive modelling (Mead 1969: 4). Furthermore, in the complex and large systems like contemporary society, cybernetic modelling Mead argued supported quality of decision making as it allowed decision makers to improve on their 'large number of breakdown in thinking' (Mead 1969: 5). Scholars suggested cybernetics in a purposive system allowed accurate prediction for decision making (von Foerster 1969). Systematic communicative capability seemingly improves the quality of decision making. A systemic approach like cybernetics was thought to be able to allow better social control and regulation of society (Buckley 1968). These scholars may have been ahead of their time for anticipating the potential of contemporary devices like computers that potentially allow for decision support systems to inform decision making in open collaborative government (Parycek and Edelmann 2013: 109, 127, 148, 186).

The use of computers in the cybernetics field influenced, along with other issues, political science. The computer, from its information generating power, was anticipated to potentially allow breakthroughs. Cybernetics was initially assuming that the development of computing power would lead to creation of artificial intelligence and autonomous systems (von Foerster 1969). Like the biological organisms, these complex human-made cybernetic systems were to become self-aware and adaptive to their environment. These may have been speculative to some extent as modelling human behaviours and values is complex. For purposive systems, cybernetics investigates 'machines which behave as if they had goals' and 'such machines can simulate goal-seeking behaviour through the feedback principle' (Buckley 1968: 74).

In the social and political sciences, cybernetics system's feedback mechanism was a metaphor to emulate control and the regulation of social and political systems, respectively. The mechanism provided a systemic way to study the relationship from inputs and outputs between the system's processes, for example, to study organisational power in social sciences (Parsons 1956, 1969), and political life for equilibrium and support in political sciences (Easton 1971). In making use of interdisciplinary concepts and general system theory, cybernetics arguably provided potential for new ways of thinking about systems. Complex systems like society, for example, were seemingly made up of both organised as well as disorganised systems' behaviours (Mead 1969: 1–11).

Cybernetics assumed that, from the emergent informational power of computers, new thinking could be brought to bear on social issues to create purposive, self-regulating systems, like living organisms studied in the field of biology (von Foerster 1969). This created new ideas for some scientists (Buckley 1968). In political science, a novel behaviouralism paradigm, using positivism (Easton 1971), aimed to be predictive like cybernetics. Cybernetics' goals as discussed by von Foerster (1969) were for models to predict human, computers, and human and computers' behaviours. Easton assumed that from the informational capacity of computers, he could use factual evidence to deduce political relationships within what he called a political system.

Cybernetics promotes this notion of self-regulating and self-directing social systems (Buckley 1968) known as the cybernetic goal for autopoiesis (self-sustaining system) from which more capable technology for systems like artificial intelligence results (Kaplan 2016; Gordon 2011). This led some scientists to explore the idea of control of behaviours through the 'conscious promotion of social goals' (Buckley 1968: 385). Easton is one of the few American scientists, as explained below, to explore this issue through the use of modern systems analysis (Buckley 1968: 385).

2.1 Systems Theory and Political Theorising: Eastonian Value Free Empiricism

In political science, Easton's positivist based systems concept was an ambitious attempt to develop a unifying concept like Newton's laws in physics for natural sciences (Easton 1966a: 13). Using a positivist approach, Easton intended to develop a grand theory as in the natural sciences.

Through his systems theory, Easton sought a unifying, law-like concept for political science to generalise political phenomena (Easton 1985). To create empirical theory like the natural sciences, Easton promoted his systems analysis approach to investigate political life, claiming it would shift political science's field of methodology and philosophy (Easton 1965a). Systems analysis, he would initially claim, was a value neutral and a positivist approach. But it was a position he would be forced to relinquish down the track (Easton 1971). Due to the presence of varying values, political events are complex to model as some of the values are based on a belief systems that are changing according to time. For Easton, the attempt to quantify these values objectively required that people's beliefs or worldviews were excluded (Miller 1971). This made Easton's positivism lens problematic when analysing political phenomena.

In politics, Easton was a reformist who began from the idea that 'the bulk of social science is being transformed from an enterprise pursuing the traditional approaches to one seeking to model itself upon the methodology of the natural sciences' (Easton 1957a: 111). Easton suggested that systems analysis allowed political systems to be potentially classed in a similar manner to the chemical elements in the periodic table. This classification metaphor for the political system, he posited, allowed clearly articulated properties (variables) that could define what class a democracy belonged to for democratic theorising (Easton 1957b). In his political system framework, Easton argued for *a fortiori* research like those in natural sciences (Easton 1971: 60), to allow a grand theory to emerge.

Following the natural sciences paradigm at the time, Easton was using a positivist approach to develop abstract systematic theory (Easton 1985), that would be law-like with predictive capacity through modelling. The new movement of post-behaviouralism, that displaced behaviouralism, was partly due to a shift towards an applied research focus seeking to solve actual political problems. There was a need to develop solutions from observed political facts but the facts from reality were '*a posteriori*' (Von Mises 1951: 274), and as we discuss below they cannot be mathematically modelled through systems analysis. In seeking a value free model, the fact that Easton required the political scientist to decide what to define as the political system from what was of interest was a value judgement for the scientist and it introduced a degree of bias. Furthermore, Easton mentioned that his approach 'could set forth ['*a priori*'] all the different types of authority relationships in which the members exercise authority in a political system' (Easton 1957b: 4). This would imply the relationships between members in his political system were static over time, now a reasonable assumption is that system changed over time and that change would shape the relationships accordingly in the system. Easton mentioned equilibrium and use of variables to define stability, but the equilibrium states for his political system in practice would shift when the regime changed and new variables emerged (Easton 1971: 268).

There were also associated changes like the prominent acknowledgement of the philosophy of science itself which impacted on understandings of the validity of both qualitative and quantitative research methodologies (Kincaid 1996). The new tendency was for solving actual problems in society, like philosophically to explore what was the best form of political system rather than abstract theorising about political life for a generic political system's survival. Actual problems in society were sometimes solved through speculation that led to discoveries for what constituted a best solution at that time in that context. A positivist perspective was potentially limiting if one was to produce evidence from causal facts only. Interpretation of those facts might lead to a new insight, but it could also limit human imagination and intuition that allowed the exploration of speculative solutions. In the twentieth century, a view was taken of science and technology that its impact on society also mattered; for example atomic fission was good when it was controlled to produce electricity safely but bad when it destroyed through an atomic bomb. People have values which could not be understood solely through logic and reason alone and these values shaped their worldviews and behaviours in politics.

Research needed to justify its practical use in terms of solving actual problems; for example, from a system's perspective it was to investigate a democracy's survival. Democracy was assumed as the best form of governance, but for Easton's theory democracy was a 'second order' problem to solve (Miller 1971; Easton 1965a). Easton explained democracy raised too many questions that 'we have not found it necessary to address ourselves' (Easton 1965a: 481). He seemed to have left this task for others though he was seeking through systems a unifying theory.

This is despite Aristotle's view that a good form should be the goal in governance, which this book posits democracy ought to be. Easton was concerned about the laissez-faire attitude of liberalism (Gunnell 2013: 196), but he also seemed to have difficulty in relating democratic values to his empirical system framework. It did not have a democratic bias as was highlighted by Birch (1993: 221). Easton was also accused of shifting his position to a constructivist one in the mid-1960s. He suggested that a system was 'any aggregate of interactions that we choose to identify may be said to form a system' (Miller 1971: 219). This shift in perspective from Easton may have been a recognition that his theory was not able to solve actual political crises that were the new imperative of the post-behaviouralism movement towards a more applied political science (Miller 1971: 207–208).

Easton's conceptual framework on the political system was too abstract to solve actual problems. Easton's conceptual framework presented confusions due inter alia to inconsistencies that emerged in his later writings as he adapted his works (Miller 1971). For Miller, there were changes in the field given '[p]owerful intellectual currents have arisen in the twentieth century to deny that the scientific method can obtain knowledge, particularly knowledge about man and society, that is generally reliable' (Miller 1971: 209). These changes were necessary as a scientific approach like those of Easton had seemingly not produced the solutions to the political crises of the time that Easton himself had identified with to create his theory described under mood (Easton 1971: 3–31).

Easton (1985: 133) also argued that political science was about policy making. For him, it was 'the study of the way in which decisions for a society are made and considered binding most of the time by most of the people' (Easton 1971: 133). He also clarified his positivist use of values as those things that were valued by the society. This seems like a utilitarian stance rather than an ethical one. His attempt to explain values was seen by some authors as his learning curve about politics (Miller 1971). Miller, using a value theory lens, suggested that Easton was seemingly exploring politics and his inconsistencies in his writings were from his learning over his two decades of scholarly contributions which though Easton attempted to adapt and change it remained unsatisfactory. Easton's system theory was something that did not meet the value theory lens that Miller applied (Miller 1971).

Miller argued that political change must 'make an intelligent distinction between beneficial and harmful changes, it must be guided by an understanding of the good and just regime' (Miller 1971: 234–235), and society may choose a good political system for example like a democracy. Easton seemingly had a generic political system in mind and this was hard to contextualise to a democracy, for example. However, Easton was not exploring values for good and bad governance or democratic values for his political system. He was interested in the survival of the political system and a generic one was adequate for that goal. Easton's clarification of his use of values constrained it to policy development but this value creation was applicable to any type of governance system (Easton 1971). In his definition of systems theory for policy making for the whole of society, Easton included institutions to support his concept of allocation of values in a system environment. He developed a method of analysis he called systems analysis to investigate political life of any governance system (Easton 1965a). This technique was to use variables that could be measured and from the data analysis allow predictions for a system's behaviour. Easton's detailed methodology is described in his work *The Political System*. In 1971, he updated *The Political System* to include changes from a paradigm shift in his field to post-behaviouralism, but his stance was still positivist.

To discuss his contribution, this book refers to the updated version of his work *The Political System* (Easton 1971). Unlike some critics of his work, for example (Miller 1971), I believe it is normal that when a ground breaking change is introduced it is reasonable that an original work needs to be improved from the new knowledge emerging in the field. In the era after the initial version of *The Political System* emerged, there were many changes in the field of how scientific research was conducted. These changes were within the philosophy of science itself (Rosenberg 2013; Ladyman 2012; Kincaid 1996; Curd and Cover 1998; Chalmers 2013; Caws 1965). Easton could not remain immune to the new paradigm shift taking place within this field. He was trying to improve research for political science through his new scientific framework. Being pragmatic, he adapted his work which he acknowledged as necessary for a 'search for a value-free or ethically neutral social science [research] has been short-lived' as he concluded that all research were built on value premises from the process of selecting the problem to the data collection step and also its interpretation (Easton 1957a: 113).

Easton repositioned through an acknowledgement that science could be both qualitative and quantitative, and that there was room for behaviouralism as a newer method that he had been introducing to build on the work of the traditionalist (Easton 1957a). He admitted there was a shift from a positivist approach in science, stating '[o]ther criticisms of science were directed at its positivistic claims that behavioural research was value free', and in fact this value neutrality was not objective as it served the 'purpose of the establishment' (Easton 1985: 142). He added that the attack based on this understanding of science broadened when Kuhn suggested that 'all science, natural as well as social, is essentially an irrational process' (Easton 1985: 143). Easton did not support this view of science and he explained that his position was still scientifically oriented (Easton 1985), and he supported empirical research. However, the post behavioural era was just the beginning of an array of new methodologies for research, which he acknowledged was shaping the positivist area of research.

To solve actual political problems, the research field is even richer now from methods that have become acceptable for scientific research. So, from a dearth, which Easton complained about to justify adopting his systems approach in politics, we now have a variety of methods to conduct research in the field. Easton mentioned, for example, Max Weber's '(*verstehen*)' or interpretive method common in contemporary political research as an exemplar of new methods which along with the 'earlier impressionistic methods' now 'have even regained some plausibility' (Easton 1985: 143). To evaluate Easton's contribution further, we explore how his systems theory came about and what it added to the field of politics. There were times when Easton despaired about the lack of reason in society as is explained below.

2.2 Systems Theory and Political Science: Towards Eastonian System Worldview

Easton reviewed the progress of political theory made over time and commented on the richness of theories in the post-1945 era which he claimed was the 'major turning point in the history of political science' from what had occurred 'over the previous 2000 years' (Easton 1966a: 1). He felt that politics should look externally and he identified some key theories that he felt contributed to decision making in politics (Easton 1966a). He also felt that a classification of theories like grand, partial and low-level theories might help the field with a grand theory being law-like, thus providing generalisation to explain political system behaviour both past, present and potentially in future like Newton's laws in physics.

In his view, Aristotle was 'the first significant scholar in the West to treat fact seriously' (Easton 1971: 7). In that sense, Aristotle is the first system engineer to design six forms of governance but then based on moral values he differentiated the good forms from the bad forms. This differentiated the constructivism inherent in Aristotle's work in comparison with Easton's value free empiricism that was limited to facts. Easton deplored speculation from those researchers that did not use facts.

According to him, scholars like Hobbes and Spinoza created their 'elaborate doctrines' from speculative axioms that were built from 'casual empirical observations' rather than from scientific observations like he assumed Newton had conducted (Easton 1971: 8). He added that it was fortunate that these speculative doctrines were a close model of reality. This would seem to mean that intuition and inductive methods were not recommended. Easton also mentioned that utilitarianism as a concept in the eighteenth century used facts and not 'a priori knowledge' for him Bentham observed and then deduced that men desire happiness (Easton 1971: 10).

Another speculative example, which he disparaged, was Rousseau's definition of primitive man as an argument 'against the growth of technology and the ubiquity of reason' (Easton 1971: 10). In so doing, Easton seemed to be defending a need for quantifiable means to measure political phenomena, or empirical theorising in politics. However, this positivist stance also opened him to critique about the extent to which political phenomena could be measurable. Norris though, as mentioned in Chapter 2, used Easton's systems concept of 'support' to measure democratic deficit.

The nineteenth century was the era of scientific rationality according to Easton (1971: 10–15). There was a significant move to systemic observations to analyse contemporary behaviour in society for theory creation. According to Easton, the key figures of the time were Comte, Marx, and Spencer for their study of society which excluded intuitive premises. Easton regarded Comte as a trendsetting positivist scholar and the creator of 'a closed scientific system' (Easton 1971: 11). The shared scholarly belief then was that science will solve any social issue, moral or causal. It could integrate both values and facts. The assumption was that the world was rational not chaotic, and men were rational creatures that could create a 'positive (scientific) polity', as Spencer, Comte, and Marx searched for 'laws of development and interactions' (Easton 1971: 12). For Easton, their research developed transformative theories that had a massive impact. Comte explained the evolution cycle of society as moving from religious to metaphysical and finally positive, while Marx's cycle was from 'primitive communism' to feudalism, capitalism, and socialism (Easton 1971: 12).

Easton was exploring political system's life cycle analogy to comment on various theories. He was using a biological cycle where an organism grew, was stressed, and then aged and died. His system analysis was focused on the life cycle of his political system or how a system persisted. Easton's views of political life seemingly aligned with the possibilities that emerged with systems theory when seeking law-like theorising which Newton achieved in science. The life processes of a system were its fundamental functions without which the system died, these processes he posited were the central issue of political theory (Easton 1966a: 143). System was the starting point of analysis for theory as an open adaptive system (Easton 1966a: 144). This was a view he reflected through his strong critique at Easton (1972) of exchange theory. Easton viewed exchange theory as a subset of systems theory. It showed his belief for systems theory as an all-inclusive theory, and in its empiricism potential, for his stance was for scientific theorising which he promoted. He was seeking through his attempt a new kind of empirical theory that 'seeks to illuminate the whole domain of political interaction' (Easton 1966a: 2).

Scientific positivism was an age of optimism, but when Easton reflected on his era, he was pessimistic for he noted the scientific method is no longer associated with social good and there was a 'growing disillusionment' (Easton 1971: 5). His pessimism was apparent in his work *The Political System* where he critiqued positively other scholars like Pareto's and Sorokin's claim that 'social change fluctuates aimlessly' (Easton 1971: 17). An implication from Easton was that non-scientific methods were failing to deliver social improvement outcomes. He was against the belief that society could not be improved through science, and he consistently abhorred speculation in research that was irrational and non-scientific. Easton projected his faith in science as a means to find political solution through system theorising. He was concerned with the way human beings resorted to violence to fix stress in society and equally concerned with the rise of some of the doctrines to justify such violence.

Easton's system study was a means to an end and that was for defusing stress in society through empirical theorising. He was seemingly building up a case of desperate times that science needs to address to avoid chaos as an outcome from stress build up in society, he even linked Marx's theory to violence. Similarly, his view of Sorel's work was that from social stress it seemingly ended in violent outcomes (Easton 1971: 18). Easton found such contributions scientifically incomplete as they failed to explain the cause for change in society, like from his perspective regime failure in society resulting from stress.

Easton opposed doctrines as they aroused emotions and laid their authority upon tradition (Easton 1971: 19). For example, he found that an increased theological membership was linked to an irrational behaviour observed from a loss of faith that occurred when theology was displaced by scientific reasoning in the natural and social sciences (Easton 1971: 19). Human behaviour in his view was such that hegemonic beliefs like theology could not be displaced even by facts or empirical theorising that brings stability and control. In fact it would seem that his experiences of people who were controlling the interdisciplinary research may also have been a target of such criticism as 'much of the effort was under the control of individuals such as Mortimer Adler, Richard McKeon, Yves Simon, and those attached to the newly created Committee on Social Thought chaired by John Nef, which had close ties to the theologically oriented *Review of Politics* and theorists such as Voegelin' (Gunnell 2013: 194). To Easton a desire for spirituality created through an appeal to faith led back to theology, a path that for him led research away from reason to 'revelations' (Easton 1971: 19).

This desperation with irrationality in human behaviour allowed him to comment that 'democracy too ought to avail itself of blind faith of which people seemed capable' as the 'flight is from political order' (Easton 1971: 20). Easton was suggesting that democracy could outbid its competitor if it were to appeal to passion. He was berating moral values and the sceptical scientists in politics for losing faith in reason, for he added 'especially moral ones, the critics of scientific method have insisted upon a need for a revival of an emotional attachment to high spiritual ideals' (Easton 1971: 19). He believed that there was a 'concerted attack against use of scientific method' (Easton 1971: 15) and that it was leading to a crisis in political science. He felt concerned about the loss of belief in reason and the norm and alternatives that were being perpetuated in political research. Even the political scientist's knowledge creation techniques failed when he scrutinised them for their reliability. He contested the argument offered which was that social laws unlike physical laws are not as enduring. This for him was seemingly an excuse for not exploring for scientific answers. Easton expounded that scientific methods like behaviouralism would allow empirical understanding of democracy by creating 'middle range theories' in political science (Easton 1985: 138). He provides a listing of the 'major tenets' of the differences between the traditional and behavioural research at Easton (1965b: 7).

Easton assumed research in political science could only discover something that was true for a particular time and place (Easton 1971: 35). Political theories were limited to generalise as 'political science has been reluctant to inquire into such a theory' (Easton 1971: 65). With this assumption, any relationships that were discovered in political science were true for the human behaviours observed at that time. This is reasonable as human behaviour changes with time, given the changing nature of the culture and society humans are embedded in. Therefore, it would seem reasonable to assume their values are not constant.

Easton reflected on the limitations of cross-cultural theories for generalisation being produced in research noting that 'after the sanguine expectations of the nineteenth century' it appeared that 'scientific reason excludes from its scope and skill the discovery of ultimate values a society ought to pursue' (Easton 1971: 35). This was perhaps a potent reason for Easton to embrace a less positivist stance. But his arguments in *The Political System* were that science 'dependence on mathematics was not accidental' (Easton 1971: 9), it delineated the different paradigm that he proposed with his systems theory. However, even with mathematical modelling it would seem that human behaviour seems too complex to solve for his systems analysis. This would limit predictive models to be developed from accurately simulated political or human behaviours due to indefinable variables.

For Easton though, political theories must be based on facts. He argued against an unsubstantiated extension of facts like when using 'historical positivism' which he treated as speculative at Easton (1971: 74). This idea sums up Easton's search for a scientific way to conduct research that is positivist. Easton critiqued the narrowness of the theories that emerge from historicism, for example like elite theory, which focuses on state and power, calling them 'narrow-gauge' theories (Easton 1971: 121). His scientific bent would even bring him to suggest a democratic cataloguing arrangement that could support his generic political system to find what properties defined the causes for a political system to change from one type of regime to another (Easton 1957b).

Easton assumed that if somehow political systems were to be classified then there would be specific variables developed to define those systems' (regimes) properties. Now it is possible that a grouping could be used to accurately define something, like elements in chemical tables, but if the model or element being described is not static, then this task would become very complex. Easton did not seem daunted by this prospect though. With the advent of computers (cybernetics), perhaps he assumed that a systemic solution was possible. This classification need for Easton was due to his belief that the traditional methods were stifling creation of new ideas. He described this concern with the bias of traditionalist methods in their being continuously projected as some superior skills from the past to shape the present and future (Easton 1957a). This critique explains his need to innovate and would seem a laudable venture to embark on theoretically. One of the causes that he identified was traditionalist research that was stifling new thinking from the way students were being trained to develop research skills. Traditionalism was about recreating an existing paradigm. To those trained in traditionalism, his systems theory challenge to a hegemonic view may have seemed very radical at that time. However, Easton was not the average conformist, his training was in the social welfare system and his teachers, one who was anti-capitalist, and another an expert in communication theory, seem to have shaped his thinking (Gunnell 2013: 192).

Easton berated the educational system as being responsible for perpetuating a myth which he believed the empirical methods that he was proposing would dispel and he even proposed a curriculum to teach it (Easton 1966b). His concerns were about bias in perpetuating a given paradigm that remained unchallenged. He argued a continued hegemonic paradigm eliminated new thinking. This hegemony then shaped the type of theorising that emerged for it was not creating new thinking that could, for example, relate policy making from people's inputs to a system's survival. He also expanded on a need for holism like a systems approach for studying what is now associated with policy making in a political system. This policy-making process he described as the authoritative allocation of values for society (Easton 1966b). Values which were linked to demands which he presumed could be measured from studying stress and its impact on his unit of analysis the political system. Easton was visionary as he captured the essence of communication on the core issue of policy making and the resultant stress when carried incorrectly.

3 Eastonian Contributions to Political Theory

David Easton's systems theory emerged between 1950s and 1970s to reform political science through a set of three works *The Political System*, *A Framework for Political Analysis*, and *A Systems Analysis of Political Life* (Easton 1957a, 1965a, 1965b, 1971). For Easton, 'the nineteenth century witnessed the transformation of American political science *from discourse to discipline*' (Easton et al. 1995: 132). Seemingly, Easton was the first 'to replace the state and its institutions' with the concept of 'the political system and its processes' (Birch 1993: 220). Using the idea of 'political system', Easton aimed to establish an enquiry mode that was based on pure sciences. Birch suggests that all serious studies of politics requires a knowledge of Easton's contribution, which 'was an ambitious attempt to build a large scale conceptual framework known as system theory' (Birch 1993: 220). We explore Easton's contributions to show how SE might be seen as being in the spirit of Easton's approach and indeed how it be able to address some of the problems that bedevilled his approach.

Through his extensive works, Easton provided a significant impetus to systems theory by developing a whole method of inquiry. In 1953, he introduces the idea of 'system' in politics and the idea of political life to explore changes in political systems through his work in *The Political System*. Easton's political system was for exploring political life that he argued was key to understanding regime changes from stress due to inputs that became unmet demands, this application of systems theory to politics is depicted in Fig. 1.

According to him, political science was 'the study of the authoritative allocation of values for a society' during the life of the political system (Easton 1971: 129). Political life, according to him, was 'all those varieties of activity that influence significantly the kind of authoritative policy adopted by society and the way it is put in practice' (Easton 1971: 128). He explained his conception of political life as participating in an activity that related in some way in making and executing policy for society. For

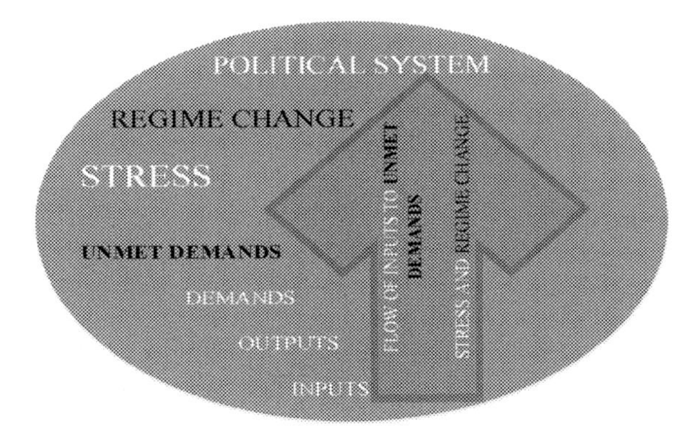

Fig. 1 The Easton model from inputs to regime change from unmet demands (Adapted from Easton [1971])

Easton, a decision was not a policy, a policy was a web of decisions and actions that allocated values (Easton 1971: 130). Easton here was drawing on the whole process of policy making from inputs to outcomes, where conflict in decision making leads to unmet demands that create stress to the system. Thus, he links these inputs that become unmet demands to stress as depicted in Fig. 1.

Stress is a trigger for decision makers to react to either change the system, a regime change, or in rare occasion if the stress was catastrophic the system died. Stress is important to understand Easton's attempt to develop a pure form of political science through what he called the common sense idea of 'political life' (Easton 1971: 126). He posited that in political science the authoritative 'policy-making process constitutes the political system' (Easton 1971: 129). Easton claimed he was searching for 'uniformities in political relationships' to create a theory (Easton 1971: 86). He differentiated between pure and applied research and he viewed his efforts as one towards pure scientific research, which was his self-appointed goal as a systematic theory for political science. He even wanted to split political science research into a 'pure political research' and an 'applied political research' (Easton 1971: 87–88). His political system allowed scholars to study the processes that started from people's inputs, which if demands remained unmet in the political system being observed, the stressed system caused regime change, this is depicted in Fig. 1.

3.1 Eastonian Systems Analysis

Systems analysis was a concept that Easton adopted from communications to explain how a political system 'itself comes into existence and changes in its basic content and structure' (Easton 1965a: 475). His systems analysis is a mode of analysis that 'tell us little about the way in which any particular type of system, such as a democracy, might persist' (Easton 1965a: 480). Communication of information he projected allowed authorities to take action so the political system persisted (Easton 1966a: 152–153); this was for any type of system, democracy or autocracy for example. Easton's key variables were 'the making and execution of decisions for society' and 'their relative frequency of acceptance as authoritative or binding by the bulk of society' (Easton 1965a: 96).

Clearly, Easton created a modern political theory, a revolution as he claimed (Easton 1965b: 3), aimed through developing his framework for a general theory but it is of limited use to democratic theorising. In the 1950s, he mentioned that though various theories were created the scientific ones did not seemingly influence politics which was apparently quite static for empirical theory (Easton 1966a). Easton (1966a: 7) applauded the scholars external to politics for creating various empirically oriented theories that explored power, decision making and 'systems analysis from communication sciences' thereby integrating other disciplines into political science.

Easton's concern with modernisation of the field saw him urging for a shift from theory which exclusively implied moral philosophy in its various forms to a modern one where one looked externally and borrowed from other fields to adapt them (Easton 1966a: 1–13). Now like Aristotle and Plato (Dunn and Harris 1997), before him, Easton believed in educating about politics. He did not propose the philosopher king concept, but recommended a close working relationship between political scientists and politicians. He wanted a paradigm change for political science students by using the systems concept to solve political problems and to innovate a systems analysis framework that created alternative strategies for political research (Easton 1966a). Systems analysis, he assumed, would link academia and politicians to develop the right synergy to solve the problems of society based on facts. In this way, he provided some promises of being a visionary as he understood perhaps the limits of his framework and the dangers of being stuck in a given paradigm.

To modernise, he brought together various concepts from interdisciplinary fields to create his positivist theory. He even developed curriculum for systems approach to be taught in the political sciences (Easton 1966b). He considered that variables from systems analysis would be found to mirror reality and this would allow a better understanding of the world as is. A modern theory, he argued, must innovate through creation of new information from logical operations that 'extend the horizons of our understanding and explanation' (Easton 1966a: 2). He claimed that modern political theory, by exploring political behaviour, represented a break from the past and that political science must stop being inward looking.

For Easton, the revolution of methods and techniques that is accompanying modernisation in the sciences was 'usually described as the study of political behaviour' in political science (Easton 1966a: 2). He claimed that

> In looking at the relationship between political science and relevant theories external to it, we are reminded that in every age there have been dominant political patterns that have filtered into the basic areas of political knowledge. But these patterns become dominant, not through any mystical, ethereal force but because they are in fact borrowed by one discipline from another and adapted to the needs of the host. (Easton 1966a: 13)

An interdisciplinary approach allows breakthroughs that are green field. This insight allowed him to borrow from the biological sciences to explore political life as an open system. The list of theories he reviewed mentioned above (Easton 1966a), he claimed was not exhaustive but as external theories they allowed insight into systemic decision making. Decision-making theoretical insights, he claimed, were core aspect of politics and with the social-choice processes it could explore power as the criteria. He also offered other theoretical links, for example systems analysis, from what he believed to be a 'growing field of knowledge – the communication sciences – in which the dominant theoretical direction can be described as systems analysis'. He also acknowledged the work of Anatol Rapoport in postulating that the '"system properties" of political life' were important to study political events (Easton 1966a: 11). His attempt to open the field of politics to interdisciplinary scholars was summed up in the following 'to draw the attention to the variety of seminal ideas in theory that are available outside the normal range of political science, and that can serve to stimulate the discovery of additional patterns of analysis by political science itself' (Easton 1966a: 11). Interdisciplinary integration can be quite complex, with systems analysis he introduced, what he termed, a key concept—equilibrium theory.

3.2 Eastonian Systems Equilibrium: Input/Output and Support Stress

Easton (1966a: 143–154) mentioned equilibrium theory in his revised essay on systems analysis. In the essay, he was considering the persistence of a political system in a world which might be stable or changing. At this stage, he was exploring equilibrium theory assuming it was the crucial aspect to allow a system's persistence. The analogy he used for explaining persistence was the life processes of a political system. He claimed that it was crucial to analyse the life processes and the responses from the system to find the equilibrium that allowed a political system to persist. He assumed the persistence of a political system from its behaviour as an open system was analogous to an adaptive living entity. The system sustained itself from its ability to manage the conversion of the inputs and outputs for ongoing support. Political life to him was a complex set of processes, and

> Although I shall end by arguing that it is useful to interpret political life as a complex set of processes through which certain kinds of inputs are converted into certain type of outputs we may call authoritative policies, decisions, and implementing actions, it is useful at the outset to take a somewhat simpler approach. We may begin by viewing political life as a system of behaviour embedded in an environment to the influences of which the political system is exposed and in turn reacts. (Easton 1966a: 144)

Political system as a concept seemed abstract at the time to his peers. Easton's goal was to simplify, but with a system of behaviour as political life he only created further complexity. He then brought in both dynamic and static equilibrium to mean stability. Easton was faced with meaning issues, an argument he advanced was that '*equilibrium* and *stability* are usually assumed to mean the same thing' linguistically (Easton 1966a: 145). It was assumed that a stable political system was in equilibrium. If disturbed, the equilibrium caused the system to return to its original state of stability, this was position 'O' as shown in Fig. 2. Easton borrowed and developed the concept of strain from engineering whereby the state of the system may not behave in an elastic manner (returns to original state after being stressed) at the time of disturbance (stress forces change and strains the element) (Silver et al. 2013: 45).

The analogy was that when a political system is disturbed (changed) from its equilibrium state, it would tend to return to a new state of equilibrium which was not the original one (if strained outside its elastic limit, for example plastic deformation occurred). However, it may so happen the system adapted (adaptive system concept) and attained a new state of behaviour or equilibrium due to it being strained, the point 'S' became a new regime. There was a point 'A' (yield limit- maximum change) where too much stress from changes would overstrain the system to fail or perish at point 'P'. To sum up, the assumption was that stress led to

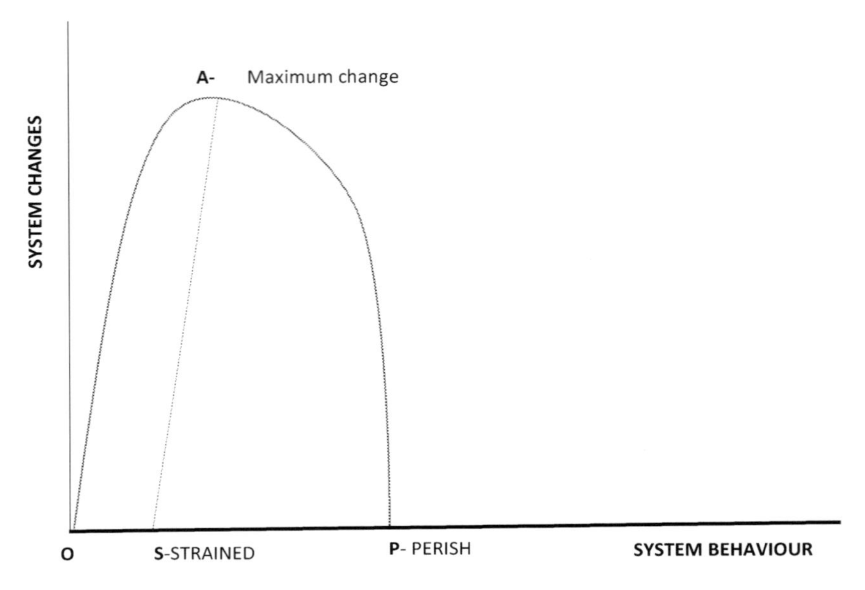

Fig. 2 The engineering analogy for political system stress and regimes change (Adapted from Silver et al. [2013])

regime change as the political system moved from one equilibrium point to another.

The case of the political system does not exactly follow the engineering stress-strain curves analogy shown above but does simulate it to some extent. The system can also transform adapting to survive. According to Easton, this made it unique and he used the analogy of an open system (a living organism from biology) that learned from inputs to survive by adapting itself. The strained position became the new point of equilibrium and so the system persisted. This would imply that an event that could destroy a political system would have to be very unique (catastrophic). Easton called this ability of the political system to survive as self-sustenance, it was seen in biological entities which adapt themselves to their environment to survive. His idea was that at a higher conceptual level a political entity could change to a different regime (democracy to autocracy) and survive. One should not read into this that he supported an autocracy.

A regime could change from a democracy to an autocracy (tyranny) for example, but both would still have a political system in place. Easton argued that by studying the political system from this perspective, one could find the conditions that might allow a democracy for example to survive. He argued that the democratic political system could adapt, it could make 'efforts-limited only by the variety of human skills, resources, and ingenuity-to control, modify, or fundamentally change the environment or the system or both' (Easton 1966a: 147). The outcome for a democracy to persist could be through a system that was able to adapt by defusing the stress being caused by the changes. Easton conceded, in rare occasions a system could perish if the changes were catastrophic. It would not matter whether the regime was democratic or autocratic when such an event (perish) happened from his theoretical perspective. This stand allowed Easton to explore power and political members' intent towards the political system.

Now there were some issues that Easton had identified with the equilibrium theory. One was that the intent of the members in the political system was to return the system to its initial state of equilibrium, to make stability a priority. Thus, members of a democracy must be willing to return the system to a democracy. Secondly, the system was assumed to use the same path to return to that initial state and that this return to the initial state was without resistance. However, Easton (1966a: 146) also stated that members of a political system might wish to destroy the previous equilibrium or even keep it in a state of disequilibrium, for example when they wanted to keep themselves in power through creating fear of external dangers or internal threats. Easton brought the analogy of engineering elasticity of a system by stating that 'it is a primary characteristic of all systems that they are able to adopt a wide range of actions of a positive, constructive, and innovative sort for warding off or absorbing any forces of displacement' (Easton 1966a: 146), and he posited that 'a system has the capacity for creative and constructive regulation for disturbances'. These were two issues he intended to explain through the systems analysis theory. As to address the stress, he brought in the concept of feedback for managing equilibrium so his system could persist.

The feedback mechanism was an important part of Easton's framework to allow the system to react to persist. With this concept borrowed from cybernetics, he provided a 360 degrees feedback mechanism where inputs and outputs could be measured to allow the system to survive by taking timely corrective action, if stressed. The decision-making process became an ongoing cycle in the loop. Overall, Easton had attempted to integrate the scientific concepts of his time but at times there seemed to have been a hasty integration that has created more holes than answered questions.

The criteria for the political life that he analysed had its objective that it allowed the system to persist despite the strains placed on it by unmet demands. The systemic constraints caused stress to which the system responded or adapted by identifying the stress through its feedback mechanism. Feedback was a control mechanism and initially applied by Easton in this sense as he mentioned to develop a systems approach in politics he had borrowed from cybernetics and other technological fields like communications (Easton 1965a, 1965b, 1971). So, we would assume that the system Easton had in mind is mechanistic such that the variables, when identified, had an upper and lower limit that for equilibrium would allow appropriate actions to be triggered for the system to respond for its survival.

Such mechanical control, through computers for example where if it was made autonomous, when implemented could deliver or perpetuate an autocratic governance system. It could become strictly rules based and potentially tyrannical in nature. The people could be constrained to the point of system equilibrium before their unmet demand became identified by its regulating mechanism as a threat to the system's survival. A threat which would trigger an automated response for the political system to react for its survival. This may be an undesirable outcome as a tyranny can keep persisting through appropriate responses. In a democracy, an assumption is that the representative will accommodate the people's demands or lose support but a ruler by optimising the regulatory mechanism could potentially maintain a status quo.

Easton himself explained that he provided the 'skeletal outline' for systems analysis and that it posed more question than it answered like

> What precisely is the nature of the influences acting upon the political system? How are they communicated to a system? In what ways, if any, have systems typically sought to cope with stress? What kinds of feedback processes must exist in any system to acquire and exploit the potential for acting so as to ameliorate these conditions of stress? How do different types of systems- modern or developing, democratic or authoritarian- differ with regard to their types of inputs, outputs, and internal conversion and feedback processes? What effects do these differences have upon the capacity of the system to persist in face of stress? (Easton 1966a: 154)

Easton presumed to have answered these questions in his two works *A Framework for Political Analysis* and *A Systems Analysis for Political Life*. Seemingly, there is more work ahead before a systems theory becomes well integrated into politics. But the work of Easton has created a significant leap in that long journey, for it opened up an alternative way of thinking about the political (Easton 1966a). As a conceptual framework, it allowed analysis for phenomena that might be of interest to a scientist. However, as deficiencies are uncovered with the behavioural movement, in the late 1960s, change is on its way. This shift leads Easton to acknowledge in September 1969 that an era of post-behaviouralism is the new norm (Easton 1971, 1985). There have been critiques to Easton's positivist stance.

4 A Critique of Easton's Work

For Easton, the systems theory approach served to differentiate analysis from historicism. As part of this approach, it rejected treating the state as a key explanatory concept, seemingly due to its previous association to historicism (Easton 1981). Easton aimed to develop a scientific approach to research in politics. However, this delinking from previous ways of studying politics led to inconsistencies in his approach. For example, Miller's interpretation of Easton's work pointed to the problem of how values were to be accounted for, and this went beyond the question of which values were to be given priority and how they might be embedded in the systems theory approach. Miller also pointed to the values that a political scientist might bring to the research process in the sense that the researcher's values affected the research process. In a general way, Easton was aware of this (Easton 1971), as it had been acknowledged as a limitation with positivism in scientific research (Von Mises 1951; Ball 2007).

4.1 Inconsistency to Understanding Values

Values shape a person's world views and are contextual; thus limiting predictive modelling accuracy. Miller observes that 'the values which shape thinking and colour observation are said to vary in an inexplicable way from one epoch, culture, or society to another: Men think and perceive differently at different times and places because their values are different' (Miller 1971: 210). The post-behavioural revolution that Easton identified to supersede behaviouralism was due to behaviouralism missing the contributions of moral values. Historicists agree that no set of values transcends time and remain valid in all contexts (Miller 1971: 211). Values provide a degree of subjectivity and 'thus cannot have the status of final or objective knowledge' (Miller 1971: 211). Some form of positivism accommodates subjective views however observed reality limits logical interpretation through mathematical modelling for the 'rules of logic correspond as an approximation to certain facts of everyday life' (Von Mises 1951: 5). Human values could not be accurately measured using variables.

4.2 *Inconsistency to Other Political Science Theory*

Easton's works used two variables and these were critiqued by scholars for their contribution to theory in political science. Reflections on his inconsistencies with variables were from Berlinski (1976: 122–127) where Easton's contributions to systems analysis were examined using the control theory lens. Berlinski argued that even if variables were to be created to develop an empirical model for the political system, it would not be able to model the system's behaviour accurately. Thus mathematically, the model provided by Easton for inputs and outputs would have no predictive capability to explain exact behaviour of the political system (Berlinski 1976: 112–154). Even though Easton advocated use of interdisciplinary fields to develop his systems theory approach, it cannot be assumed that biological or cybernetics (engineering) assumed the same notion of system as Easton did in his theorising. The metaphor that Easton used would therefore create inconsistencies. It would also require too many variables to be able to accurately reflect the system it wanted to predict. However, Easton's works should not be seen in isolation though.

Miller's review creates inconsistency with terminology in Easton's works. Using value theory, Miller raised issues, like ambiguity of terms, for example system, society, value and political (Miller 1971). This type of critique was normal as Easton similarly critiqued in his review of other people's works, like the exchange theory at Easton (1972), to promote his work over the exchange theory's contribution. However, Easton to his benefit was bringing in a new thinking in politics that he believed was comprehensive like system without a link to ideological values. He mentioned in his works of his very strong faith in humans and their ability to use science to fix the issues emerging in politics from ideologies (Miller

1971; Gunnell 2013; Easton et al. 1995, 2002; Easton 1965a, 1965b, 1966a, 1971, 1975).

In the second edition of *The Political System*, Easton (1971) responded to some of his critics. Easton reminded his critics that he was not after 'a new science but of an appropriate contemporary strategy for science' in politics. In this light, the mathematical lens of Berlinski (1976) control theory for system analysis was not what Easton was proposing. Berlinski's comment may be correct in the context of control theory but this was not what Easton was proposing to apply to systems analysis. Systems analysis has been used in other field like engineering (Antill and Wood-Harper 1985), this was also criticised by Berlinski (1976) as mathematically inaccurate.

4.3 Inconsistency to Adapt Communication Field

Now, if systems analysis was borrowed from the communications field and adapted, it was not adapted in such a way where it focused on the communication process (inputs and outputs) that was needed to improve the political system's concept. For example, the authorities in Easton's systems analysis were not informed about how to improve the governing processes. An attempt was made towards bringing in inputs and outputs towards creating relationship between systems. But again with a complexity generated from lack of clearly articulating what a political system should be (Miller 1971), the ambiguity it created reduced its analytical value. Different systems (cultures) react differently even though they could be democratic in nature.

4.4 Inconsistencies with Parsimony and Abstraction

Another critique against Easton was his ambiguous use of the system construct which created confusion about the framework he promoted. This reduced the political value of his work accordingly (Miller 1971), but this is not unusual given democracy itself cannot be accurately defined. Though the systems concept was hard to comprehend for some at the time of its appearance in politics, systems as a term is extensively used now.

A key problem, however, concerns the level of abstraction at which Easton's approach operated in terms of connecting it with the actual circuits of power, including that of the state and its various institutions and processes. This seemed one of the major drawback in the 1970s for Easton. With the return to prominence of the state in political analysis, Easton regarded this as a threat as he felt that behind such a move were concepts that were adaptable even to the ideology of Marx which challenged his liberal views. A focus on the state was a move away from what he regarded was a scientifically oriented analytical framework. An alternative might have been to explain how the state could be conceptualised as a system and it might have reduced some of the abstractness, but it would have complicated his framework. So, Easton seemed to trade off the degree of abstractness for the simplification that it provided for his theorising. An example of his simplification was his reducing his definition of what was a political system to two major variables which according to him were sufficient as 'all other variables may be considered non-essential or incidental' (Easton 1965a: 96).

4.5 Inconsistency to Applied Political Science

A key problem with Easton's approach was that he assumed what he called reform theories which was from applied science was not a goal for his systematic theory. In creating his idea of a political system (Easton 1971: 96), he missed the intent of an applied science which was to provide solutions. Easton berates the focus of political science in developing applied solutions (Easton 1971: 78–97), as a revamping activity. A revamping activity for him meant it was used to collate facts to 'suggest some projected reform of the political process or structure' (Easton 1971: 79). This imposed some limitations on what political science could be and could deliver. Easton even felt threatened by the concept of the state and some of his peers' views on theorising, as we explain below. In politics, the purist scientific stance limited the idea of systems theory as projected by Easton for democracy. His arguments for a lack of limitations for his systems theory actually limited his theorising as it could not explain democracy (Birch 1993). In fact, I argue that an applied science like a Systems Engineering approach is a transformative systems approach that has the ability to design a democracy as decided by the people using technology.

Despite these criticisms, Easton is mentioned by some as one of the few political scientists 'to use modern systems analysis seriously and extensively as the basis of a model of political process as an input-output, feedback-controlled, goal seeking system' (Buckley 1968: 385). One of the goals to be sought through systems analysis is a democracy perhaps where people decide, but this was not Easton's focus. The rulers' decision-making issue noted by Birch is but one of the key democratic problem with Easton's works. However, this is an issue that democratic systems in practice are yet to solve as well, for even though ICTs allows extensive reach to potentially engage every citizen, a virtual *polis* is yet to emerge. Those for terminology raised by Miller (1971), we assume that Easton's works and its evolutionary nature contributed to such nascent issues. This is not uncommon in new fields, for example like the ICTs where a proliferation of e-terms are noted (Crespo et al. 2013: 2). Easton saw the potential of his work to add to the systems thinking which the emergent technology like computers and now ICTs could extend on.

4.6 Inconsistency to Engineering

To explain stress to the system, Easton mentioned that not all disturbances caused strain to the system, some disturbances might be favourable to the system, but he failed to articulate clearly in what context, he did not explain if his political system had an elastic limit for democracy, for example. Once he moved to the 'two distinctive properties – the allocations of values for a society and the relative frequency of compliance with them – are the *essential variables* of political life' (Easton 1966a: 148), as Miller argued, Easton created inconsistencies. He alluded to structural engineering when he posited that if the essential variables were pushed beyond what we might designate as their critical range mentioned above, Fig. 2, stress occurred. He failed to mention what degree of stress he was referring to or the elasticity limits within the system which he was alluding to earlier to explain stable state equilibrium theory. He concluded that the severity of the stresses caused a system to collapse and that the political system disappeared, yet at the same time he treated the resilience of the system to persist under stress, these are what I call within its elastic and plastic ranges. A system might exist even if it is strained (deformed), when it is stressed to what is alluded in engineering as the plastic range.

In engineering, beyond the plastic range the system can reach a critical load, yield limit such that it deformed, and the deformation continued even though the load was reduced. The system keeps deforming and subsequently fails even under lower stresses. In trying to adapt this analogy from structural engineering (Silver et al. 2013), Easton complicates by using a decision's acceptability as the criteria to measure deformation within the system. He was neglecting a key ingredient, the lack of support from citizens. Citizens' tolerance of a regime might have a limit and this limit was not clearly defined, also for example a revolution would not fit in his framework if it was an opportunistic one by a military coup that became civil war.

Easton mentioned that the life processes of the political system allowed it to adapt, as the informed authorities had the ability to react whether 'they desire or are compelled to do so' (Buckley 1968: 428–436). What had started as an elegant system with inputs and outputs crossing boundaries that defined the political system gets lost in this overly complicated additions like respond or adapt (Miller 1971: 200). Easton seemingly left open the failure limits or to define how one could identify when critical range was being reached and breached. This made it questionable if this could be applied in every circumstance to develop a generic political theory which was his goal. The degree of stress also created more questions than it answered, if the system was self-perpetuating then did every political system have a point of no return when stress even if reduced would continue the deformation of the system to lead to failure. Regimes might change as mentioned by Miller but Easton did not clearly articulate when this happened in his framework (Miller 1971). The concept of critical limits between regimes would have assisted scientists to measure or even create criteria to measure to develop better forms of governance using a systems approach.

Easton's systems theory allowed an incorporation of external ideas from engineering and the other sciences to enrich the tools available to scholars for pursuing their own search for answers. But one would tend to agree with Miller that the theory seems incomplete to deliver a grand theory like Newton's laws in physics where the applicable constraints for the laws to work are known. Though undoubtedly, Easton did provide a significant body of work to incorporate a systems approach in politics. Perhaps a focus on democracy rather than a generic system would have provided a more pragmatic body of works as the critical range of democracy could then have been defined, thus leading to a significant scientific body of work in politics for democratising and managing a democracy to persist. Democratic limits which if defined would lead engineers to build rather than design the system democracy.

Easton also relaxed the idea of a political system to mean 'any set of variables regardless of the degree of interrelationship among them' (Easton 1966a: 147). According to him, the set, any set, selected to be a system had to be an interesting one in the sense that it explained some aspect of human behaviour significant to us. This potentially makes the number of variables for a set infinite in a system. Furthermore, he divided the environment of the political system into two, one he called intra-social and the other extra-social. This was where he started to develop significant complexity which with the idea of 'values that are authoritatively allocated' reduced the clarity of his theoretical position (Miller 1971). It made the political system ambiguous by adding too much detail at times and too little at others. Undoubtedly, his contribution at the time when it emerged was significant as it was novel, and his intent was laudable, that was an attempt to improve the state of affairs within the field. In focusing on generalising the political system, Easton might have created too much complexity and failed to simplify which was one of his goal, for as he stated 'it is useful at the outset to take a somewhat simpler approach'(Easton 1966a: 144). A path he abandoned as he expanded his theory for systems analysis to cover every possible regime.

The term system, which is very common in engineering, was something new at the time and apt to create some inconsistencies. Furthermore, Easton mentions that though there has been a rich array of theories, currently there also seemed to be extensive overlap and duplication in the field. Easton's positivist argument was that scholars in the field seemed to have lost a sense of direction which existed before due to alluding to external factors rather than finding the dependent and independent variables to explain (Easton et al. 2002). A view is that Easton's scaffold was to be used to develop further and better research for he understood he was starting a process that could evolve. This scaffold that he created was built on by Norris to investigate democratic deficit as an independent variable. It is acknowledged by SE which is also a form of systems thinking but from engineering as mentioned in the rest of the book.

5 Relation Between Re-Emergence of State and Eastonian System Worldview

According to contemporary scholars, in Easton's theory 'the actual text had not been fully and carefully analysed' (Gunnell 2013: 198); and, thus it is hard to understand. Easton seemed to have created significant resistance amongst certain groups of scholars who were perpetuating what he perceived as the hegemonic view which had not produced improvements to allow for new thinking. His paradigm shift created the requisite space in the field to allow emergent scholars to create new theory. However, when the state re-emerged within political debates Easton was concerned about scholarly inquiry reverting to the old ways. The issue of the re-emergence of the state according to Easton was problematic to his theory.

Easton expressed his concern that his theory which provided systems analysis to study politics was under siege by the re-emergence of what historicism used to study (Easton 1981). With SE, the state can be reduced into smaller systems which can be aggregated to the nation-state function like an e-democracy, but this reductionism was not something that Easton could accommodate from his theorising. He saw with the state an emergent risk of a reversal to the old ways of conducting research in political science. He was pragmatic though about his own theoretical contributions. We explain this SE reductionism in the next chapter.

Easton (1985: 141–142) reflected on the era that was known as the 'post behavioral revolution' where he accepted that human behaviour was too complex to model, and that the number of variables made the discovery of regularities unpredictable (Doebelin 1980). Easton also shifted his views to accept that his earlier stance of a value free positivist framework was not feasible as even science was not value free from the observer's perspective. Also importantly in the assumption of value free, there was implicit acceptance of the status quo whereby the ideology which shapes the research is that of 'bourgeois liberalism' in 'the existing power structure' (Easton 1985: 142). This could be seen as a response to the earlier critiques of scholars like Miller on his conception of values (Miller 1971). However, Easton was also flagging the danger that a status quo may lead to maintaining elitism, which has been already entrenched within a political system through scholars like Schumpeter (Schumpeter 2010).

A value free stance was false objectivity as it allowed research to withdraw within an ivory tower which resulted in that it did not actively develop solutions to pressing social problems. This allowed problems to fester, and with attacks on scientific methods as practised, according to Easton, it broadened the debates of the basis of epistemology and ontology in social research to create a reassessment of the positivist scientific method that was prevailing during the 1950s to 1960s behavioural era. This paradigm shift also brought about new methods approved by Easton like the 'interpretive method (*verstehen*)' that was developed by Max Weber (Easton 1985: 143). Easton lamented the dilution of the scientific method as an incoming change. Such research, according to him, has resulted in a purposeless political science to solve problems in society.

Easton defended his positivist value stance as he argued that in an actual liberal democracy there was a disconnect in values from what was and what ought to be. If this discrepancy from what is and what ought to be could be measured, then his method would assist a democracy to emerge based on scientific grounds rather than speculation. Once the democratic process was defined, scientific measurement could develop capability measures devised to monitor democratic capability. Process capability is routinely measured in manufacturing to improve quality and reduce wastage. Processes are not capable to deliver a democracy for example, technically waste resources. Given such a democratic process is yet to emerge, Easton may have been ahead of his time, though his framework is no less significant from a lack of such a process.

Easton's systems thinking opened new avenues of inquiry that challenged scholarly thinking about whether democracy as practiced is the right form of governance (Easton 1971). A study of the way policy is formulated in representative democracy might show that the process has been highjacked by vested interests to maintain a status quo. Representative democracy projects itself to be democratising, but instead the existing system actually perpetuates elitism by restricting the people from access to resources through effective gatekeeping by the political system. Such a political system can persist by insidiously perpetuating myths about the complexity of the problem rather than using scientific tools and methods from other disciplines to solve the problem through a systems approach. Easton's systems approach was to challenge and create space for a new thinking in politics. In political theory, Easton's systems theory is a precursor of SE. A commonality is that both refer to systems, but with some differences from the way it is applied. SE is about designing new or existing systems which it creates or upgrades to solve a need, a process that is explained in the next two chapters.

6 Conclusion

Easton's systems theory in political science was a radical approach to political analysis aimed at explaining every regime as an empirical system. Easton intended to avoid metaphysical speculation that was grounded in abstract things like values which he regarded as subjective. Using observable facts, he aimed to apply scientific rigour from systems analysis to create variables to predict systems behaviour. From cybernetics, Easton adapted communications as a key aspect to identify stress from policy demands as he assumed a political system's stress threatened political life of the governance system. The system scaffold that emerged was to allow new thinking to emerge to analyse a political system's behaviour. In neglecting to factor that human interpretations from values do shape a political system, the Easton's approach became limited for good government. As in his pursuit of a simplified positivist model, which was a generic political system, Easton's theory neglected democracy leaving it to others. The systems theory approach had potential, but Easton's lack of focus on democratic practice in its own right and secondly his approach's inability to predict the emerging and increasingly capable information technology reduced its application to contemporary problems.

However, systems thinking in other fields evolved, and one form like SE is used in many engineering ways to design and manage both complexity and large systems (USAF 2010; Eisner 2011; INCOSE 2007). SE allows complex designs to be created using the systems approach, and with a System of Systems (SoS) approach allows autonomic operation of the subsystems within the larger systems (Eisner 2011). SE overcomes a potential shortcoming with Easton's irreducible unit, the political system.

SE provides flexibility to study systems in smaller groups which may or may not be autonomous but add to the function of the overall system, like a system democracy. A democracy exists through values that are democratic, but these values are shaped by the people who contribute these through the SE techniques. SE develops specifications for things that are valued by the people through its techniques like Requirements. When Easton mentions values, he was referring to things that were valued by the people rather than democratic norms. In this sense, Easton developed a system that could be coercive as he was only concerned with the life of the system rather than a purposive governance system geared to deliver the good life through the nation-state for example. Easton may have intended to reflect democratic values but his theorising focus on system persistence through stress missed this democratic goal.

SE's multidisciplinary approach has some similarity with Easton's attempt to integrate other fields with his systems theory in politics. However, there are some differences. Easton assumed the system exists and when it changes from one form to another it is due to stress. He was not purposively designing a system to deliver a required function for its life, like for example a democracy that democratises. Rather, Easton was only interested to observe the system's behaviour through studying the various causal interactions of the political system to predict its behaviour as it went through its life cycle. Though Easton hinted at constructivism referring to 'constructive systems', he was not keen to pursue this avenue (Easton 1965a: 30–31). His generic political system was aimed at a fact-based positivist construct. Democracy and its values seemed complicated to fit into a fact-based purely positivist paradigm. However, as we saw earlier measurement for democracy performance through democratic deficit and democratic quality for e-democracy have been developed by scholars (Norris 2011; Beetham 2012; Kneuer 2016). Easton seems to have paved a new way of thinking about quantifying democracy.

For defining his political system, Easton by proposing that it was only the scientist who decides what was interesting versus a trivial system deviated according to his peers into abstract explanations. Explanations were critiqued for creating too much ambiguity to be useful to political theory as mentioned earlier by both Miller (1971) and Birch (1993: 222). Easton's positivist theory along with others of a similar nature got embroiled in debates which led to an intellectual war between those who wanted to root themselves into values for good government which they argue democracy embodies and those disputing values as restrictive to progress (Dryzek et al. 2008). As an alternative paradigm in political theory, Easton's view can be interpreted as to what governance could become. However, to shape the democracy Easton's political system required a means to incorporate the people's input. To design operational systems for the military, SE discovers and interprets the end user's need in an environment of technological growth and of changing human values like ecology in mind for the life of the system. It is in this dynamic context that the SE process is considered for designing an operational e-democracy system (Blanchard and Fabrycky 1998).

SE expands on the work that Easton has contributed to. SE explores a bottom-up approach that uses people's inputs towards a customised form of democracy of the people. People's acceptability of the system is the point of equilibrium for democracy, the requirements for this equilibrium is shaped by SE's techniques. With the ICTs, SE provides a democratisation capability. This is through its Requirements technique that ensures that people's needs are identified and thus aggregated for support by the majority (legitimised). The role of ICTs in this process is to communicate and aggregate the peoples' needs, this we posit is critical for large systems or *polis* to build democratic capability (people's voice is heard to shape policy decisions). SE's unique life cycle design process is for both the now and a system's future. It is to a detailed discussion of SE that we now turn.

References

Antill, L., and T. Wood-Harper. 1985. *Systems Analysis: Made Simple Computerbooks*. London: Heinemann.

Ball, T. 2007. Political Theory and Political Science: Can This Marriage Be Saved? *Theoria: A Journal of Social & Political Theory* 54 (113): 1–22.

Beetham, D. 2012. Defining and Identifying a Democratic Deficit. In *Imperfect Democracies the Democratic Deficit in Canada and the United States*, ed. R. Simeon and P.T. Lenard. Vancouver and Toronto: UBC Press.

Berlinski, D. 1976. *On Systems Analysis: An Essay Concerning the Limitations of Some Mathematical Methods in the Social, Political, and Biological Sciences*. Cambridge: MIT Press.

Birch, A.H. 1993. *Concepts and Theories of Modern Democracy*, 2nd ed. London: Routledge.

Blanchard, B.S., and W.J. Fabrycky. 1998. *Systems Engineering and Analysis*, 3rd ed. Upper Saddle River, NJ: Prentice Hall.

Buckley, W. 1968. *Modern Systems Research for the Behaviorial Scientist: A Sourcebook*. Chicago: Aldine Publishing Company.

Caws, P. 1965. *The Philosophy of Science: A Systematic Account*. Princeton, NJ: Van Nostrand.

Chalmers, A. 2013. *What Is This Thing Called Science?* St Lucia: University of Queensland Press. https://ebookcentral-proquest-com.ezproxy.newcastle.edu.au/lib/newcastle/reader.action?docID=1181566&query=. Consulted 28 January 2018.

Crespo, Rubén González, Oscar Sanjuán Martínez, José Manuel Saiz Alvarez, Juan Manuel Cueva Lovelle, B. Cristina Pelayo García-Bustelo, and Patricia Ordoñez de Pablos. 2013. Design of an Open Platform for Collective Voting through EDNI on the Internet. In *E-Procurement Management for Successful Electronic Government Systems*. Hershey: Information Science Reference (an imprint of IGI Global).

Curd, M., and J.A. Cover (eds.). 1998. *Philosophy of Science: The Central Issues*, 1st ed. New York: W. W. Norton.

Dewan, E.M. (ed.). 1969. *Cybernetics and the Management of Large Systems: Proceedings of the Second Annual Symposium of the American Society for Cybernetics*. New York: Spartan Books.

Doebelin, E.O. 1980. *System Modeling and Response: Theoretical and Experimental Approaches*. New York: Wiley.

Drack, M., and D. Pouvreau. 2015. On the History of Ludwig von Bertalanffy's "General Systemology", and on Its Relationship to Cybernetics—Part III: Convergences and Divergences. *International Journal of General Systems* 44 (5): 523–571.

Dryzek, J.S., B. Honig, and A. Phillips (eds.). 2008. *The Oxford Handbook of Political Theory*. Oxford: Oxford University Press.

Dunn, J., and I. Harris (eds.). 1997. *Aristotle Volume II*. Great Political Thinkers, 2. Cheltenham, UK and Lyme, NH: Edward Elgar.

Easton, D. 1957a. Traditional and Behavioral Research in American Political Science. *Administrative Science Quarterly* 2 (1): 110–115.

Easton, D. 1957b. Classification of Political Systems. *American Behavioral Scientist* 1 (2): 3–4.

Easton, D. 1965a. *A Systems Analysis of Political Life*. New York: Wiley.

Easton, D. 1965b. *A Framework for Political Analysis*. Prentice-Hall Contemporary Political Theory Series. Englewood Cliffs, NJ: Prentice-Hall.

Easton, D. 1966a. *Varieties of Political Theory*. Prentice-Hall Contemporary Political Theory Series. Englewood Cliffs, NJ: Prentice-Hall.

Easton, D. 1966b. *A Systems Approach to Political Life*. Lafayette, IN: Purdue University. http://www.eric.ed.gov/contentdelivery/servlet/ERICServlet?accno=ED013997. Consulted 28 January 2018.

Easton, D. 1971. *The Political System: An Inquiry into the State of Political Science*, 2nd ed. New York: Alfred A. Knopf.

Easton, D. 1972. Some Limits of Exchange Theory in Politics. *Sociological Inquiry* 42 (3–4): 129–148.

Easton, D. 1975. A Re-assessment of the Concept of Political Support. *British Journal of Political Science* 5 (4): 435–457.

Easton, D. 1981. The Political System Besieged by the State. *Political Theory* 9 (3): 303–325.

Easton, D. 1985. Political Science in the United States: Past and Present. *International Political Science Review* 6 (1): 133–152.

Easton, D., J.G. Gunnell, and M.B. Stein. 1995. Democracy as a Regime Type and the Development of Political Science. In *Regime and Discipline: Democracy and the Development of Political Science*, ed. D. Easton, J.G. Gunnell, and M.B. Stein. Ann Arbor: University of Michigan Press.

Easton, D., J.G. Gunnell, and L. Graziano (eds.). 2002. *The Development of Political Science: A Comparative Survey*. London and New York: Routledge.

Eisner, H. (2011). Systems Engineering: Building Successful Systems. *Synthesis Lectures on Engineering* 6 (2): 1–139.

Gordon, B.M. (ed.). 2011. *Artificial Intelligence: Approaches, Tools, and Applications*. Hauppauge, NY: Nova Science Publishers.

Gunnell, J.G. 2013. The Reconstitution of Political Theory: David Easton, Behavioralism, and the Long Road to System. *Journal of the History of the Behavioral Sciences* 49 (2): 190–210.

INCOSE, I.C.O.S.E. 2007. *Systems Engineering Vision 2020*. San Diego, CA: International Council on Systems Engineering.

Kaplan, J. 2016. *Artificial Intelligence: What Everyone Needs to Know*. New York: Oxford University Press.

Kincaid, H. 1996. *Philosophical Foundations of the Social Sciences: Analyzing Controversies in Social Research*. Cambridge, UK and New York, NY: Cambridge University Press.

Kneuer, M. 2016. E-Democracy: A New Challenge for Measuring Democracy. *International Political Science Review* 37 (5): 666–678.

Ladyman, J. 2012. *Understanding Philosophy of Science*. Hoboken: Routledge.

Luhmann, N. 1995. *Social Systems*. Stanford: Stanford University Press.

Mead, M. 1969. Cybernetics of Cybernetics. In *Purposive Systems: Proceedings of the First Annual Symposium of the American Society for Cybernetics*, ed. H. von Foerster. New York: Spartan Books.

Miller, E.F. 1971. David Easton's Political Theory. *The Political Science Reviewer* 1: 184–235.

Norris, P. 2011. *Democratic Deficit: Critical Citizens Revisited*. New York: Cambridge University Press.

Parsons, T. 1956. Suggestions for a Sociological Approach to the Theory of Organizations–I. *Administrative Science Quarterly* 1: 63–85.

Parsons, T. 1969. *Politics and Social Structure*. New York: Free Press.

Parycek, P., and N. Edelmann. 2013. CeDEM13 Proceedings of the International Conference for E-Democracy and Open Government. Paper Presented to Conference for E-Democracy and Open Government, Danube University Krems, Austria.

Rosenberg, A. 2013. *Philosophy of Science a Contemporary Introduction*. Routledge Contemporary Introductions to Philosophy. Florence: Taylor & Francis.

Schumpeter, J.A. 2010. *Capitalism, Socialism and Democracy*, 1st ed. Florence: Taylor & Francis.

Silver, P., W. McLean, and P. Evans. 2013. *Structural Engineering for Architects a Handbook*. London, UK: Laurence King Publishing.

USAF. 2010. *Air Force Systems Engineering Assessment Model*. Air Force Center for Systems Engineering, Air Force Institute of Technology: Secretary of Air Force, Acquisition. http://www.afit.edu/cse/. Consulted 8 September 2013.

von Bertalanffy, L. 1950. An Outline of General System Theory. *The British Journal for the Philosophy of Science* 1: 134–165.

von Foerster, H. 1969. *Purposive Systems: Proceedings of the First Annual Symposium of the American Society for Cybernetics*. New York: Spartan Books.

Von Mises, R. 1951. *Positivism: A Study in Human Understanding*. Cambridge: Harvard University Press.

CHAPTER 5

Understanding Systems Engineering

1 Introduction

SE is a structured process where staged decision making occurs around an end user's need to deliver technical solutions. SE is understood as an engineering discipline whose design approach applies systems thinking to combine interdisciplinary fields' activities in a purposive manner to build a system (solution) (Eisner 2011; Department of Defense 1993, 2008; Blanchard and Fabrycky 1998). SE designs develop various relationships from a set of processes that, through the system SE creates, constrains the interacting processes to deliver a desired outcome which meets an end user's need. SE is coupled with new and emerging technologies, which creates opportunities for new and improved systems being developed for humans' use (Blanchard and Fabrycky 1998: 17). SE's design success criteria are that it must meet the end user's need initially and also during the life cycle of the system.

SE can be seen as a complex process that combines the required disciplines together (input) so that they interact with each other to create the knowledge, hardware and software in a specific way to deliver a desired function (output). To develop a solution, SE combines technical and managerial systems to make it work together and communicate with each other. There are three key elements in the SE process: a function, a system where the function will exist and the end user who needs the function to be designed. The relationship between function, system and the end user

S. Bungsraz, *Operationalising e-Democracy through a System Engineering Approach in Mauritius and Australia*,
https://doi.org/10.1007/978-981-15-1777-8_5

has feedback, which is a key to linking each of the elements to the others in different ways during the SE design and operational stages. These links are discussed below and are within a SE process systems framework. However, they are dynamic in the sense that they can be customised to suit a project like e-democracy.

SE integrates a multi-lens understanding of the world around us from input of various professionals to then create the variety of conceptual and physical systems required by human beings (Blanchard and Fabrycky 1998). The SE systems can be conceptual like a body of knowledge understood as a particular organisation of ideas, for example this book, or as a physical entity where it has a specific form like a car or plane. SE's conceptual stage can be seen as an apperception process where, as a mental process, an idea is integrated into a body of ideas, which means with SE a single idea drives the rest of the functions. Prior to building a solution, SE uses the main idea to initiate a conceptual design of the whole solution process that is developed through a conceptual roadmap. This roadmap as a concept of the whole solution starts from needs identification through to system design and a system's retirement at the end of its operational life. This holistic life cycle approach is discussed below.

During the SE design process, a conceptual system is the precursor for an operational system. The first step in knowledge creation is to define what the conceptual system is required to solve. This is done using a structured requirement process that is also discussed below. Requirements discover the need as an idea that is communicated by the end user to an engineer. This idea (need) is then used to develop a body of ideas from a group of professionals as a set of requirements that define the solution. SE governs the set of activities to generate the solution ensuring that all the requirements converge to the key idea, which is the ultimate requirement in a process that we will discuss below.

The solution as a conceptual or physical system can be simple or complex. For SE, it represents a set of processes that are organised through a structured approach to become a relationship that makes the system deliver a specified function as long as the system exists. SE as applied by the book could be equated to a knowledge management system to translate a need (from the people) into a desired outcome (for the people). In practice, the SE process uses a life cycle approach to design and manage complex systems like a health system or a defence system made from discrete elements (hospitals or military aircraft) that are made to work purposively such that when the components or sub-systems are fitted at the highest level, health care or air defence, a function is delivered to meet a health capability or defence capability (goal). The SE approach enables a birth to death or life cycle approach to system development; this whole of life approach is explained below. SE's life cycle approach starts from an idea (need identification) to the capability development (solution) including its retirement when it becomes obsolete; this life cycle process is discussed further below through the Waterfall model. A core to SE's design philosophy is its cradle to death approach.

Some describe SE as a process that can be defined either as 'an iterative process of top-down synthesis, development, and operation of a real-world system that satisfies, in a near-optimal manner, the full range of requirements for the system' or alternatively as a process involving a 'methodical, disciplined approach for the design, realization, technical management, operations, and retirement of a system' as the life cycle of the system from its inception to its disposal (Eisner 2011: 1). In this context, the system or product is defined as 'a construct or collection of different elements' that work synergistically together to produce results not obtainable by the constituent elements aggregated on their own. Meaning the collective as a system itself produces purposive outcomes which are different from the sum of outputs of each element of the system on their own (Eisner 2011: 1). That is, SE can develop from simple interactions at a subsystem level which, when they are transformed in a purposive manner, fit the component to make the organised collective deliver very complex outcomes (functions) at the higher system level. This hierarchal aggregations of inputs can create new systems with new capabilities (Blanchard and Fabrycky 1998). The SE design process develops this orderliness through a system's specification. Specification aims to deliver at the system level a function that is self-sufficient so that it meets an identified need for the life of the system. Though SE finds extensive use in Defence, it is as a means to develop solutions to complex problems; for example, Configuration Management technique discussed below is used by non-military to manage costs and obsolescence for projects during their life cycle. The System of System (SoS) construct from SE allows the development of capabilities created by integrating autonomous systems to meet the need for managing emergencies, natural disasters and large bush fires as mentioned below.

2 Emergence of Systems Engineering

System engineering (SE) has its root in systemology and was developed during the 1940s to respond to military need for a means to understand a system in operation (Blanchard and Fabrycky 1998: 10). The need in the military was to develop large and complex systems. Systems deliver specific functions, roles or missions for the military, developing these systems and operating them required interdisciplinary expertise and knowledge. This military interdisciplinary need to solve cross-disciplinary problems gave rise to an initial form of interdisciplinary fields' cooperation to develop systemic solutions, which was termed as systemology. Derived from general system theory, systemology was applied by US Department of Defense from its need to understand a host of common processes in military operation across its systems, and it developed, for example, operations research.

This initial research to find what was common in military processes with systemology led to the creation of a body of knowledge that allowed other disciplines to emerge like SE (Blanchard and Fabrycky 1998). SE has an emphasis on technical integration for solving a need using interdisciplinary teams. Integration of many disciplines, engineering and non-engineering allowed SE to build large systems from component level to create complex functions at the supra-system level. The military required complex functions which could not have been built without the interdisciplinary know-how, like building an aircraft from engineering and pilot's ideas (conceptual) to its retirement (post operational life). Such a complex engineering enterprise requires various specialist and generalist fields to work together. SE as a methodology was developed to facilitate design of large and complex technical systems for the military weapon systems (USAF 2010).

To implement a systems thinking in design, SE guidance is provided by standards developed by the military. The Mil-Std-[Military Standard] 499B, Military Standard Systems Engineering, was the last standard drafted by the US Department of Defense; it was superseding Mil-Std-498 which as new technology emerged became inadequate (Department of Defense 1993). The need for an engineering governance design utility for the military across its different forces air, ground and sea was the common thread that resulted in the development of the SE methodology. SE's initial guidance from synthesis of the military need created the requisite knowledge to develop a standard that was prescriptive, Mil-Std-498. It intended to standardise a command and control centric operation. It was both for performance and also to provide effective control over decision making in a complex environment. However, in the face of the growing technological innovation from a fast-evolving digital technology, a need for balance emerged and the self-imposed constraints quickly became obsolete from the complexity of a technology that was itself becoming more capable. The prescriptive constraints had to be removed to allow the flexibility that a dynamic capability such as technology required to innovate to create novel technologies. To exploit the emergent technologies, a paradigm shift needed to occur.

The paradigm shift resulted in a subsequent standard which was the superseding standard Mil-Std-499B that was for guidance. It was recommended for best practices so as not to restrict innovation like its predecessor Mil-Std-498. SE itself had to develop further as SE's link to technology required a degree of flexibility to accommodate the pace of technological change. This technology relationship of SE was one of the foundational requirements for SE's initial standard but that had to be improved upon further given technology's importance to US military to maintain its capability. Defence needed capabilities which required systems to work together to deliver very complex operational outcomes that cutting-edge technology would provide. This meant a reduction of the control function within Defence to provide the empowerment that technology needed to innovate. Information technology introduced a new era from a culture like a paper-based system which was becoming obsolete to that of the communication capabilities of the fast-evolving digital technology like the Internet and Intranet. Also, to provide a technological edge, it was realised that with integration more capable technology modules could be quickly configured by SE for new roles that emerged from the evolving operating environment, which was also being shaped by digital technology. This required SE as a technique to manage interfaces between the modules, so the new roles could be delivered while using technology that at times was still at a developmental stage or old and therefore unique. This mixed technological environment created its own demands for flexibility as a SE approach that was mechanistic or paper-based quickly became an operational constraint. Developing technology about which knowledge was yet to be acquired given a lack of experience with such technology needed a new culture of more risk-taking. Implicit is that SE had to be dynamic if it was to remain a technique used to enable improvement as it had to adapt the technology developed through the multidisciplinary teams' collective efforts to meet Defence needs.

The obsolescence of technology as new ones emerged and also the complexity as nascent technology became more capable plus Defence's dependency on technology required a convergence through a standard that was not constrictive. The SE technique that was created to deliver practical outcomes as technology became more capable through the end user (Defence) exploiting the new capabilities of technology meant a less prescriptive process. In this dynamic technological environment, SE grew to be embedded into the culture of the organisation with only a guidance as recommendation from the end user (Defence). The military in turn had to improve decision making so it accommodated technology's need for growth and innovation. Thus, the SE process itself grew as it was realised that its link to technology required such growth to accommodate the emerging complexity. It had to manage from a highly integrated and increasingly complicated but technologically capable system. For Defence, there was no alternative. If it was to deliver outcomes that were going to be more capable, then its SE process itself had to keep improving. Defence processes and practices had to change. SE had to become more efficient and flexible. In taking or looking to adapt the SE approach from its origin, there is no reason to expect that SE will retain the hallmark of its origin. As when SE is removed from its military context, the options exist to contribute to the renovations and transmutation of existing representative democracy. This is because SE is becoming more able and complex. As it evolved, it generated complex SE models like System of Systems (SoS). Some of the technologies had very short life cycles as new more capable technologies were discovered, and they became obsolete. SoS was used to manage these complex transitions of the old and the new advanced technologies for Defence. Now, this SE feature is attractive to an e-democracy that cannot find a means to remain relevant in the context of growing ICTs, which is becoming more capable and also more integrated in society.

As the integrative capability of technology grew, it was becoming more able and complex, SE had to evolve, and it adopted concept like SoS. SE which started as a means for decision making in the military through a paper-based command and control approach that fundamentally required transparency for decisions needed to change. As technology became more capable and complex, the SE outputs mandated by the military adapted. Its requirement for a development activity mentioned a 'decision data base' to document and organise data from engineering efforts so an audit trail from requirements to agreed solution could be traced (Department of Defense 1993: 9–10). The step to make decision making transparent through technology has flow on to its empowerment capacity. Decision making through the databases became traceable and thus accountable. A move to a database made many of the previous activities in the previous (paper-based) standard obsolete. Use of technology allowed further virtual departmental (systems) integration from new digital capabilities like Internet and Intranet. At its core, it was the transfer of information between systems or communication between systems that were becoming automated. This informative aspect of the system to be integrated required new concept like a SoS. At the same time, there was a move to human managing higher functions while the lower functions were digitised using the capability of digital technology. A degree of empowerment took place from automation that was not feasible prior to digitisation.

In this technology-driven context, the SE concept like SoS grew and it is now a specialised subfield of SE exploring interface integration of complex systems. This SoS concept is equated to a federation of systems 'in situations where there is little central command and control like authority and power' (Brooks and Sage 2006: 264). This empowerment has been possible due to adaptability of the SE concept to use technology that was emergent and more capable. However, importantly from Defence's capability requirement, there was a need for accommodating high-risk emergent technology to be fielded and operationalised. Using the SE framework, this emergent technology could be fielded and also upgraded as more was learnt about their properties. Some of which emerged only after the technology's operationalisation given that the new technology's behaviour was unknown until applied by Defence. The trade-off for Defence, as an end user, was a release of its command and control through accommodating the technologies capacity to do more. This process of use of state-of-the-art technology to innovate though common in Defence is a process which society is experiencing now from digital technology which seems to be penetrating the commercial sector and society, but stagnating in the governance area.

3 Systems Engineering as End User's Need Centricity

The system's need that the end user defines is the central feature in SE (Blanchard and Fabrycky 1998: 21). The Systems Engineering Standard Mil-Std-499D, from which both military and industry's current SE process applications are derived, has the end user clearly at the centre of the system being designed. The following simplified schematic as shown in Fig. 1 depicts this end user-driven process, and it is the key strength that this book draws from. As depicted in the schematic, the SE design is outwards from the end user to the primary system functions, system elements and then the system itself.

Now, the SE developed system has to provide a solution to meet the end user's need. When designed, the system will be made up of processes, products and people (personnel) that deliver the function (system e-democracy) derived from the end users' need. That is, SE as an engineering process allows the development of a system that even when it would be built from scratch is centred on the end user. The book though is mindful that it does not have to build something from an idea as a representative democracy already exists. However, it is the core of the SE

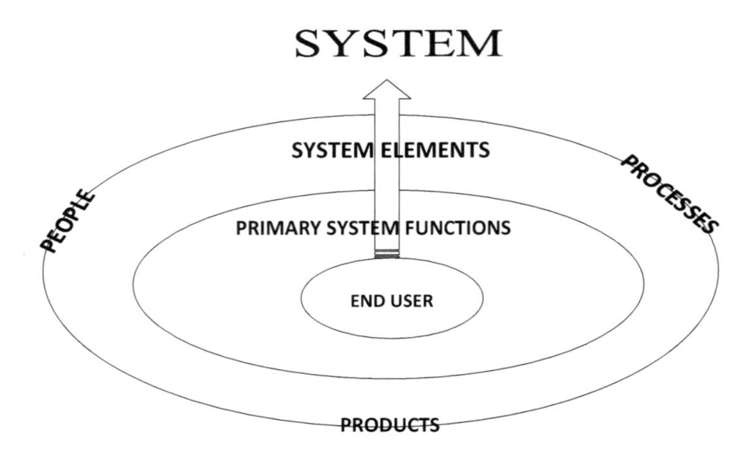

Fig. 1 SE process is people centric (end user) (Adapted from Department of Defense [1993: 1])

process centricity that even when a legacy system (representative democracy) needs an upgrade, the change process is people centred. That is, the function of a system e-democracy through a SE process is people dependent for its decision making. Using the concept of the 'centrality of end user in SE', the book argues for a paradigm shift to embrace technology to empower people (end user) for decision making we expand further on this idea below.

The key aspect depicted in Fig. 1 is the centricity of the SE process where the end user as the decision maker shapes the system as a whole. This is the core of this research that the people (end user) must have decision-making power in a democracy. Figure 1 also shows that SE is holistic in its approach as it ensures that all likely aspects of a system are considered through requirements that are to be met through the multidisciplinary team and integrated into a system as a whole. Unlike a process that is focused on efficiency through study of repetitive activities to achieve high-quality outputs with minimum cost and time, the SE process is a discovery process. This discovery process involves the end user needs which must be identified to be solved. The need is analysed as the first step whereby the SE process discovers what has to be solved in terms of requirements, like an e-democracy that involves people in the decision making at the highest (or policy) level. Once the need identification step is complete, then the SE process embarks on the actual stage of developing more detailed requirements for a solution to the problem; this step also involves the end user consultation. SE's end user centricity as mentioned above is maintained through the whole development process for a solution that gets operationalised. This is a fundamental link to the end user that SE created as a structured engagement to ensure from previous lessons learnt that a valid or successful solution is developed and maintained (USAF 2010). The end user is in control of the design being involved in every stage of the system development from inception to its operationalisation. Figure 2 illustrates the design flow for an initial solution.

SE is a sequential process flow to develop the highest-level function of an e-democracy as defined by the users of e-democracy in the form of requirements. If for these users democracy is people's rule, this function is then the top-level requirement for the SE process (Grady 2006; Blanchard and Fabrycky 1998). In a simple design, the SE process starts from a conceptual design and it then proceeds to a preliminary design before the initial form is developed as shown from its flows in Fig. 2. This

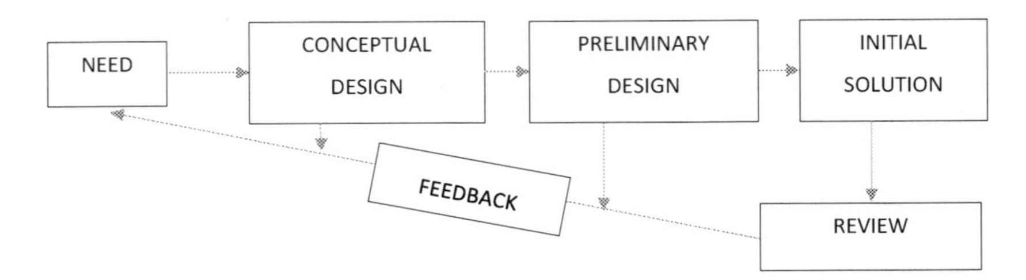

Fig. 2 SE design cycle to develop an initial solution (Adapted from Blanchard and Fabrycky [1998])

interim form is reviewed, and if changes are required by the end user, new requirements emerge to capture the intent of these changes so the system or product is improved upon using feedback. The feedback loop allows this change in requirements to be reviewed and justified against the originating requirements in consultation with the end user. An adapted SE process cycle to design a solution is shown in Fig. 2.

The conceptual design is the first stage that is initiated to validate a need, and at each step of the design process, SE seeks feedback. Feedback is to inform the end user and seek agreement. The consultative approach ensures there are no surprises when the solution emerges from SE's design as the end user is part of the solution process. Using an iterative process, as shown from the arrows in Fig. 2, the SE process develops a solution (initial product). Each iteration leads back to the need which is used to validate the product and changes to it. The user's need is the centre of the design process (see Fig. 1). This need can only be changed by the end user as decision makers not by those engineers involved in the designing process.

At preliminary design, the ultimate requirement (the need) is decomposed into system element (CIs) and functional (FCIs) requirements, which are discussed in detail later below. These requirements define what the engineer understands about the need of the end user, and they are used to manufacture and test the product required by the end user. The initial product when developed is reviewed by the end user and the designer to validate the need. The review mechanism is a decision point used to ensure that the product (e-democracy) meets the end user need (people's rule). If the need is not met, new requirements for change are developed to amend the design so a modified product results that moves closer to meeting the end user's need.

For an abstract idea like e-democracy to satisfy the need of people's rule, this iteration can go on for many years. For the conceptual design, SE first must prove the need for e-democracy exists; this is a feasibility of the concept. To build a successful system, SE does not design without a needs analysis as the product or system is speculative given that there are no requirements for such a system. Thus, for SE ensuring a need exist improves the success of the designed solution when realised, as mentioned above, in practice the realisation of the e-democracy with SE is anticipated to be an ongoing developmental task towards system e-democracy. The interim solution development shown in the above schematic (see Fig. 2) provides the basis for further refinement as the system is refined towards its improved versions in a structured SE manner. The structured approach brings a sense of order as the product is realised through each iteration in the cycle referred above in Fig. 2. Feedback to the end user provides a means to iterate to develop an improved solution.

A final product evolves from the interim solution in an iterative manner going through many cycles of refinements as requirements change to new ones or are amended. With SE, requirements development is a critical process applicable to the whole SE cycle. Therefore, a system development that applies the SE process must have in place a Requirements process for its conceptual design, as 'regardless of the type and complexity of the system, there is a conceptual design requirement (i.e. to include requirements analysis)', mentions Blanchard and Fabrycky (1998: 25).

This advice for a Requirements process being core is consistent with the other system engineering scholarships like the military Systems Engineering Standard Mil-Std-499D at Department of Defense (1993) from which most of the SE practices are derived. This process remains a criterion for success even when SE evolved to new concept like SoS at Department of Defense (2008). The Requirements process is still crucial and occurs at the start even when the SE process is customised (USAF 2010). For example, Requirements process is a core process at USAF (2010) which is an assessment model of SE processes. It is also a core process at USAF (2013) which provides a handbook for its systems engineers to develop communication devices like communication satellites used in information networks for both military and commercial application like the Global Positioning System (GPS). For SE, Requirements is a crucial process to define what to build.

As noted, to design a system SE always starts with a conceptual design first. For an interim or simple solution, Fig. 2 shows a simplified design life cycle to develop an initial model of a system e-democracy. It is to be noted that this is only the first phase of the SE design process towards a mature solution, a total capability solution, which is the mature system explained below. A key aspect of a conceptual design is to allow technological opportunities to be explored to solve the problem (Department of Defense 1993). For this technological interface, translating the problem to something accurate like a requirement is necessary so it clearly articulates the problem. Requirements definition as a core process is discussed in more detail later in the book; we explore the SE process life cycle further through one of the SE models that are tailored and used to develop communications solutions (USAF 2013).

With SE, there is a learning post the design phase and this is through the performance of the system in its environment. The top-level function, Functional Configuration Item (FCI), specification provides a governance role to the rest of the functions hierarchy created from the lower components', Configuration Items (CIs), interactions. At the design stage, the top-level FCI provides the focus of efforts so the system does not fall into chaos during design that is it ensures that components or CIs achieve the overall goal of the system's specification. However, some of the assumptions made during design may not be how the system actually performs during its operation as not all the issues to affect the performance could have been anticipated by the designers when the system is actually embedded in the social context it operates within. The knowledge of the actual performance provides new insights about the relationships some of which would be emergent. These emergent relationships thus require changes to cope with the unexpected behaviours of the system if they undermine the top-level performance (FCI). The feedback loop described above, in Fig. 2, allows corrective action to maintain system performance in the stage when the system is operationalised. These changes are anticipated by SE and discussed below. The changes to performance are managed by the Configuration Management (CM) technique, while Requirements from engineering addresses the limitations in performance normally through an upgrade. An upgrade in this context is through new requirements to address the performance issue uncovered during operation.

SE concepts also allow reductionism and expansionism, like with a SoS approach that we expand on later. This hierarchy of system of SE is mentioned as allowing a supra-system to deliver complex outcomes that are dependent upon the subsystem's units yet not a direct (purely mathematical) aggregate of each subunits inputs (Mesarović 1964). A key SE issue here is the purposiveness of the SoS approach. When aggregating using a top-down approach, the specifications which are generic become requirements that are more specific and a bottom-up approach from the simple data inputs from elements of the system that are transformed at each stage to create the complex function (designed role of an e-democracy). We expand on this relationship between systems to create the supra-system SoS that is governed by SE later in the book. In other words, an SE design aims to align elements around a need (role) that the top function of the system must meet (e-democracy that democratises in our case). The need (role) is the soul of the system, or its design reason.

4 Some Key Terminology in Systems Engineering and Its Application to E-Democracy

Now, before we progress further, first we explain some key concepts that will be used in the book in the fundamental processes of the key techniques of Requirements and CM. This discussion is applied below from a bottom-up approach, which are CIs that create FCIs which in turn deliver systemic functions. Some other key terms which are used by Requirements, system and Configuration processes are explained below. These terms which explain SE will be applied in Chapter 6. These key SE terms adapt SE techniques to conceptualise the system e-democracy solution discussed in the rest of the book.

4.1 Configuration Item: An Elemental System in Systems Engineering

The term '"configuration item" in the SE process refers to an aggregation of work products that are designated for Configuration Management [CM] and treated as a single entity' within the process (USAF 2010: 19). This concept allows Defence to select the level of control that it applies to a product's attributes (Department of Defense 1992, 1997). These attributes are the functional and physical requirements that make up the specification for the product. Though every component will have specifications, not every component in a system is worthy of being a Configuration Item (CI). CIs are useful to decompose the system or product into smaller elements. A CI is an item that requires formal release meaning there is control of the information defining the item which requires end user acceptance. This information identification must be understood by all stakeholders; the information set is therefore allocated a unique identifier through which the CI's set is associated with management. The CI's level of control depends on factors like safety, regulations, unique properties, or requirements and also the level of delegation approved by the end user (Department of Defense 1997). The item's attributes are managed in house or delegated with a level of oversight from the end user. Using the CI identifier, the changes to systemic performance, for example, can be addressed either through decision making by Defence as end user or a contractor who will consult with Defence for upgrades to the CI prior to incorporation of the change.

CI is a key to the process of CM as it is where management and control starts. It manages every item for both software and hardware from the highest level to the items at the lowest level of decomposition. Even though an item is being sourced from a contractor by Defence, yet the contractor does not have control authority over the configuration of the item, Defence retains decision control as the end user. The configuration is documented by Defence or the contractor. If data is managed by a contractor, then it must provide access to Defence to audit its data management activities. These data checks are through regular audits (baselining) mentioned in configuration documents like Configuration Management Plans (CMPs). Configuration baselines are established regularly, and these baselines are checked by Defence during the audit process. These audits verify whether the CIs conform as well as perform as mentioned in their documentation. A list of CIs makes up the system or product; with computers' databases capability, this list can be every single item that makes up the product as decided by the end user. CIs allow reductionism as well as aggregation into specific systematic configurations.

SE uses the CI concept for managing what SE designs and that the end user decision is important to control when designing and operating the product. CI is a means to provide backward and forward traceability as a system design evolves or changes during the design phase and later during its operational phase. It provides a means to associate and trace changes that are incorporated and which shift the system from one product baseline to another.

In digital environment, each CI that is created has a unique identifier assigned to it. The unique identifier is used to describe and link the attributes of the CI. During the life of the product, the CI concept provides a means for a history against each CI to be recorded against the unique identifier (USAF 2010). At any time, the status of CIs is traceable from the records. CIs allow the integrity of system baselines to be maintained. Baselines are the snapshot of a CI status in time; importantly, it allows an audit trail to ensure that the system is actually what it is. A CI is verifiable by auditing the CI's requirements listed in the product's specification (USAF 2010). Verification is a key aspect to ensure that the product developed meets the needs of the end user and its configuration managed for the life of the system.

CIs also allow delegation; it can be used to identify who is responsible for its ongoing management during the life of the product. Once the CIs are listed for a product, then for each CI management responsibilities are allocated against it. This allows identification of the organisation or individuals responsible for every CIs of a system for the product's life cycle (USAF 2010; Department of Defense 1992, 1993). Any changes to the CI would involve these stakeholders during the life cycle of the product. Nowadays, this information is recorded using a relational database. As the system evolves, the system and related changes are managed through the database. The records ensure traceability and accountability; ongoing audits are conducted to confirm that the baseline reflects the records against the CI.

CIs are used as a means to collect information about the system that ensures the system's integrity in both design and operation. It is the most useful concept to manage the complex systems that are developed to deliver outcomes. Databases are effective tools to assist in these informational activities for the end user to assure itself about the system; some typical activities are shown in Fig. 3.

The above schematic, Fig. 3, is derived from the USAF System Engineering Assessment Model (SEAM); it uses CI to conduct an audit as mentioned for Configuration control to identify, inform and check the design of a system (USAF 2010). CI as a control element forms part of the CM technique mentioned below. Identification is a unique way to validate each CI in the system, while verification is as mentioned above from the information set that will uniquely define what the CI is. The information which is requirements allows CIs to be uniquely identified and managed as a set called specification, and this specification is essential especially when tracing the information that is describing the system. Information is crucial for both management and description of the item

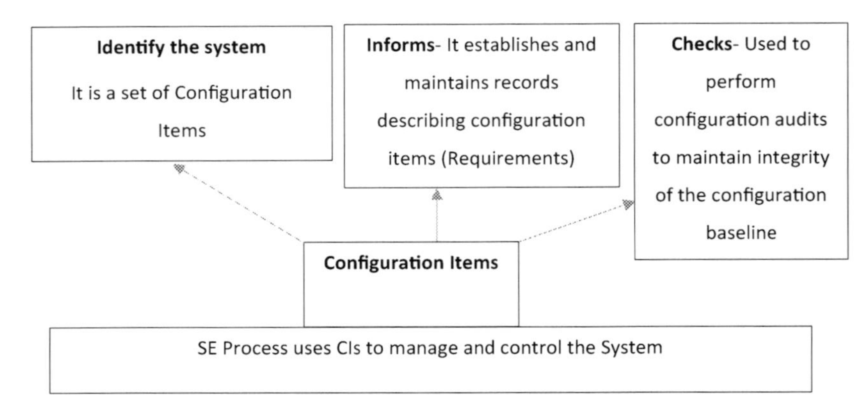

Fig. 3 Use of CIs as a control- to identify, inform and check (Adapted from USAF [2010])

in terms of what the system is or its configuration at a given time. The CIs attributes will be a set of requirements or specification.

To inform records are created and kept against each CIs. CIs allow a set of records to be grouped together so the CIs baseline can be managed using CM technique. Requirements are the fundamental records that CIs are linked to. These fundamental records define what is to be built. Checks are audits that ensure that what is being described is what actually exists; the checks ensure integrity of the system at any phase of the product's life. The checks are against requirements that define each CI both conceptually and physically. Audits ensure that requirements are not overlooked, like orphaned requirements derived during the conceptual design phase but overlooked at delivery. Checks assure the end user that what was specified has been delivered.

CI allows the key links mentioned earlier between requirements, specifications and the product to manage control of the product's evolution during its life cycle in both design and operational phases. A list of CIs can be either large or small depending on the degree of control one decides to have on the system and changes to it; for example, change can be managed to a component level of the aircraft like screws and bolts or subsystems level like wings and power plants. During the life of the system, control of CIs is through the CM process. CIs are associated with different issues that need to have ongoing management; this could be, for example, lack of performance, obsolescence of parts, new technology and a changing environment.

The decision to support management of a CI is from the end user and developer who are part of the design process. Costs of management, as well as the need for that level of management, will drive the level of control to be exercised for an item. This decision occurs when the solution is transferred to the end user. There is need, for example, to ensure safety when using the designed solution; this is dictated by risks generated from use of the solution. Every technology or system delivering a function entails residual risk that must be transferred to the end user for acceptance. As an example, due to safety and risk of failure of the items, for an aircraft system's safe operation within the context of its environment, a high degree of control is needed to component level for the aircraft as compared to a car. A component like a screw failure for an aircraft when flying may be a catastrophic event, while a similar screw failing in a car may not incur a similar event. The end user decision to control is dependent on the risk and consequences of the product failing when operating. This tends to be a judgement call based on experience managing similar risks from previous or similar system failures in that operating environment.

Importantly, the idea implied here is decision control for CM by an end user is customisable based on risk acceptance by the end user. The list of CIs under management will be identified and approved by a Configuration Control Board and mentioned in a Configuration Management Plan (CMP). Now, the level of control is customisable through the CI construct based on the product in operation. The level of control will be decided by the end user who accepts the risk entailed by the solution being delivered. CIs are therefore crucial to ensure that the end user understands what is being delivered and accepts that an appropriate support system is in place for its safe use when delivered for operation. Importantly, the end user retains control of changes to the CI through a Configuration control process mentioned in the CMP. Changes to CI must be approved by the end user unless the CI management is delegated as mentioned in the CMP. CIs and CMPs authorise a level of empowerment for decision making to manage the system's performance.

4.2 Functional Configuration Items and Performance

Functional Configuration Items (FCIs) allow an initial description of a system so a functional baseline can be developed for its functions or performance. FCI is defined through functional specifications required to operate the system. In this book, we use this construct to link the top-level need in a hierarchical manner that allows the defining of functions that a system or subsystem must deliver from its lower-level components or CIs. FCIs are used to differentiate the complex from the elemental or CIs in the book. The functional breakdown starts from a top-level need and it breaks down into lower-level functions until all activities of the system for development, production, operation and support are identified for its life cycle. FCI is a conceptual construct used to manage a system during design. It informs about functional specifications in a system that links with what is needed as one set of requirements or baseline to a system's function, like lift.

FCIs allow the grouping of lower-level elements in similar functions in terms of CI, like a left wing and a right wing. FCIs are characteristics of a function by a system or subsystem whereby the performance of the FCI can be measured as described in its functional specification, lift in right and left wings. CIs are linked to the FCIs which are higher aggregates of the CIs behaving in a specific way constrained by the way they interact due to design. Like CIs, the FCIs are a construct, FCI though is to identify higher outputs like functions from outputs of the interacting CIs. These are used to describe performance; FCIs are aggregates from CI outputs that are transformed to deliver a function. The book separates the elemental system CI from the complex functions FCIs.

The idea of FCI is a concept we use to allow the identification of the function that is delivered by the system to meet a need. This need could be complex, and it may require many lower-level functions interacting in certain ways, like an aircraft needs wings to create lift and a power plant to create the thrust which generates lift, so the system aircraft can fly the highest function being sought. Now, a function refers to a relationship that must exist so that a specific or discrete action that is necessary to achieve a given objective can happen. Top-level functional specification is used to decompose to more detailed lower-level specification. Decomposition is from the top-down, from the higher level function (need—primary FCI) to lower-level functions (subsystems FCIs) and then to group of entities (CIs) specification. Functional specification defines performance of the FCI and these are measurable, like the thrust from an engine's performance or wings lifting force.

In the book, we use these two levels to differentiate between CIs which we treat as elemental and FCIs which we treat as aggregates in a defined way designed to create a desired outcome, a function. FCIs are CIs but fitted or aggregated in a specific way. CIs create performance through a wing that is made up of rivets, sheets metals and formers arranged in a specific way, but FCIs are the function lift of the wing; the wing delivers the lifting performance. For example, there exists a relationship between maintenance and ongoing operation which is that for the system to perform an operation to achieve its design outcome—the maintenance operation must take place. For conducting the operation, there may be a maintenance action that is necessary to restore the system to its normal operation state (Blanchard and Fabrycky 1998: 62). Maintaining as function assures that this relationship between maintenance and ongoing operation allows, for example, the aircraft to perform as designed. This functional baseline is assured during a functional configuration audit when the system is operationalised. Maintenance activities are then designed to allow the aircraft to perform at the functional baseline during its operational life.

The FCI at the higher level is linked to every FCI at the lower level on which it is dependent; these, in turn, are linked with every CI at the lower level on which they are dependent. This hierarchical relationship when arranged in a given manner allows the need at the highest FCI to be met; it also differentiates between the low-level interaction of the CIs and the more complex interaction of the FCIs. Now, it should be noted that at the highest level the need is defined in a vague manner but as we drill down the need becomes defined in more specific terms with the lowest CIs being quite accurate in their description; this accuracy allows components to be manufactured and then aggregated (fitted) in specified manners to deliver a means to solve a problem or need. For example, if the FCI performance is noted to degrade, then a maintenance action could restore the FCI's performance. This maintenance function would either be carried on the dependent FCIs or CIs.

In the schematic example shown in Fig. 4, from the bottom up, CI-1 and CI-2 interact to deliver a function FCI-A and similarly CI-3 and CI-4 interact to deliver a function FCI-B. Then, FCI-A and FCI-B are necessary to deliver the need at the top-level (system) FCI-NEED. This idea can allow complexity to be broken down into manageable functions that work together to deliver a higher function; for example, the wing

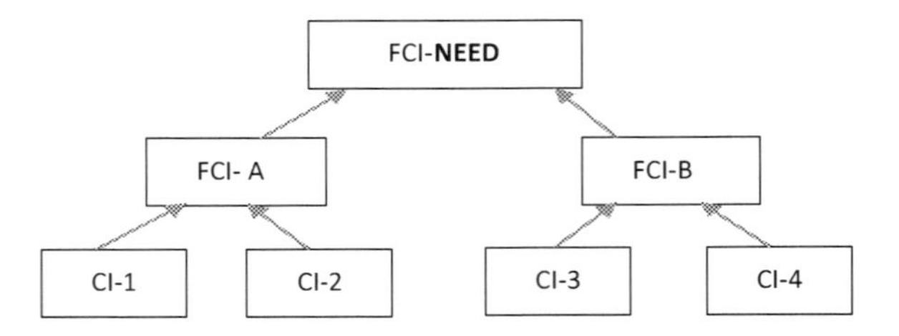

Fig. 4 Relationship between components (CIs) and functions (FCIs) (Adapted from Department of Defense [1992])

creates lift (function FCI-A) for an aircraft and the power plant provides thrust (function FCI-B) to the aircraft.

At the aircraft level, both the wing and the power plant are necessary to make the aircraft fly (highest function the need). The wing can be built separately from the CIs as long as it meets the lift requirements (FCI) for the aircraft, and the power plant (FCI) can be autonomous as long as when it is attached (constraint) it interfaces with the wing to provide the thrust (FCI) required by the aircraft. Thus, the FCI modularity allows a complex device like an aircraft (a system) to be built from less complex subsystems like the wing (FCI-A) and power plant (FCI-B) that both in turn can be decomposed further into components called CIs (CI-1, CI-2, CI-3, CI-4). The FCIs baseline also allows control of changes to ensure that the function or performance of the system does not degrade so that the flying need of the aircraft is not undermined. Restoring performance is through a separate function called maintenance identified during design, post the product being designed and operationalised.

The FCI concept as mentioned above allows the SE process to manage system performance after the functional baseline is defined so that it does not degrade during the operational life of the system. It also allows the interface issues to be managed as changes to one function increased power from power plant has an impact on the wing's attributes. These impacts can be mitigated by ensuring changes are compatible. These attributes are derived using the Requirements process (Grady 2006; Blanchard and Fabrycky 1998). Changes to the function and its attributes or specification for the aircraft (system) are managed through Configuration Management techniques. Functional checks are through audits or routine maintenance checks.

4.3 System of Systems

A SoS approach modernises to make SE evolve as it brings about the idea of a federated system. This evolution was necessary from the lessons learnt when dealing with increasing complexity of a more capable technology. For example, the USAF recommends that a SoS is 'a critical concept for the development of the new and improvement of the old systems' (USAF 2013: 4). Thus, to either replace or upgrade a system, the concept of SoS is applied to analyse the components of the system needing upgrade. This upgrade could be due to obsolescence resulting in interface issues arising between systems which the SoS is made of or to improve a system's performance. The SoS concept allows many systems to be considered together to understand how they might interact in a meaningful way to deliver an upgraded desired function. In such cases, it is a means to break down a complex system into smaller systems or subsystems to investigate the interface issue and allow the upgrade. For example, if some part of the system that is operating is not delivering the desired output that is required for the system to perform, it makes a case for an upgrade for that part. It may not be necessary to make a change that creates a totally new system; a performance improvement through an upgrade can be achieved through the modification of part of the larger system. A SoS approach allows the complex system to be reduced into smaller sections so that aspects of the system are upgraded to improve its overall performance.

SoS allows very large systems to be integrated into a complex supra-system delivering desired outcomes to address the need. Each SoS system through interface contributes to another system through specific outputs that are then used or transformed by the other system to deliver a different outcome. Thus, a computer can transform signals sent via a digital communication system into images displayed on a computer system's screen. Both the computer system and the digital communication system work autonomously of each other. Through SoS engineers break down the complex modern world into systems that interface with other systems and deliver a desired function. In a technological environment, SoS allows a larger collective or system to persist and operate as autonomous systems from each other (Department of Defense 2008). For example, in this book the political system operates within the larger nation-state system; this SoS concept allows the book to focus on specific areas of the system nation-state that it needs to address during its life cycle using the SE technique to upgrade the legacy system. We show that the life cycle approach that SE applies to deliver solutions (systems or a set of requirements to build a system) is compatible with a decision-making cycle for the representative system.

4.4 Primitive Statement Conversion to Specification

The essence of a requirement statement is a raw statement. Now, if a Specification is stripped away, with the style and formatting reduced like from reducing a paragraph to the essence of the paragraph, the resulting statement is a sentence that is called a raw or primitive requirement statement. A primitive requirement is an essential attribute of a system like a FCI or a component CI of the system (Grady 2006: 49). It is made up of a controlled attribute like weight for an aircraft and a relation which could be equal, less than or greater than some value, for example, 'Weight less or equal to 23,600 pounds'. The controlled attribute is used to limit the weight of an aircraft to be designed to 23,600 lb; however, at the raw stage, it is not adequate. This attribute will need to be clearly communicated to many departments, so it must provide some further clarification.

The next step after defining this attribute is to document it into a Concept Requirements List (CRL). The CRL identifies and lists the requirements for the system that could be either done on paper or using a database for a larger system. One can combine several of the primitive statements to form a very simple requirements document for an item which places the overall limit 23,600 lb on the design of the whole aircraft. A CRL will be a listing of such raw statements from engineers. This statement even though it is at an early stage of development of the design nonetheless provides all the teams involved in the design with a controlled factor for the design. Now, this aircraft limit weight will need to be communicated to every engineering field, so there is a need to add some more information about what weight it is about and also other additional information so it can support detailed design and procurement activity. The statement is therefore given a subject 'Aircraft' and turned into a complete sentence which is, for example, 'Aircraft empty weight must be less or equal to 23,600 pounds'. Basically, the engineer's raw statement is expressed more clearly into what will be or become part of a specification when the CRL is translated into a format that is used by all to design the aircraft.

The removal of ambiguity is an essential part of the Requirements process. The design engineer is creative; he or she is specifying with the minimum words what they have in mind through the raw statement elicited from a need of an end user who wants an aircraft. The transformation to a specification is so that the rest of the organisation understand what the engineer meant, many primitive statements would have a value attached like '23,600 pounds' that clearly articulates limits. This value aspect of control limits is the quantitative aspects of Requirements. However, there are qualitative statements that must be turned into requirements like 'existential requirements' (Grady 2006: 51). These existential requirements for SE humanise, as it gives an element or system attributes that personalise which is seemingly missing in some designs. SE humanises its requirements as it integrates the human aspect into the design of the system. Often there has been accusation even in politics that 'of its view of the individual; [where] the human being comes to be understood in terms of ownership and commodities – as property' (Carter 2017: 19). SE's centricity to human needs avoids this trap from ensuring that existential requirements are part of its design consideration. Its subjective interpretation of human existence in the world is through human experience. For example, through environmental requirements which are a routine part in its design considerations, the degree of pollution is minimised. These pollutants could be aircraft noise and greenhouse gases impact of its engines or disposal impact, for example, for the recycling of materials requirements to be used during design. Substitutions for heavy metals that impact human health, for example nickel and chromium, in new aircraft are substituted with environmentally friendly alternatives; such qualitative consideration for sustainability is part of the design process.

Requirements and design activities are to consider that if there are tasks that are unnecessarily complex or that may result in frequent or critical errors, such tasks are eliminated if they 'require extensive cognitive, physical, or sensory skills' (Department of Defense 1993: 13). In SE, this human element is an implicit design philosophy as SE creates designs that are used by humans and they are reflective of human needs and values. To ensure that human factors are part of the design, further guidance for those applying SE is provided through additional requirements that may be generated to satisfy end user objectives (Department of Defense 1993: 15). Such requirements could be at higher levels like subsystems FCI or the primary system level (aircraft) where it becomes necessary for environmental operating considerations, noise and other pollutants as mentioned above.

4.5 *Agenda*

The Agenda in the context of the e-democracy identifies the scope of the deliverables from the people; therefore, any change to that scope should be with consultation from the people. Some of the causes for scope creep are poorly defined and poorly understood requirements; the complexity of a project and also failure to understand the end users' expectations right at the start creates creep during design (Kerzner 2017: 60). This reinforces the need for good requirements. If one is using only high-level requirements like a vague policy or the Agenda which is ambiguous in a political system, then one can expect scope creep when the design delivery gets to a detailed level of the work breakdown. Ambiguity may lead to delivering the wrong solution, what the people (end user) never wanted. Given that scope creep cannot be stopped, then Configuration Management processes ensure that changes to scope are transparent and undergo a defined approval process for scope changes. An Agenda under CM allows changes to be transparent and accountable.

4.6 *Scope Creep*

Scope creep is an extension of what had been agreed at the start of a design process to what is actually designed during the design process; this causes design to fail. This difference is visible when checks between the actual system and an engineering Agenda or agreed Specification are conducted. Scope creep could be equated to an extension of a decision-making scope through changes in Specification. It is a certainty that in every system design project, there is scope creep. This is due to a natural desire to continuously enhance the project's requirements as the project's deliverables are being developed and then new knowledge and requirements emerge. Scope creep creates growth in the project's scope by increasing the initial Agenda.

When dealing with end users, it is important to clearly articulate the scope right at the start of the design process. There should be a clear understanding about what is being agreed to or the set of requirements that scope the boundary for what must be delivered or an agreed Agenda.

To make decisions in Defence, it was important to have a means to define the scope of the project. From lessons learnt, there was significant budget blowout from scope creep which resulted in an undesirable outcome of products that were not approved for end user operations. In worst-case scenarios, there was non-delivery of a solution due to the increasing scope from additional design changes that were decided during the development phase. This phenomenon of a creeping or evolving requirement that was being added on after decisions had been made was termed scope creep. Scope creep has seen many projects fail due to incremental changes that sometimes resulted in a useless product due to significant deviation from what was specified and what ended up being delivered. A means to control such a growth was necessary, so decision making could have a degree of control and oversight or configuration control during design.

Economic risks, for example, arise from scope creep; it arises when a decision has already been made and costed but then if additional changes are made from what was decided it creates creep. In the case of a governance system, this scope creep could make changes unaffordable; for example, if an Agenda has been voted in an election and then a representative decided to do things that were not part of the Agenda, that is a case of scope creep. Scope creep would undermine the legitimacy of the original scope that has been approved by the end user, the people. CM with change control allows a structured decision-making approach that controls scope creep. Change control is also to maintain the performance of the system through approving changes that are valid; we expand on this CM process below. First, we discuss scope creep and its relations to an Agenda as well as some implication from a technology perspective for e-democracy for this phenomenon of scope creep. The Agenda is an analogy for the Specification or at a less refined level the CRL. In the case of an e-democracy to be discussed later, a set of requirements from the people is an Agenda for government.

Many projects have become distressed due to scope creep resulting in poor performance on delivery. It is therefore essential to get all stakeholders to agree to what must be delivered. An Agenda to scope the project creates consensus between end user and the project as a contractual agreement prior to embarking on the delivery of the project. Scoping of the project is conducted and agreed to right at the beginning of the project. One way to address scope creep identified by Defence was through the process of CM (Department of Defense 1992). We expand on this technique below. What is certain is that human nature is such that scope creep is apt to occur when designing large and complex systems. Thus, by setting a CCB, the scope creep can be managed but not stopped. If an Agenda is set for delivery, for example, then the Agenda should not be changed without consultation from the stakeholders. Apart from scope creep for a SE design, in a technological context, another area of weakness is technology itself and its obsolescence management is crucial through effective control.

4.7 Configuration Management Plan

A CMP is the means to document the processes and procedures for an organisation to manage decision making for the system. This master document lists every sub-plans of stakeholders like contractors as well as government agencies including the manufacturers of the components that make the system. All those who have some responsibility and empowerment (through delegated authority) in the management of the data and subsystem or components attributes that make the supra-system are listed in the plan against the CIs or FCIs that they manage. The CMP defines the roles and responsibilities of each stakeholder of FCIs and CIs that are involved in the support and delivery of the function of the system that is being managed. The CMP document is used to define and delegate management decision making to stakeholders who are involved in configuration control.

For complex systems, like a fleet of aircraft, for example, there is a hierarchy of CMPs for systems that are made of multiple systems like a SoS. For example, a FCI like an engine, its maintenance and its upgrade could be delegated to a given organisation with its own engine CMP, this arrangement could change over time and it is then reflected in the weapon system CMP, which is a master CMP for the aircraft fleet. A master CMP is a dynamic document that can be updated by the owning organisation responsible for the overall management of the capability, like Defence which maintains control over the aircraft fleet system. The CMP ensures that authority to make changes to the system is clearly defined in terms of scope and CIs and FCIs responsibilities. A CCB is the key body that implements and assures the intent of the CMP. The board has oversight over the whole system's capability and makes final decisions about changes to be incorporated into the system based on justification for those changes. The board's role could be assumed to be that of a parliamentary head in the case of a representative democracy where the master CMP is a Constitution. The plan also defines activities like audits that ensure the integrity of the product or system during its life cycle. At each audit, a baseline is struck, and this baseline is a snapshot of the configuration of the system at a given point in time.

4.8 Baselining

Baselining activities are necessary to ensure the integrity of the system as they are a means to audit the system to ensure that the described system in its documents matches the actual system. A baseline is a specific configuration at a certain timeline; it is a reference point for the system. Baselines provide traceability as with audits the system configuration can be verified from the documents (specification) that describe the system and these are managed through CM techniques. Baselines are important to manage the performance of a system. Over time a system undergoes change, and as these changes are incorporated, the baseline will shift to new baselines. Thus, baseline is an assurance activity but also baselines are a means to trace the history of changes made to a system over time. They are important for managing the configuration of an item over time. Baselines need to be verified to assure if the baseline meets the description in the documents which are maintained through the CM process. If when carrying a check, if the audited *as is* (actual) configuration is assured for the end user, then the baseline is validated. Baselines are checked against the approved documents that describe a CI to validate functional or physical characteristics of the system.

At the conceptual design stage, the approved tier-one documents are its specifications and requirements. However, specifications which are tier-one documents grow during the operational stage to include like operational and maintenance documents, like drawings of the physical system, for example. A baseline assures that only approved and current documents are being used by the organisations to support design and operational activities for the CIs that make up the system. It is a way to improve trustworthiness of the SE process that the CI and FCI are reflecting what their documents describe; this is important for accountability. Changes can be traced for authorisation at each baseline, but more importantly the status of the changes to be implemented can be assured through this baselining activity. Assurance through baseline checks is a key activity enforced by a Chief Engineer during the life of a system.

4.9 Validation and Trustworthiness

Validation is the act of assuring that the requirements are correctly specified and agreed by the end user. For example, the end user may decide that a validation be carried out to assure that the system will perform as intended in the operating environment. The validation goal is to ensure that the system performs, for example, safely for its intended use in the environment where it will operate, and through the validation process, it demonstrates the system's operational capabilities. Validation is conducted early in the design of the system and continuously during the systems' life, and can be applied to all aspects of the system, its support, maintenance and training (USAF 2010: 69). The USAF has even developed a framework from industry best practices for SE processes assessment to ensure that the SE processes being followed conform to as a minimum to best practices level. The intent of the validation activity is for continuous improvement as the engineers assessing the processes raise a report that will require corrective action for any deficient process identified during the validation activity. But importantly it gives the end user the trust they need about a validated system's performance.

5 Requirements Process

Requirements is an overall process which involves an analysis of the need; it is used by engineers from both a top-down and bottom-up approach. It is an eliciting activity to discover and translate the need into design specifications through engaging with the end user. Requirements is what links the need to the system design. The top-down approach ensures that the function (highest level specification) for which the design is being developed is broken down (decomposed) into subfunctions at a lower level like mid-level specification and then further decomposed to still lower levels that are discrete specifications to create the design hierarchy. This is a synthetic view from the designer's interpretation of the end user's need.

The bottom-up approach starts from the specific components design that is aggregated at each level to the top function; this activity elicits from the design team the elements that will build the system. SE uses requirements to firstly understand and then develop the desired relationship between the system's components that are made to interact in a given configuration to create a desired function at each level. This function is developed through configuring either a new system or an improved system (an upgrade) in an iterative manner. SE translates the information from the end user to create the requisite knowledge to design a desired product or service (outcome). The translation process is to communicate an end user's need to a multidisciplinary team; we provide an example below. This SE design activity is conducted through a cyclic approach where the team synthesises, analyses and then evaluates requirements to assure (validation) the end user need is met.

SE's iterative process to create the Requirements process starts with inputs from the end user; this need capturing activity can be from telephone conversations, emails, faxes, or interviews and meetings where the need is elicited. Now, this need, if designing complex artefacts like an aircraft for a given role or a hospital system to provide health services, may lead to potentially thousands of requirements. Normally, a database is used to capture these requirements that are created from the need. At the conceptual stage starting from the need, a process of analysis is undertaken to break down the need into subfunctions; this top-down approach is through a requirements analysis process. In this step, the requirements are decomposed as the system is broken down into modules which are designed to interact to create the function (need). We explain this below; this development process links the need to subfunctions and to specific components in a hierarchical manner using CIs as the building blocks for requirements analysis. This functional breakdown would result in a functional baseline or a hypothetical hierarchy of what the delivery of the need would be in terms of functions.

Requirements analysis is conducted as an approach that starts from the top moving downwards at various levels of details for the functional decomposition (Department of Defense 1985, 1993, 2003). Requirements analysis is a bridge between the end user requirements and the system requirements, a bridge that allows solutions in the form of system functions to be generated. The end user's need, requirements and objective are analysed to determine the system-specific performance and functional characteristics. The level of detail depends on whether the SE process analyses a system, the top-level FCI, or subsystem like a lower FCI, or a lower component like the CI.

In the preliminary design stage, for example, a system's functional requirements are applicable for exploring the feasibility of developing the actual e-democracy. The system FCI to CI hierarchy mentioned above is depicted in Fig. 5 where a decomposed view of the relationship that

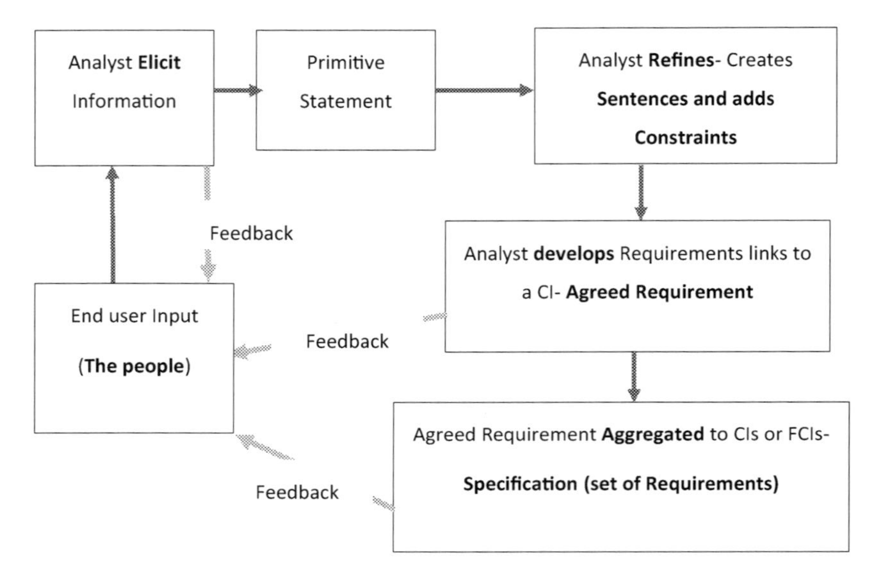

Fig. 5 Creating a Requirement and transforming to a Specification (Adapted from Grady [2006])

emerges from the preceding processes comes together. It displays how the design aim at the top has links to the description of requirements at each level. These requirements must be sufficient to define each CI or FCI characteristics using requirement analysis (Blanchard and Fabrycky 1998: 49). The functional requirements at each level are informed by creating a specification for the need.

To develop a specification, the essential requirements making the specification are aggregated through creating a set of requirements which are generated as mentioned below in the Requirements eliciting process. The aggregating is conducted through a process of synthesis where input requirements that consist of performance criteria, functional and interface criteria are grouped logically to be translated into solutions (specification) to satisfy the inputs (Department of Defense 1993). Now, requirements are generated through an eliciting process whereby the analyst interacts with the end user to define and then iteratively refines the need further. The design team through a Chief Engineer and a design engineer is also involved as discussed below. This eliciting process below is an iterative process that is crucial to the success of the design initially and during its life cycle. A system is defined by Requirements, and these drive the whole design process from its conceptualisation to its operationalisation. The analysis requires an eliciting activity where an analyst engages with the end user to extract needs and expectations mindful of the constraints and interface requirements from the people; this is conducted in a structured manner for SE (USAF 2010: 46). The analysis of Requirements as a process takes the people's 'needs, expectations, constraints, and interface requirements', translates it and collects it to define the system or product that is needed during the system's life cycle (USAF 2010: 46).

The first step to developing the requirements is to elicit 'a primitive statement' (Grady 2006: 48). This is an organised method for writing a statement to describe an essential point or individual attribute of an element or CI. A primitive statement example is provided above. The set of requirements from this step is very basic and it is aggregated to make a list of statements that are elicited from the end user. This list forms a set known as 'CRL- Concept Requirements List' that can be a paper-based list for a simple system or computerised like for a complex aircraft system (Grady 2006: 48). It is important for the SE analyst to understand this need as 'this effort will not be successful without more information about the [end user] and their expectations' (Grady 2006: 86). Developing the requirements is an iterative process and bi-directional between end user and analyst to elicit the requisite information; Fig. 5 below shows an iterative process to create a Specification from a raw or primitive statement.

The CRL as a list of primitive statements provides the starting point for further development to a specification as each statement is converted in 'one or more complete sentences in a language familiar to the [end user]' (Grady 2006: 48). This is done by adding a subject that is the component or CI and then a verb and a sentence ending as mentioned above. A paragraph that includes the title and number for identification is added; this would complete the requirement statement (Grady 2006). A set of completed requirements against a CI or FCI will form the basis for a specification of a CI or FCI description. Additional information is provided to specify or clarify under the heading 'Additional Data' (Grady 2006: 50).

Completed requirements are prioritised by engaging with the end user. The analyst liaises with the people to seek clarification about the adequacy of the requirement in describing the need and the timing for its delivery. Any identified constraints like interface issues are noted and recorded against the requirement. It is important that the people are informed about it and agree with the requirements. The aim is to create accurate requirements that are complete for describing all the attributes of a need from the people. This accuracy is fundamental to the success of the design. The structured approach through specific formats was used in a paper-based system mentioned above; these have now been integrated into databases.

With the advent of digital technology, a database is used to record the requirements; for example, those used by the Department of Defense are called CoRE and DOORS (USAF 2010: 48). These are databases that specifically provide requirements traceability for the life of the system. CoRE, for example, is a specialised software, it assists the analyst with developing requirements, and it is used in complex designs when developing new system, like C130 J software requirements (Faulk et al. 1994). DOORS, for example, is a database that is more friendly to use as per the Defence Science and Technology Organisation (DSTO) report for requirements management, as DOORS can work in most Windows-based computers (Cant et al. 2006). DOORS reduces compatibility issues for information sharing between computers, a key interface issue when using digital technology in a SoS environment. The SoS has led the US Defence to develop JEDMICS which specifies its interface needs for information systems compatibility. This JEDMICS initiative is an ongoing adaptability of Defence to accommodate technology and a dynamic SE through SoS. This is a move to a networked distributed data management communication interface guidance where technology is becoming more capable to exchange data between databases.

DOORS, for example, is preferred over CoRE and is used for large SoS environments since 'One of the major features of DOORS is its ability to handle links', and so the database provides a means for multiple users simultaneously to access and exchange the data (Cant et al. 2006: 8). As mentioned above, requirements are refined through an iterative process engaging a multidisciplinary team with the end user and every requirement that links to CIs. The complexity of multiple users makes a database for traceability an effective means to manage the links between the CIs and FCIs as well as the specifications and individual requirements. These links evolve as the specifications and requirements go through their refinement iteration. As a data repository, networked databases keep and manage the CIs and FCIs history. The history provides traceability for Configuration Management as the decision making can be traced. Teamcenter is another database which manages the links between requirements for the whole life cycle of the system from requirement generation to system retirement (Briggs and Sampson 2006). These databases are tools to improve the requirements management process to create agreed requirements from the end user.

An outcome of the SE iterative consultative process with the end user (people) as the decision maker is an agreed specification. This is derived to articulate the need of the end user in a form that could be delivered by engineers. The agreed specification defines the product that is to be created. As mentioned earlier for SE, the feedback to and from the end user continues during the life of the product or service. SE is a structured process where end user needs and wants are transformed and specified for agreed delivery for the life cycle of the system. If any changes are made to the delivered product or service that does not match the agreed specification, the end user is informed through the feedback mechanism as a request for change.

Issues may arise in practice, for example, when a specified product or service from the end user may not be feasible as per the agreed specification. This feasibility issue may be discovered during the design phase or post design when more becomes known about the system's performance due to its operation. Thus, through the feedback system, a renegotiation of specifications may become necessary between the end user and the designer. There may be trade-offs needed which would require changes to the agreed specification. These changes are in consultation with the end user (the people). The end user validates the new specification to ensure that they agree that what is being changed is acceptable. Validation with the end user also ensures that the specification or full agreed set of requirements has been met when the product is delivered. With validation, inconsistencies emerge from any agreed requirements that are left out. Validation is the bi-directional traceability activity between what was specified and what is delivered; it assures the end user.

Negotiations with the end users allow the resolution of any inconsistencies; in an e-democracy, this will reduce stress within the system. Thus, SE is a process to inform and also reform. To meet a need, it continuously creates and maintains the system by consulting with the people about their requirements through the life cycle of the system. SE generates a consultative culture that is people centric—see Fig. 1. The people may know a better way, all that is required is for them to be consulted so their requirements are elicited to improve. As mentioned above, the process of developing raw statements that are then developed into specifications could be used to create an Agenda. The CRL is the analogy of an unrefined Agenda.

6 Configuration Management as Governance

Configuration Management is an overarching process that ensures that what was specified is actually what is delivered based on the data it manages. This is a core governance process for decision making within the framework of SE. CM is a technique that Defence uses which it had to develop as CM is 'a process for establishing and maintaining consistency of a product's performance, functional and physical attributes with its requirements, design and operational information throughout its life' (Department of Defense 1997: 1–3). CM started for establishing configurations of delivered products and controlling the changes to them, or scope creep, but it has evolved from lessons learnt in practice to optimise the outcome for the end user. CM is a means for maintaining accurate and valid data to retain knowledge and history (i.e. lessons learned) so one can improve the systems being managed. The process supports informed decision making in Defence by providing a process for defining a product through CIs and FCIs, and it is also a basis for empowerment through a CMP or Configuration Management Plan.

The CM process starts its journey as the standard Mil-Std-973 which was mandated by Defence; the standard was a burden in that it added significant cost, and also, in a fast-changing technological environment where innovations needed to be the norm to maintain cutting edge capability, it was seemingly suppressing innovation. This led to a review and a change to a new guidance in the form of the military handbook, Mil-HDBK-61. Similarly, to the SE standard mentioned earlier in an environment where information technology was rapidly advancing, configuration practices moved to a best practice guidance. The Mil-Std-973 paper-based prescriptive data control standard became obsolete in a digital environment where Defence was using integrated databases to exchange information (Department of Defense 1997: i). With a more capable technology like databases to manage data, there was a new need for flexibility to innovate; standardisation from a paper-based paradigm was too rigid for such an environment.

Defence still needed to retain control of changes to its capabilities though, and CM process as discussed remains a structured means to incorporate changes to upgrade the system during its life cycle (Department of Defense 1992, 1997; Lacy 2010). CM techniques ensure that the changes are successfully implemented due to a requirement for change that may be initiated due to obsolescence, cost or a performance need or all three (USAF 2010; Blanchard and Fabrycky 1998). The integrity of an agreed specification between the end user and the contractor (as the designer) must be managed though. The CM technique monitors the integrity of an agreed baseline by managing the information about the changes to the agreed specifications that define the baseline. Baselines cannot be changed without authorisation from the end user, Defence or the people in an e-democracy.

CM practices have also evolved as Defence has adapted its practices to ensure that it kept pace with a changing environment from ICTs. There have been changes to Defence processes from advances in information technology that had a significant impact on how it did things; technology challenged the *status quo*. As the information exchanges between systems could be done through digital media, a whole new opportunity emerged for information management, data transfer and sharing through integrated and distributed databases. It impacted the way information was used to conduct CM during a systems' life cycle (Department of Defense 1997: i). In this distributed environment, initiatives like Contractor Integrated Technical Information Services (CITIS) and Continuous Acquisition Life Cycle Support (CALS) emerged for describing the technology interface required for electronic access and transfer of data between contractor and government.

The end user Defence did not itself manage the data for products being designed for it. The advent of digital technology created a more flexible SE process where databases are used in a networked manner to provide the required information to manage the product during its life cycle (USAF 2010, 2013; Department of Defense 2008). The government as the end user still retained control over decision making but it had to empower some activities. The information, for example, was stored where it is most economical to store it and there it was managed for the whole life cycle of the system (Department of Defense 1997: 2–5). The CM process required the end user (Defence) to agree to any change proposal to the original specification prior to the change being incorporated into the system. This process flow when adapted provides a degree of control to the people as the end user depicted below. For example, with the CM process in place, a change initiator in Fig. 6 cannot make changes without the end user's agreement to the proposed change if using CM principles. As when an old agreed specification is superseded, however, because it has been approved by the people, no change can be made to that agreed specification until the new specification as a proposed change has been approved by the people. This flow of change control by the end user is depicted in Fig. 6.

CM maintains the records of the specifications for the CIs and FCIs. When changes are required, it adds to the records for the CIs and FCIs in the databases like CoRE or DOORS mentioned earlier. Any specification change with a CM process is traceable to the change initiator and the change approver. This transparency ensures that every change made to a specification is traceable to both the need for that change and the history of changes that are made against the CIs or FCIs. This SE standard at Department of Defense (1992) was necessary for an environment where the US military interfaced with various suppliers that delivered a capability in a networked environment.

When it changed its processes from paper-based to information technology, a new guidance at Department of Defense (1993) was provided to interact more effectively with its departments and suppliers. It provided a means for the various interfaces to operate in a flexible manner using JEDMICS as the interface as mentioned above. The SE process tailoring was allowed to enable use of digital technology and CM moved from a standard at Department of Defense (1992) to a guidance at Department of

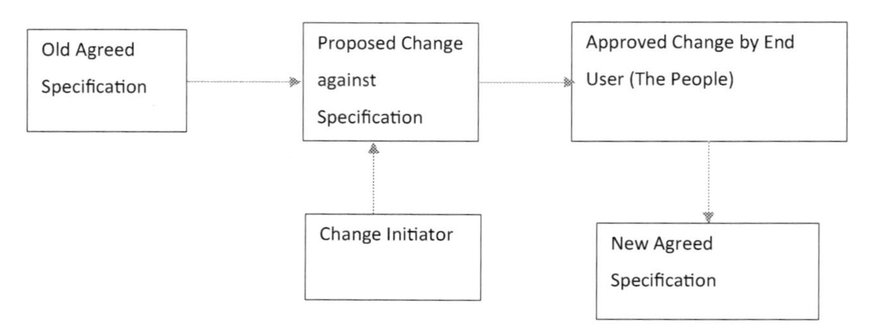

Fig. 6 Change to an agreed specification (Adapted from Department of Defense [1992])

Defense (1997). The flexibility to guidelines rather than standards allows SE to improve and use technology to innovate its techniques. SE is based on best practices as mentioned at USAF (2010: 3) and as these practices change it adapts.

The process describing the change management system is documented in a CMP (Lacy 2010; Department of Defense 1997). This documented plan ensures that changes are decided at the appropriate level as it defines the change process followed by all organisations that have responsibility to manage the CIs. However, more importantly, it also identifies and mentions the level of delegation of authority to make changes to the system. It documents the empowerment to manage the system's evolution. The end user (the government) can delegate to the supplier, for example, a level of authority to make and propose changes reflected in a CMP (Department of Defense 1997).

Another important aspect of the CM process is that impact assessment about interface issues becomes visible and the addressing of the issues assures the overall integrity of the system. A change in one CI may affect other CIs or FCIs. CM is the disciplined management of changes through a structured process that is transparent to all stakeholders. The following information is a typical adapted change request from USAF (2010); the information for a change request may consist of firstly identifying the CI or FCI that is affected. A reason for the change being sought, this could be performance or safety related, for example. Then, an impact of the change is provided in terms of how it affects existing CIs it interfaces with; the cost of implementing the change and degree of urgency for the change to be implemented is quantified. This change request is a proposal to allow informed decision making about the change by the end user who is presented with the above information seeking approval for change. The information mentioned above is not exhaustive. The end user may require more information from the proposed change before a decision is reached. This criterion to inform about the change proposal is so that the end user assesses the impact of implementing or not implementing the change.

For ease of management, a relational database is used to track the status of the baseline. It identifies as a minimum every CIs and FCIs for every change that has been requested and its progress, if incorporated or pending. The data set allows the aggregation of all CIs and FCIs and records every specification that is linked to these items. A Configuration system allows planning of the upgrades (changes) required to maintain a capability as well as scopes the future work outstanding against the capability. Use of a database allows records against individual or groups of CIs to be efficiently traced (USAF 2013).

There is a need for CIs history to analyse changes to a baseline or configuration of a system. Baselines show a system evolving from where it is to where it needs to be with every change accounted for against CIs. The previous baseline is the old system prior to the change; on incorporation of the change, a new baseline is created or the new system, Fig. 7. The progress from one baseline to another shows the evolution of a system. The history of change incorporation can be captured; databases are effective in the recording of such baselines. Systems are not static, they evolve with time, but decisions made to make it evolve can be traced using CM. CM provides an accountable framework to understand how the new system came into being.

CM is a crucial part of SE as a holistic technique that maintains an existing capability by continuously augmenting it to meet current end user needs. When CIs become obsolete during design or operation of the system, they can be replaced in a structured manner to maintain or augment the capability. The system's highest function is monitored so it does not degrade when changes occur. CIs are a modular approach to create or transform into a new system while maintaining the system's ultimate requirement, the need to fly in our previous example. A system requires changes to continue operating for some CIs become obsolete and are replaced with new CIs. Some CIs become irrelevant as the operating environment changes or new regulations make them redundant. CM integrates these changes in a systemic manner.

Changes can be from the operating environment or internally from the system; all these changes against the CIs are managed using CM techniques. This structured approach allows a high level of control on the evolution of the system. It informs every stakeholder in a transparent

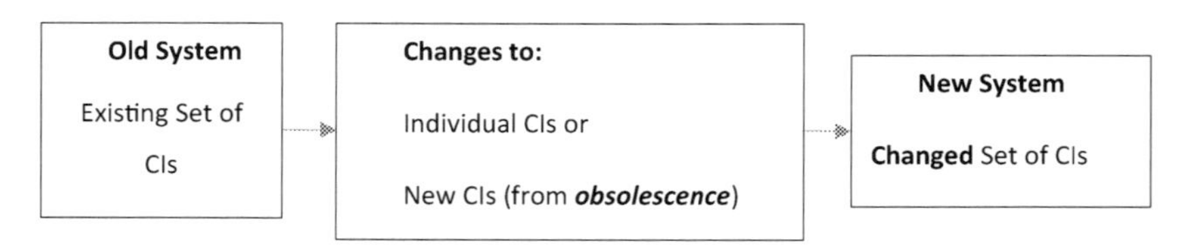

Fig. 7 Evolving system (Adapted from Department of Defense [1992])

manner for the need for a specific change. Importantly, the change to an agreed specification as shown above must be approved by the end users who were identified against the CI as explained above—see Fig. 6. A CM process ensures that configuration information is available for all phases of the life cycle of the product to support 'Metrics to assess achievement of objectives and foster process improvement' (Department of Defense 1997: 1–2). CM is a transparent and accountable process as history of an item at CI and higher level FCI can be traced both forward for those changes that are pending and in the past for those changes that have been incorporated. CM could be seen as the decision-making framework in a federated environment. CM also assures end user need for baseline validation through cyclical audits as mentioned above. A CMP would document this auditing activity.

7 Systems Engineering Life Cycle Design

SE design for the life cycle 'goes beyond the product life cycle' (Blanchard and Fabrycky 1998: 21). A more detailed design cycle to operationalise a system is shown below in Fig. 8, called the Waterfall model. It is adapted to provide an overview of the steps for implementing a product life cycle design for an e-democracy. A mature operational system e-democracy is created from the model depicted below. The mature operational system is

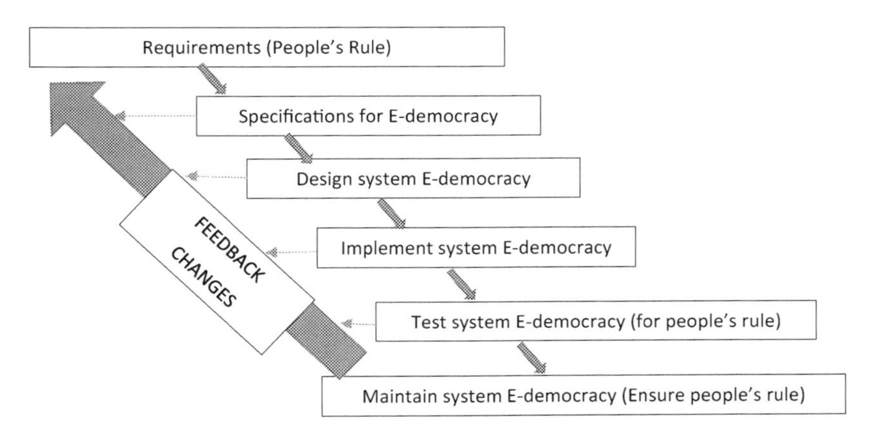

Fig. 8 Conceptual design of the system e-democracy (Adapted from Blanchard and Fabrycky [1998])

from applying a SE design process for the whole e-democracy life cycle. Note the below is not the only model, it is shown here to provide an example of SE in the life cycle of a design that starts from identifying a need to its ongoing support mentioned above. The model below brings a new feature that is e-democracy's maintenance (persistence) stage when it is operationalised. This model depicts the life cycle links for the SE design, from a people's need to a system e-democracy through an iterative process. Feedback is crucial during the whole life cycle of the system. In the context of a technological environment where new technologies that are becoming more capable, the below flow depicts a way an operational e-democracy may be conceptualised for design. It is anticipated that as changes will have to be made given that an e-democracy does not exist, this design will be evolutionary with frequent consultation between the people as end user and the design activity, which may be a contractor tasked to create the system e-democracy or the government itself. The below is roadmap to conceptualise the whole system life cycle from an idea to an operational solution.

A key process element noted here is that all changes at any of the SE stages depicted above like Specification, design of the system, implementation of the system, testing of the system for validation and finally, maintenance of the system are linked to the core process of Requirements which is at the beginning of the cycle. Control of requirements through CM ensures that the people remain the decision makers during the whole life cycle. A CMP mentioned above allows such control to the people, as the CCB becomes a collective as a society. Delegation through the CI list in the CMP allows empowerment by the people to delegates, representatives in an e-democracy system.

Now, SE should not be seen only as a mechanistic process as there may be such a tendency given its link to technology. SE is not about efficiency, but it is a whole new way of thinking about how to solve problems for society. SE should be about a structured way to think about a problem and its outcome which is an operational solution. The governance that SE provides through its processes allows innovative solutions to emerge as multidisciplinary engagement to solve the people's needs is feasible. SE allows a high degree of flexibility as based on its application the relevant processes may be used as applicable (Blanchard and Fabrycky 1998). 'Tailoring' of the SE process is mentioned in Mil-Std-499D at Department of Defense (1993: 40). Tailoring guidance of SE is based on the product being developed; it allows customisation of the SE design process. Also, as discussed tailoring accommodates a more capable technology. However, in any customisation, the Requirements process is the core first step of the SE design process. The requirements analysis that is conducted links requirements to meet the need to the ultimate requirement or key need mentioned above (USAF 2013; Department of Defense 1993; Blanchard and Fabrycky 1998). The Requirements process when using technology, a database for example, ensures traceability of every requirement to the need from conceptual design to the operation phase. A hybrid SE process like the Waterfall model shown above allows a new form of engineering governance to be explored for political theory. Different organisations tailor SE to suit their product; a system e-democracy customised to suit the people it is designed for seems feasible through a SE design adaptation.

The SE top-down process where each function is linked to lower functions in a hierarchy keeps the designed system e-democracy on track to deliver what it requires to solve the need even after the operationalisation of the system. The SE advantage is that it is applicable beyond the system life cycle (Blanchard and Fabrycky 1998). SE is used to design requirements even for the retirement phase of the system or its disposal (Department of Defense 1993). The ongoing support functions to maintain the system are designed after the verification that the manufactured product meets the need at the implementation stage. The testing phase allows the support (maintenance) functions to be developed. The cyclic audits identified from the people in a CMP allow the performance validation of the system over time to assure the performance of the system is maintained, if not enhanced through approved upgrades.

7.1 Technological Obsolescence Risks

E-democracy would require technology. Now, with every technology, there are risks associated with the technology's obsolescence. This has been an ongoing issue with Defence whereby obsolete technology had to be replaced or upgraded using SE processes. Obsolescence creates its own problem of costs of replacement and risk to performance. However, as a new technology offered better capability, therefore, decisions had to be made about upgrade to the new system for improving the Defence system capability such that it maintained its edge. Apart from performance risk due to system degradation from obsolete technology, the obsolescence had costs associated with it and therefore a management technique was required for decision making for ongoing as well as future capability. The CM process that resulted is a structured decision-making process to ensure that the impact of the obsolete technology as well as the upgrade being considered did not have any adverse effect on the overall system's performance. A structured decision-making approach of CM allowed Defence to optimise the risks and benefits for decisions to replace or continue with an obsolete technology to be assessed in a transparent manner.

In the information era, obsolescence is normal, and as new innovations come on board, CM ensures a planned approach to replacement decisions. Technological obsolescence cannot be eliminated but it can be anticipated and a rational decision arrived at so as to have an optimal replacement outcome (Furneaux and Wade 2017). Therefore, managing changes is important and thus a documented approach is through one defined in a CMP which is discussed below. The plan is drafted to provide guidance on how the changes that are to be implemented into the system would be approved. It provides transparency as the decision-making process is defined to clearly articulate the ongoing management of Configuration change decisions in the organisations that are part of the system.

8 Conclusion

This chapter has explained what SE is and how it constitutes a system of processes that are used as a governance process in engineering to develop operational solutions. Key concepts and the techniques to which they give rise have been introduced and explained and their inter-relationships discussed. It can thus be seen that SE techniques provide a structured process to create an operational solution. It transforms both qualitative and quantitative inputs from the end user (the people) into requirements. The solution which SE designs, persists for the life of the designed system. Underpinning this discussion is the conviction that this SE systems knowledge with its key concepts of Requirements and Configuration Management processes can be applied in a political science context. In particular, the argument is that SE can enable the creation of an operational e-democracy. The next chapter gives substance to this thought experiment by exploring how SE can be applied to develop an effective and sustainable e-democracy, a democracy that democratises.

References

Blanchard, B.S., and W.J. Fabrycky. 1998. *Systems Engineering and Analysis*, 3rd ed. Upper Saddle River, NJ: Prentice Hall.

Briggs, C., and M. Sampson. 2006. *Tying Requirements to Design Artifacts*. Orlando: INCOSE. http://onlinelibrary.wiley.com/doi/10.1002/j.2334-5837.2006.tb02795.x/full. Consulted 28 January 2018.

Brooks, R.T., and A.P. Sage. 2006. System of Systems Integration and Test. *Information Knowledge Systems Management* 5 (4): 261–280.

Cant, T., J. McCarthy, and R. Stanley. 2006. *Tools for Requirements Management: A Comparison of Teleologic DOORS and the HIVE*. Edinburgh, SA: Defence Science and Technology Organisation. https://www.researchgate.net/publication/27253971_Tools_for_requirements_management_a_comparison_of_telelogic_DOORS_and_the_HiVe. Consulted 28 January 2018.

Carter, A. 2017. Reading Lessons: C.B. Macpherson's Immanent Critique. *University of Toronto Quarterly* 86 (2): 19–42.

Eisner, H. (2011). Systems Engineering: Building Successful Systems. *Synthesis Lectures on Engineering* 6 (2): 1–139.

Faulk, S. et al. 1994. *Experience Applying the Core Method to the Lockheed C-130J Software Requirements*. Gaithersburg, MD: IEEE. http://ieeexplore.ieee.org.ezproxy.newcastle.edu.au/stamp/stamp.jsp?tp=&arnumber=318472. Consulted 28 January 2018.

Furneaux, B., and M. Wade. 2017. Impediments to Information Systems Replacement: A Calculus of Discontinuance. *Journal of Management Information Systems* 34 (3): 902–932.

Grady, J.O. 2006. *System Requirements Analysis*. London, UK: Elsevier.

Kerzner, H. 2017. Introduction to Scope Creep. In *Project Management Metrics, KPIs, and Dashboards—A Guide to Measuring and Monitoring Project Performance*, 2nd ed. Hoboken: Wiley.

Lacy, S. 2010. *Configuration Management*. Swindon and Biggleswade: British Computer Society, The Turpin Distribution Services Limited.

Mesarović, M.D. 1964. *Views on General Systems Theory: Proceedings of the Second Systems Symposium at Case Institute of Technology*. Case Institute of Technology, Systems Research Center publications. New York: Wiley.

USAF. 2013. *SMC Systems Engineering Primer and Handbook*. Space and Missile System Center. EverySpec. http://everyspec.com/search_result.php?cx=partner-pub-0685247861072675%3A94rti-pv850&cof=FORID%3A10&ie=ISO-8859-1&q=SMC+Systems+Engineering+Primer+and+Handbook&sa=Search&siteurl=everyspec.com%2F&ref=&ss=20497j31595539j48. Consulted 8 February 2018.

U. S. Department of Defense. 1985. *Military Standard Specification Practices MIL-STD-490a*. Andrews Air Force Base, Washington, DC.

http://everyspec.com/MIL-STD/MIL-STD-0300-0499/MIL-STD-490A_10378/. Consulted 28 January 2018.

U. S. Department of Defense. 1992. *Military Standard Configuration Management MIL-STD-973*. Falls Church, VA. http://everyspec.com/MIL-STD/MIL-STD-0900-1099/MIL_STD_973_1146/. Consulted 28 January 2018.

U. S. Department of Defense. 1993. *Military Standard Systems Engineering MIL-STD-499b*. EverySpec. http://everyspec.com/MIL-STD/MIL-STD-0300-0499/MIL-STD-499B_DRAFT_24AUG1993_21855/. Consulted 28 January 2018.

U. S. Department of Defense. 1997. *Military Handbook Configuration Management Guidance MIL-HDBK-61*. Falls Church, VA. http://everyspec.com/MIL-HDBK/MIL-HDBK-0001-0099/MIL-HDBK-61_11531/. Consulted 28 January 2018.

U.S. Department of Defense. 2003. *Department of Defense Standard Practice Defense and Program-Unique Specifications and Format MIL-STD-961e*. Fort Belvoir, VA: Defense Standardization Program Office. http://everyspec.com/MIL-STD/MIL-STD-0900-1099/MIL-STD-961E_11343/. Consulted 28 January 2018.

U.S. Department of Defense. 2008. *Systems Engineering Guide for Systems of Systems*. Office of the Deputy Under Secretary of Defense for Acquisition and Technology, Systems and Software Engineering. Systems Engineering Guide for Systems of Systems, Version 1.0. Washington, DC: ODUSD(A&T). http://www.everyspec.com/. Consulted 28 January 2018.

USAF. 2010. *Air Force Systems Engineering Assessment Model*. Air Force Center for Systems Engineering, Air Force Institute of Technology: Secretary of Air Force, Acquisition. http://www.afit.edu/cse/. Consulted 8 September 2013.

CHAPTER 6

Applying Systems Engineering to Create an E-Democracy

1 Introduction

This chapter takes up the task of bringing the thought experiment to fruition, of discussing how SE can be applied in a political context to develop a workable e-democracy. The approach thus far has combined experience and knowledge from engineering and system theory with the aim of addressing successfully some of the problems with representative democracy discussed in Chapter 2. The argument is that SE is capable of delivering a technical system that can enable people's rule. This chapter draws on the SE principles identified in the previous chapter and thinks through how it might be applied to an existing representative democracy, what might be termed as a 'legacy system', to produce an e-democracy. The FCI and CI concepts are adapted to a hypothetical e-democracy design in the context of creating an operational e-democracy. As has been argued in the previous chapter, SE's technical constructs simplifies through CIs, but it also delivers complex functions through FCIs like a decision-making function through its adapting of the SE's Requirements process, with Configuration Management (CM) providing the governance element.

These core processes in an e-democracy aim to shift control to the people. SE enables this goal to subordinate all the functions that a government system delivers during the life of the regime to the will of the people. Basically, SE arranges the CIs and FCIs in a systematic manner

S. Bungsraz, *Operationalising e-Democracy through a System Engineering Approach in Mauritius and Australia*,
https://doi.org/10.1007/978-981-15-1777-8_6

that can make it possible for people to have meaningful inputs into decision making and in so doing can develop a democracy that democratises.

2 Systems Engineering Advances Decision Making and Control of the People

'In general, the complexity of systems desired by societies is increasing', note Blanchard and Fabrycky (1998: 13). With this trend, as new technologies become available both new and the existing system become 'augmented' (Blanchard and Fabrycky 1998: 13), and the need for SE is anticipated to expand in a technologically advanced society. This societal need is from an ongoing desire for better systems and accordingly it creates 'an ever-changing set of requirements' (Blanchard and Fabrycky 1998: 13). Therefore, identifying the true needs to develop the real requirements is a technological challenge. SE as explained in the previous chapter allowed the end user to retain control of decision making with the SE process for the life cycle of the system. Here, the system is the e-democracy solution, which is designed by the people for the people using FCI and CI constructs.

Now, if we make an analogy with the end users being the people, the SE process becomes the people's process as it is designed to meet their need to have a say in decision making. This is a modernised democracy that retains its traditional ideological intent for people's say in decision making. An assumption is that to communicate their interests, every person would have access to this process, and that the government would have institutions established for citizens consultation to occur. This is an interface issue between people and the government institutions. ICTs' reach can make this access for the people an effective one. Before exploring an actual polyarchy to seek an institution that may suit the democratisation purpose in the next chapter, Fig. 1 depicts such a process from a conceptual sense.

In Fig. 1, people are consulted and this generates information which a systems engineer synthesises to create requirements using the Requirements process. These requirements are then collected to form a specification or Agenda for the government to deliver on. Each constituency provides their Consolidated Requirements List as a specification or Constituent Agenda. This Agenda is then voted by the people which makes it binding. Through cyclical audits progress against each requirement

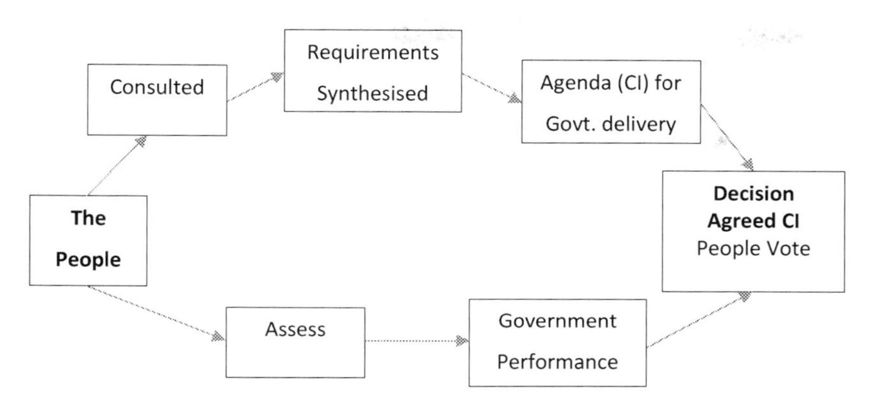

Fig. 1 SE process for decision making in an existing system

becomes a means to assess the government performance. Issues of non-performance are then raised through the representatives for clarification. This flow of inputs from the people allows the people to have a say in decision making. Each constituency as a CI becomes part of the larger system of the nation-state. Data in the form of raw statements or primitive statements is captured from each constituency with each representative becoming the custodian of those requirements. The CM technique, as a people's governance process, is used to manage the requirements set from each constituency from its creation by the system engineer to when it becomes after an election, an agreed requirements set or Constituency's Agenda. CM technique ensures each representative as the people's delegate consults with constituents to make changes to the people agreed requirements. Changes to an agreed requirement need a change proposal to be raised to discuss the merit of the change. This change proposal would require costing, a risk assessment, and importantly a justification for change. This flow for a legacy system, shown in Fig. 1, uses cyclical voting to start the process of people's engagement in decision making for agreed requirements. The idea is that the Requirements and CM process empower people to take control of the key decision-making tool in the form of an Agenda. However, as technology is applied in the political system the cycle can become an ongoing one through ICTs whereby any change or Constituency's Agenda (CIs) can be debated and ratified by the people. Now we expand on the concept of applying CIs and FCIs to develop the idea of a system e-democracy which uses the same process of Requirements and CM for the Constituency's Agenda but at a higher level for FCIs.

3 Systems E-Democracy as Functional Configuration Item (FCI)

In a SE approach, interacting CIs allow a different way to view e-democracy. CIs are considered as components that make up a system e-democracy or the state in our case. A set of subsystems or FCIs in the SE framework makes up a functional political system (e-democracy), the primary or higher FCI which is the e-democracy at state level. The information (data) is the essence of the interface between them. As mentioned above, each constituency as a CI (or FCI—if an autonomous system) interacts with each other at the parliamentary level which is the primary FCI. This construct allows the state to be broken down into CIs and FCIs, with various degrees of autonomous decision making.

A common link between systems, subsystems and components is that they (CIs or FCIs) exchange data. The data allows the system to interact and as a whole to contribute to deliver purposive outcomes by the functional CI (e-democracy) or primary FCI. CIs are essential modules that are components of the primary FCI which is the e-democracy. CIs and FCIs, as the building blocks of the state, exchange information for the political system in degrees of complexity from primitive statements to specifications for deliverables. However, extraction of information and value from this data interchange needs a transformative process. This transformative process is Requirements which adapted translates people's input into Specification or policies in an e-democracy at the state level. This transformative process or Requirements process is explained in Chapter 5. Requirements is the core process that elicits people's inputs in each constituency or CI. As already explained in Chapter 5, the data capture or information eliciting process from the people (end user) is through Requirements, this process at the CI level is embedded in what we equate to a constituency. Data capture can be digitised using the breadth of the Internet as a networked system between CIs, FCIs and the primary FCI or the e-democracy. The networked system of CIs and FCIs can be structured in different ways.

Through SE's SoS approach, SE extends systems theory to create a democratising governance system by making decision making dispersive (from the unitary CI) yet holistic (upwards to the primary FCI). This disperses power to individuals yet concentrates it at the top level by delegation from the people. Now with SE even though the power is concentrated through delegation, the people as the end user still retain control, like the US military end user mentioned earlier. At the FCI level, the inputs from the citizen level CI are used to specify deliverables for the primary FCI level or nation-state. These deliverables are the demands that are mentioned by David Easton but not clearly articulated to flow from the people as his critiques like Birch contends (Birch 1993: 220). SE builds from the ground up with the people's inputs. In doing so, SE identifies a role for the ICTs to assist the SE techniques by creating access for engaging the people for inputs for specifying deliverables. ICTs systemic purpose is to enable capture of the data whereby SE's analysis transforms it into requirements. Inputs are collected systemically by involving the citizens and using Requirements transformed into specifications for delivery. This citizens' inputs for data eliciting for primitive requirements can be conducted on-line and then followed up by the systems engineer (analyst) who clarifies and transforms the primitive statement into a requirement using the Requirements process.

SE in a system e-democracy has ICTs facilitating data capture against CIs and its interchange between FCIs to identify what the people really want at nation-state level. An assumption is the representative is willing and able to use SE techniques or has the support in place like analysts to generate requirements through the Requirements process. The purpose of the data capture is to communicate the good life outcome from the citizens to those that govern (representative) or the policy gatekeepers as defined by Easton (1971). The data, in the system e-democracy, flows in the direction mentioned by David Easton, it is from the people to the institutional representative. The CIs are a means to aggregate the data but without the will from citizens to engage and the institutional representatives to facilitate this activity, the SE process will not work. There are alternatives for automating the SE process to educate and train.

Alternatively, data capture using ICTs can be done through sensors. Changes in the digital field may allow such data capture automatically by governments through on-line activities. These can supplement existing data to allow human needs to be better understood by analysing patterns emerging from the data collected through sensing activities. Sensed data could allow current and future human requirements to be anticipated and perhaps modelled even to an individual level. For delivery of good life outcome requirements, SE techniques allow trends to be identified from requirements. As when similar requirements emerge over time, it would indicate for example that either there are delivery failures or increased demands. SE techniques of transforming needs and wants into requirements can identify both current and future requirements to allow the governance system (e-democracy or Functional CI) to plan ahead in terms of anticipated needs. This provides a degree of foresight. This foresight is for advance planning not speculation. It needs to be validated by engaging with the people, this is a means for use of the emergent capability from ICTs.

This data capture to allow foresight elevates SE's requirements analysis techniques into the areas of contingencies, or an e-democracy that anticipates its citizens' future needs and plans and prepares for that future. This is a leap forward where technology is used to improve government to continuously meet citizen's goals for a good life. Having future requirements insights, governments can organise the now in anticipation of the then functional requirements. A key aspect is the willingness of the people to allow such data capture by the state for the SE analyst. Eliciting data from a face-to-face interaction is the better means as found by the e-democracy experiments carried in twelve of the OECD countries mentioned in Chapter 3. Face-to-face interactions supplements on-line engagement between government through the analysts and the people.

This eliciting process to convert the data from the people into specifications or policy is to reduce systemic stress as mentioned by David Easton in Chapter 4. But importantly, it allows learning for both the people and the institution eliciting the data. The eliciting process is akin to a pilot case where a new system is being operationalised to create knowledge. The knowledge creation is from experimenting and transforming, and then improving from the learnt experience how an Agenda is developed from the inputs from the people where all citizens participate. The intent is to stabilise the political form through understanding the people's needs. Eliciting needs makes the system evolve to refine the technique of decision making for every citizen towards developing a national Agenda.

In the modern state, this stability has been pushed to a political party's survival. Each party in its attempt to survive exhibits distorting behaviours as they try to outbid the other parties with promises eventually leading to exaggeration. Leading to a cycle of hopes that are being dashed due to unrealistic bidding for votes. This emergent behaviour is from the competitive system, from representative democracy. Party programs causes stress to the system as they are exaggerated to win votes. It is anticipated that behaviours will change as control shifts to the people as depicted in Fig. 1 where the voting becomes requirements that are aggregated into specifications. Specifications become deliverables mandated by citizens to their state.

The SE processes define factual (from the people) rather than professed deliverables (promises proposed by the representatives and their parties). As a collective, the stress from unmet demands would become transparent with SE to every citizen. This transparency leads to new collective behaviour where representatives are made accountable for their decision making. The deliberations in Parliaments become focused due to the specificity of what must be delivered to the collective by the few as representatives. The SE top-down approach to check for unmet requirements makes demands visible. If a database is used, any unmet demands can be easily traced from its links to requirements and the need generating the requirement. Such relational databases applications were introduced earlier, like DOORS and Teamcenter for examples. An e-democracy with a networked database system could manage every need for the whole state when powered by a SE governance process. A suggested collective goal for the SE process at the state level is the good life for the whole of society. This is achievable through the SE Requirements process developing the good life specification.

4 Upgrading an Existing System to an E-Democracy

In practice, governance forms already exist. They can be treated like legacy systems in engineering that must be upgraded by SE to a democracy where people decide. However, when the SE process is applied to an existing legacy form for example liberal democracy or other forms mentioned by Held (1983), in such cases the SE process uses the SoS approach to lead the existing form (legacy system) to a new one. The new form is a system e-democracy with an ultimate requirement of people's rule. The old forms will ideologically constrain the new form where e-democracy interfaces with the old form. However, once the SE process is in place a culture change occurs both from technology as it becomes more capable and the people as they learn to use the process. SoS allows many types of systems to come together to deliver a common outcome, for example the three independent emergency services such as fire, police and hospital coming together under the SE governance system for an emergency event, mentioned in Nielsen et al. (2015: 18).

SoS allows SE to design a conceptual system or a system e-democracy building upon the legacy system as a system that it interfaces with to upgrade to people's rule. A systems approach simplifies as SE design concentrates on the ultimate goal the new system must deliver for an end user for the whole life cycle as the SE process explained in Chapter 5. The SE processes have been successfully trialled and applied in complex technological environments which an e-democracy will operate in. The system that SE design ensures it fits within the environment where it will operate. These external requirements are part of its design consideration (Blanchard and Fabrycky 1998). Additionally, SE can modularise the problem and use a multidisciplinary team to seek a solution. SE aggregates and synthesises from the simple which is one component to the whole system which is complex. SE governance allows multidisciplinary inputs to solve a problem.

Importantly, the SE process is consultative with inputs every time from the end user during the life of the system. The upgrade activity for the legacy system is treated by SE as a set of interface requirements that must be addressed during design, so the designed system fits into its operational context. Each operational context (different nation-state) has the same SE process but different solutions. That is the solution is dependent on the citizens from each nation-state and the solution would reflect this contextual customisation. With a SE system approach, the goal of people's rule for the conceptual system e-democracy does not deviate during its design life cycle. The SE process drives the need or ultimate requirement and applies it to both the design and operational phases for the life cycle of the system e-democracy. Through requirements SE for example designs an aircraft which delivers multiple roles, the key issue here is the requirements. We discuss this design process below.

Due to design constraints, the first solution is normally a compromise as the SE process works within the available resources and that places constraints on its designed solution. The solution is also constrained by the state as the operating environment of the e-democracy. For an e-democracy, ICTs are an example of a resource constraint in some states where the digital infrastructure may be inadequate. This may impose an ongoing process of incremental development to achieve people's rule using digital technology. When the development is towards something unknown or abstract, an incremental strategy of learn and evolve is adopted by SE (Blanchard and Fabrycky 1998). SE designers learn from the design process and as knowledge is created they keep improving upon the initial product to meet the design need for example as new technology becomes available and as problems that emerge during operation are known and fixed (USAF 2013; Department of Defense 1993; Blanchard and Fabrycky 1998). Technology may be created by the market and then integrated within the system to either develop new capabilities or enhance the existing capabilities. These capabilities may allow better integration using autonomous data interchange; for example, these changes are accommodated through the change process mechanism of SE, CM. As the capability of the technology grows for example the concept of Artificial Intelligence (AI) delivering technology that automates complex functions between systems through automated data exchanges, like bots mentioned earlier, these capabilities are incorporated into the system through an upgrade using CM.

The SE process is structured to allow either new upgrades or new solutions to be continuously implementing to maintain the operational requirements during a system's life cycle. This dynamic ability inherent in the SE process allows the system to modernise. This keeps the system relevant to its time and context. CM has a key role to ensure the management of obsolescence during the life cycle of the system. Advancing technology not only extends the current capability but it tends to provide opportunities for new capabilities from creating new systems. Technology has a crucial role in the operational capability of the military where SE plays a critical role as an engineering governance principle to develop systems as explained in Chapter 5.

The 'explosive growth of technology' in that last few decades has been the single largest factor which has made Systems Engineering crucial to engineering complex systems (Kossiakoff et al. 2011: 7–8). The technology has had a flow on effect even in the way engineering used to be conducted given its complexity from increased integration. Technology has the biggest impact at the conceptual level stage which is the first phase of the system development. Many solutions developed by the military have technological risks as the development of the system is based on new technology much of which is not well understood. So technology introduced in governing is anticipated to have risks, but these risks have been mitigated by SE through developing an interim solution as mentioned in Chapter 5. The interim solution is fielded through a pilot study to learn about the system when operationalised. Deficiencies that are observed from operating the system are addressed as they become known. The CM process incorporates the system upgrade requirements to the e-democracy system as improvements for obsolescence as technology becomes more capable.

Traditional engineering practices like in civilian aircraft would use known principles in design, in the military the technology is at the experimental stage and much is not known or understood therefore there is significant risks from use of the technology. One learns as the technology is implemented in the system's operating environment. Innovation produces new processes, materials and technologies which is not well understood in term of its characteristics and as these are applied new issues emerge that were not expected during design. This may affect the performance of the system and require changes that are sometimes costly down the track. Systems Engineering allows these risks which are unknown to be managed to create the mature or operational system. Understanding the risks requires a broad knowledge of the whole system which is a SE perspective. SE balances the risks through its structured process of validation where the product is tested to uncover and then address risks that are found during operation. Risk identification leads to the potential for a level of autonomous behaviour as controls are implemented that automatically monitors and addresses with appropriate responses through software during the development of the interim solution. A mature form results when the system meets its performance specification. Upgrades until this maturity is anticipated to keep evolving the interim solution.

Systems engineer designs the system to develop the level of autonomous behaviour that the system can manage on its own through FCIs. Engineers applying the SE principles create knowledge of the whole system and how it is meant to perform in the desired role from CIs to FCIs. This knowledge is then built upon during the pilot phase with a goal to improve the performance of the system when it is in operation. The holistic perspective allows the components to work together and the structured processes allow the emerging issues when operationalised to be identified and addressed in a structured manner through a change management technique like CM. New solutions are required like from development of new components through design or through new technology that are discovered. In the structured SE process, a multidisciplinary team work together to develop the emerging issues from both the hardware and software as CIs. These designed CIs require multi-engineering skills to design solutions as FCIs.

This incremental design process ensures that the system is kept on developing to ideal performance, an e-democracy that democratises. The process of SE creates superior knowledge about the system CIs and FCIs which then becomes the next design baseline for the next capability to be developed for operationalisation. This incremental and iterative process creates ever sophisticated systems that embrace technology as part of its solution suite. This evolutionary design process continues for the life of the system as it undergoes major upgrade as decided by the end user. The system relevance is maintained due to this systemic change process. The path for improvements can be traced through the CM process.

5 Adapting the Systems Engineering Framework—Technology and the Nation-State

From lessons learnt to implement ICTs for an e-democracy even the OECD mentions a need for a framework (OECD 2003). Reflecting on its own attempts to implement ICTs, the OECD noted that 'there is, at present, no methodological framework that specifically addresses how ICT can be designed and used to efficiently and effectively support information provision, consultation and participation in the policy-making life cycle' (OECD 2003: 31). The USAF to function has a need to develop and maintain the most advanced technological capability. The quest for improving its system has led the USAF, for example, to develop its core SE processes so that better systems could be successfully built for the life cycle of the system (USAF 2010). The USAF's Systems Engineering Assessment Framework (SEAM) is the closest methodological framework developed for advanced technology implementation. USAF'S SE framework is both holistic and flexible. This comprehensive framework has transparent processes. The processes can be measured to keep them effective and efficient by self-audit of the activities.

Digitisation if done incorrectly can create complex systems but these may be failures by not delivering the desired outcomes. The digitisation can also be about perceived rather than real needs. USAF set of SE processes reduces the risk of failure over the whole life cycle of the system during design and its maintenance (USAF 2010, 2013; Department of Defense 1993). Such SE processes have been iteratively built and from costly lessons learnt new processes created like CM that manages scope creep. Scope creep occurs when requirements to build a system are out of control. In the absence of a framework, scope creep increases cost that can result in failures. The costs of digitisation are significant but costlier are the political parties working for their own agendas, this is democratic scope creep from the people's perspective. A failure of processes from lack of people's control in the representative system. A scope creep must be arrested to create the systemic outcomes desired by the people.

Analysis of the set of techniques provided at USAF (2010) shows that for developing policy such that people participate, two of the USAF processes have potential to engage the people. These are its SE process of Requirements which extracts the needs and wants to create requirements, the basic information for policy making, and CM both of which are explained in Chapter 5. CM provides the management framework to augment and maintain the system over its life. This CM technique is required for managing the requirements as a set (Agenda). In this book, the Agenda is a deliverable by the political system during an election cycle. While Agenda defines the needs, the system must deliver at a certain point in time, these needs evolve and the requirements change over time. Some requirements become obsolete. A system of ongoing monitoring of these evolving changes and its incorporation is therefore needed.

For change management, the USAF used the CM techniques as explained in Chapter 5. On top of scope creep, a compounding problem for USAF was technological obsolescence during the life cycle of the system. The USAF CM technique allowed it to maintain its capability through systemic changes so that a capability was continuously enhanced, if feasible, else a new one was created to meet the new needs. As new technology comes on board, the book assumes that people's needs and wants change and the society changes. Therefore, a change management process like CM is essential for incorporating changes. These two processes together should allow the development and ongoing management of an Agenda that SE elicits from the people within a political system. The following schematic Fig. 2 shows an e-democracy where the SE processes Requirements and CM create the Agenda to give the people control in decision making. The use of ICTs is discussed below to form the virtual *demos*.

A flow from the people whose requirements are analysed is to form a set, the Concept Requirements List or raw (primitive) statement list that is managed by CM techniques as an Agenda is shown in Fig. 2. The Agenda then becomes the deliverables as the Agreed requirements at the FCI level. Every constituency or CIs provides inputs to the FCIs for creating an Agenda. Use of technology like a networked database is

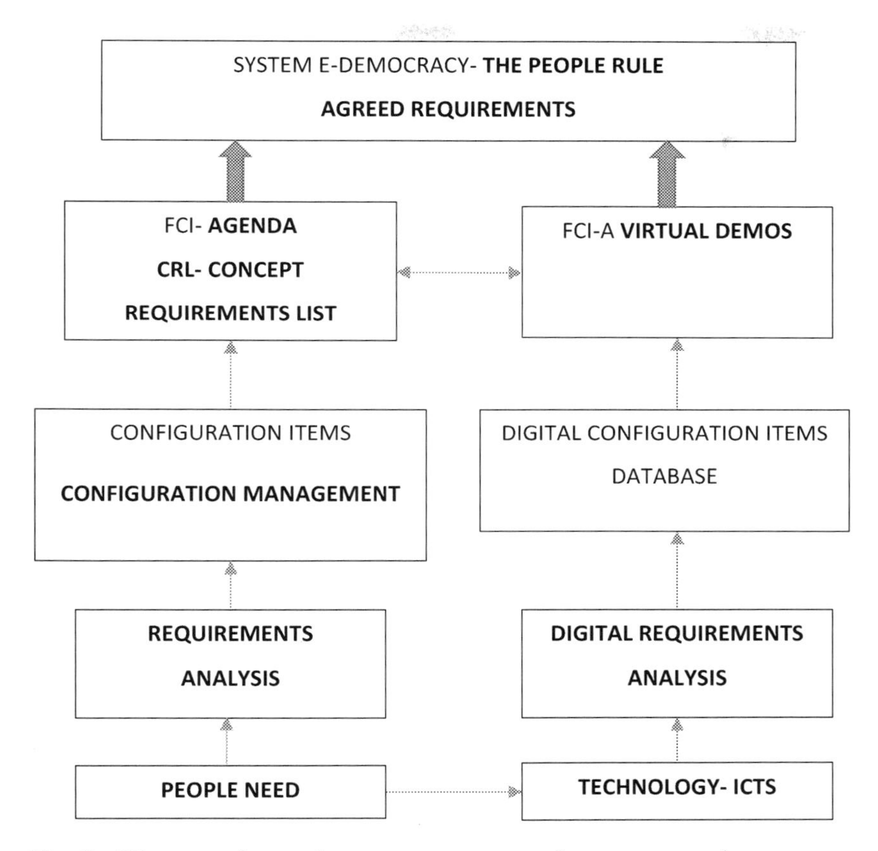

Fig. 2 SE process for need progress to system e-democracy agenda

then overlayed unto the process in parallel as shown above, this flow to develop the virtual demos is discussed below.

6 Systems Engineering Enabled *Demos* from ICTs

There might be a risk of Agenda manipulation and arena control as more powerful participants in the system exercise greater control. Lukes' point is that the more privileged and powerful participants can skew the system to advantage their interests (Lukes 2002). This is something that SE also has to minimise. Some proponents of e-democracy have suggested that a *demos* empowered by ICTs can involve more citizens effectively. A SE driven alternative would be based on an ICT enabled *demos* that would help minimise the risk of abuse of power. Some proponents have suggested e-democracy as a new form of *demos* that through the ICTs involves the people in government decision making (Hague and Loader 1999). Arguments for why such a system, the *demos*, would not work are raised by those opposing change to empower the people as discussed in Chapter 2. They are in fact restricting the idea of democracy. They argue for self-imposed limits due to the existing process which as Hindess described is a deficit by design (Hindess 2002). Such resistance was apparent for example when direct feedback was provided to politicians using websites in current representative system of government. These people's inputs did not lead to policy making, the representative argued that a 'political process is far too complex' to use these direct inputs (Papacharissi 2002: 13). This is a mindset of I have been elected, so I do not need to listen till the next cycle. A pattern of behaviour allows unaccountable terms in office. The argument from complexity is a ploy for inaction in a representative system that allows such inaction till the next cycle. SE requirements analysis process could have been applied to transform those inputs into requirements for policy making. Assuming the good life is the objective behind the individuals providing inputs through the websites, such inputs can be managed using CIs that manage complexity in large systems.

SE uses the idea of CIs to enable a digital democracy concept to create the good life through the input flow shown in Fig. 2. For the good life Agenda, SE aggregates off from the unitary concept of individuals to family to society and uses SE's CIs to create the good life requirements that inform an Easton's unitary political system. To work around the issue of complexity, the created CIs can be digitised using databases. The use of technology is a key interface issue in that the databases must be able to exchange digital information. This interface issue is highlighted and it was solved by the US DoD for its suppliers who are to provide its JEDMICS database with a standard information procedure using a shared network system (Department of Defense 1997: 1–8). Complex aircrafts are divided into CIs and FCIs to manage its functions that must be aggregated to deliver the higher function, flying specific missions. Digitisation makes this aggregation more effective than if done manually, that is in a paper-based system. With CIs in politics, SE processes use ICTs to manage the data for the good life of the interacting individual and family as organised society. Creating the CIs as digital entities to link the good life, ICTs can be used to formulate the good life for the society as explained above. Good life then is a set of aggregated needs translated into requirements by SE Requirements process mentioned above.

CIs link the various components of society with the functions of governance, informing these to create the highest level CI which is the Functional CI or e-democracy, Fig. 4 and Fig. 8 in Chapter 5 shows the links. These are applied to the Agenda generated and controlled by the people through a flow embedding SE as shown in Fig. 2 to develop the virtual *demos*. In this way, CIs are informational units that provide a means to link the people and their requirements to the State's function that delivers what they value. The CIs are informed and managed by the key SE processes Requirements and CM (USAF 2010; Eisner 2011). Parsons explored the concept of subsystems making larger system for studying power in organisations but not as CI behaving as informational units in a SoS framework (Parsons 1971). CI is a concept created by SE to manage each component that makes up the system as a collective. It can construct and deconstruct complex as well as simple systems into elements that make up the collective system. The key aspect is that each component links to contribute to the overall function of the FCI.

Now with the CIs, the deconstruction of a state into elementary components like constituencies is possible, this idea through SoS and SE's transformation into requirements was explained in Chapter 5. CIs can be aligned with constituencies where voting allows one or more representative to emerge with an Agenda for the constituency that they represent as mentioned above. A means to gather information about what such representatives have been selected to deliver is yet to be created and aggregated to the Functional CI or system e-democracy. CIs could allow this data capture which could define the scope of power delegation by people for each elected representative for delivery during their term in office. CIs would represent and generate a set of requirements from the people that could be aggregated at the FCI level. This synthesis would provide policy formulation by the people.

The people in a state also inform the representative through SE techniques to create a functional CI or democracy as a political form. This system improves upon the current one where what must be delivered is ambiguous as there is no systemic process to aggregate their requirements in a democracy. Ambiguity for what must be delivered also leads to lack of accountability. SE techniques when applied to digital democracy allow representatives to create their mandates or be given an approved mandate by the people they represent. In so doing, representatives transform to become a defender of the public will rather than party line compliance. This new engineering technique in improving their role may disrupt the electoral reengineering by parties, which Norris (2004) has identified as an issue for holding government to account. The systemic weakness of a free and fair election that is competitive allows enterprising representatives to bypass the people's will, and SE addresses that through its people centricity process that informs through people's requirements.

Now digitisation to reflect the public will from digital voting has been implemented by some states, but this according to Hilbert (2009) led to euphoric claims about an e-democracy by the proponents of direct digital democracy. Citizens' empowerment for direct decision making by e-voting is a common form of an e-democracy being proposed (Meier 2012). This e-voting has led to confusion in the literature obscuring rather than solving existing problems. A forward-looking Aristotle had predicted chaos with direct decision making from the many (Dunn and Harris 1997; Dunn 2005). With only voting as a decision-making tool, a risk of a tyranny of the majority also emerges even in a digital environment.

For consensus that is fair, inputs from people still need to be debated in a practical manner. Representatives with their approved mandates from the people could do this openly in Athenian style. The people could then access these debates through ICTs. People can review the debates to provide feedback to their representatives. Such direct feedback reduces the powerful role of gatekeepers to use information to their advantage. David Easton identified in his system theory information was a means to either change the system or maintain the system (Easton 1965a, b, 1966, 1971). This potential to use lack of transparency to one's advantage was feasible prior to the advent of the ICTs. In a democratised system, this ICT enabled feedback mechanism allows requirements to be changed and recommunicated by the people to their representatives in a transparent manner. With the ICTs this feedback may also occur directly when the debate is happening in Parliament. In this way, representatives can harness the resources which are from the peoples' input directly during parliamentary debates; thus, a virtual *demos* is created.

The people can ratify what is debated in Parliament. The SE technique of CM allows such changes to be traced for every requirement and put to the people for ratification. Any such agreed ratification process would be included in a CMP mentioned earlier. This is an alternative way for the people's involvement in decision making using the ICTs and SE techniques. With a database system, these inputs could be 'retained in networked computer system' as suggested by Grady (2006: 50) as information technology 'greatly facilitated' a primitive requirement conversion and its traceability. Digital technology allows multiple links to record every aspect of the decision. This informative aspect from technology has been enhanced for the Teamcenter database which supports the decision-making links for every aspect of the system life cycle from design to disposal (Briggs and Sampson 2006).

This collective input from CIs by SE introduces the concept of the purposive system or goal oriented democratic system. Using the CIs, ICTs can be used to create abstract things like quality of life based on its citizens' inputs that SE elicits to generate people's requirements. Additionally, a CI construct also allows various types of systems (autonomous components) called Functional CIs (FCIs) to be built through a systems of systems approach used in networked communication systems (USAF 2013). Based on the interaction and interfaces from CIs and FCIs multiple ways to form *demos* is possible. There is a degree of empowerment in a state as each FCI can focus on contributing to a pre-defined people's agenda through a SE framework. The FCI construct allows higher levels of complexity it builds functions that contribute to the higher functions for the primary FCI or e-democracy. We explore this systemic reductionism with FCI that allows both autonomous and non-autonomous subsystems to operate to provide democratic capability that empowers yet centralises.

7 The E-Democracy That Democratises with People's Inputs

To make the democratic system function in an integrated manner, systems approach allows a holistic means to decision making. As a systems approach SE considers and integrates inputs from the political, ecological, social, cultural and the psychological understandings. In doing so, SE embraces the ethical aspects of decision making. The first stage for SE is at a conceptual level to define needs which results in the requirements for the system itself, like which processes would be needed by the project.

This aspect engages both the end user and the technologies that may provide an integrated solution. At this stage, the life cycle of the solution is a key consideration. SE harnesses various disciplines to provide inputs to the solution.

When SE is applied to a problem faced by governance, a consideration of the solution is the life cycle. SE's framework emphasises on requirements being conducted methodically at the conceptual stages of developing the solution. This methodical emphasis at the conceptual stage is because SE found it has a significant impact on the effectiveness of the outcome during its life cycle (Blanchard and Fabrycky 1998). A causal relationship emerges from SE's ability to trace end user requirements to outcomes. This framework links risks from a bad decision to deliverables of the system. This risk identification of bad decisions before developing the outcome allows an opportunistic window for better satisfaction of the end user, the people in this book. In a democratic system, given satisfaction changes it is essential to have the people involved in the decision making on an ongoing basis.

SE's set of processes are meant for an ongoing system. Adapted to a democratic system, they create a goal oriented capability for a system e-democracy. It can generate requirements continuously from people's inputs. Data that are created flows through the SE processes to be translated into requirements for each CIs. The ICTs bring an enabling capability potentially to automate some of the data collecting activities. The analysis of the inputs finds what is common. It creates links between individual needs (a unitary CI component) and group needs (higher level CIs supra-components). The CIs aggregated to its highest level creates a Functional CI (FCI) that exists to process aggregated demands for the system. This functional CI could be for a constituency that may or may not be an autonomous political system. In a system e-democracy, the requirements elicited for the primary FCI are at the state level, these would be policies for the state.

To improve and create a people valued governance system with SE techniques, the book views e-democracy as a purposive democratic system. E-democracy must deliver demands from its SE augmented function. CIs are to be perceived as information aggregates that allow e-democracy to be informed to govern. Using these CIs, Systems Engineering can apply digital technology to relate empowerment for decision making to the function of the digital democracy. The functional outcomes of the primary FCI are traceable and linked to specific requirements, hence a democratic system cannot deliver outcomes that have no requirements. Requirements are what the people value. This constrains representative's scope of decision making as outcomes must be specifically related to one or more requirements. With the SE techniques, the role of ICTs is for facilitating requirements creation to inform the CIs for system e-democracy. This application of ICT by SE democratises as the people are involved to voice their inputs for representatives to deliver the requirements set (Agenda).

SE is an alternative system approach that decomposes a system into CIs yet also aggregates it to its FCI, which creates higher system function. CIs construct allows a reduction of the systems interactions to unitary level with differing complexities. The capacity of ICTs to link data and the people through the Requirements process using CIs can be aggregated through SE for creating demands.

CIs allow direct interaction at individual levels and for the complex ones at society level. The data exchanges between CIs can be captured by ICTs or manually through system analysts eliciting it from the people. SE type processes use these CI interactions to create a goal oriented system (USAF 2010). Using CM techniques, SE is capable to manage the complex issues, for both individuals (unit CIs) and groups (aggregated CIs or higher FCIs) interactions. Each constituency in a State is made up of a group of individuals and the process used by the analyst for capturing CI requirements can be digitised using SE's reductionist (FCI) and expansionist (CI) systems approach. This technique allows different thinking mode, inside out reductionist to FCI and outside in expansionist to CI.

The interaction of ICTs is democratising subject to CIs as individual needs being understood and articulated by the SE analyst. SE techniques translate the interactive data elicited from CIs through requirements analysis. This technique creates a set of requirements for and from each CI. SE can conduct requirements analysis without digital technology but then size, with too many CIs, becomes an issue. Also, the CIs may be geographically dispersed, the Internet technology allows reach to potentially every citizens to create a networked CI associated data. With a networked database, such a set of requirements when developed from the CIs concept allows relationships to emerge and be observed by the SE analyst. Aggregated to a primary FCI, for example good life policies for society, the requirements sets are mapped to the lowest CI or individuals for producing associated outcomes. The good life at FCI level is definable through these links. There is flexibility in the framework, a FCI at society level may have different outcomes from CIs at lower levels but the outcomes are both linked to Requirements process which are from the people's inputs from which the outcomes were synthesised. This is a top-down and a bottom-up approach as outcomes at the top (FCI) are related to needs from the bottom (CI).

To create the CI system's capability in the SE framework, ICTs are ensuring reach to each member of the public through its breadth. This would be critical for democratisation. With ICTs now becoming portable on smartphones, we assume that the barriers to reach and access are significantly reduced due to the portability of the technology. However, to improve system redundancy free access can be provided at various social venues. The ICTs capability though have to be implemented by the decision makers for this purpose, this is in line with the OECD ICTs implementation recommendation (OECD 2003). The representatives (politicians and senior bureaucrats) in charge of running the affairs of government must develop the required ICTs functionality and support for SE analysts. This assumes the decision makers understand ICTs and also that they will allow such a change with an intent to democratise rather than further control. ICTs capture of data can increase control in a system to have goals other than democratisation or a good life, as mentioned earlier in the Egyptian case study in Chapter 3. ICTs deployment with SE is not meant to increase control but rather to relinquish power so that people are engaged through a power share arrangement in that state. SE uncovers a technological risk from ICTs that it expands on below.

Digital technology seemingly also brings a 'disruption to political life' whereby 'decision making is unhinged' as it is misappropriated to represent other interests than those of the citizens (Wilhelm 2000: 3). A risk exist as seemingly 'emerging information and communications technologies' are benefitting the unfettered market to the detriment of the people (Wilhelm 2000: 4). A rebalancing is therefore needed to address these neo-liberal challenges. SE, as a modern system approach, is transparent and repeatable to elicit the decision-making interactions and SE constrains this market approach. SE's constraint is a set of requirements transparently elicited by SE to instil a degree of people's control over decision making in the democratic representative systems. This set of requirements (Agenda) can be baselined and managed through the CM process. Requirements as a set have been created to develop engineering specifications that create a solution from idea to products, like in the design of aircraft for example (USAF 2010; Eisner 2011), and CM techniques ensured that what was delivered conformed to the specified requirements. Similarly, SE applied to develop an Agenda scopes the deliverable for political parties as contractors to the people and this is depicted in Fig. 3.

In the absence of the SE framework, the ICTs assist the entrepreneurial representatives in achieving their goals unfettered by the current procedures of the representative system. An Agenda developed by SE scopes the people's delegation like specifications. The adapted SE process is shown in Fig. 3. Now this diagram shows a series of relationships aimed at minimising the privileging of those with more power.

This is a point that has long been understood within political science thanks to the path-breaking analysis by Lukes in his identification of "three faces of power". Here it is the second face of power identified by Lukes that is of interest. As he demonstrated, the second face of power is undetectable to inquiry, and is 'unperceived by pluralists' approach in a representative democracy (Lukes 2002: 5–6). Individual, groups or

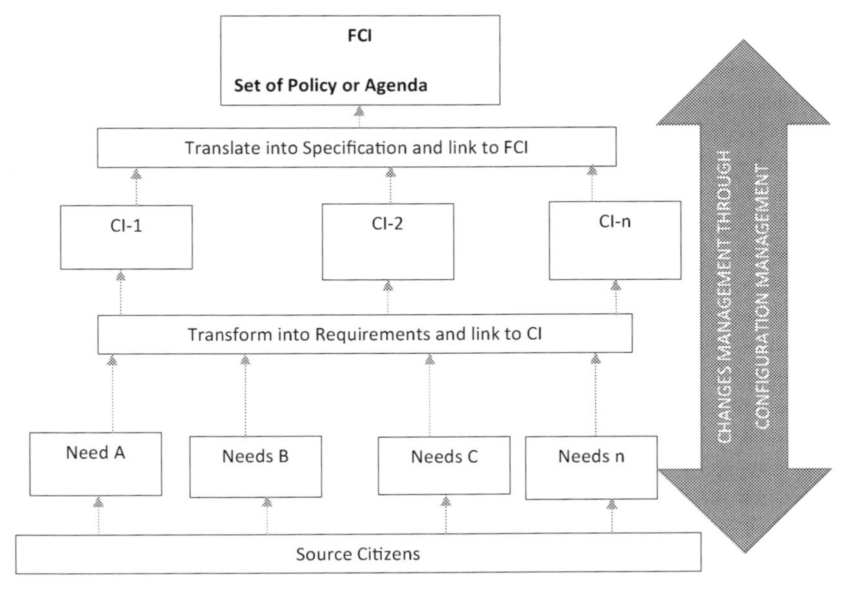

Fig. 3 A conceptual operational system e-democracy with adapted SE process

organisations exploit conflict within society, and they use bias, like for political parties right oriented or left oriented focus, this bias though eliminates some debates from public airing. In so doing, these entities exercise power from consciously and unconsciously creating or reinforcing barriers that remove or foster issues in politics. A hegemonic view amongst political parties that representative democracy as *politea* is the optimal system has the power to undermine other alternatives like an e-democracy that democratises. Political parties exercise power when operating within a hegemonic thinking as they actively defer or exclude through non-decision making for example the public debates of the idea of alternatives, as an that is untested but has potential like an e-democracy, or an evolutionary design for a democracy that democratises. Non-decision making is an insidious form of power that is hard to detect. It allows in an Internet driven era those in power to manipulate the powerless majority by covertly reducing alternatives that changes the tacit status quo. This is despite a lack of 'democraticness' in representative democracy (Wright 1994: 536).

In this sense, hegemonic views or bias are sustained, these in the form of culture or ideology become a form of class rule, like bourgeois for example. Lukes contends that a belief system, as a majoritarian view, could be maintained through the inculcating of social actors that become predisposed to execute a hegemonic view (e.g. bourgeoisie or monarchy). An issue (class rule) that the majority may be unaware of when manipulated by those in power. As an example, Lukes posits that social actors when they consent to be ruled they do so conditionally and their conditions are based on their individual limits. Now these limits could be a crisis point for the political system an equilibrium that defines the limits of representative democracy for example. In the 1970s, in capitalist democracies, the democratic limits of the capitalist political system were questioned for legitimacy. Now there were alternates like socialism or a revolution to be considered but regardless of the legitimacy crisis in capitalist democracies these choices were not exercised.

For Lukes, this lack of change to an alternative political system was from the covert and coded resistance of the neo-liberal ideas to change, instead capitalism was spread worldwide. This spread of capitalism (hegemonic) secured and maintained compliance of the subordinated masses to ideological domination from capitalism's propaganda (Lukes 2002: 9). This form of domination remains undetected and Lukes argues it perpetuates the power of those political organisations that exploit policy conflicts promoted as left and right wing in society to retain power. Conflict allows entrepreneurial parties to manipulate outcomes through effective use of power to promote their objectives.

That is Lukes' second face of power which is undetected in a pluralist system like representative democracy is used to manipulate the outcome (Lukes 2002). Due to systemic bias, a lack of decision making allows some individuals or groups to exercise power by limiting debates. Through a transparent SE process, the limits of decision making (agenda setting for policy) in the political system are defined through specifications from the *demos*. These specifications are traceable and this allows transparency of decision making which reduces to some extant Lukes' undetectable second face of power for exploitation by groups or individuals or organisations. With a transparent Agenda set by the *demos*, SE opens a new avenue for the potential enhancement of accountability for decision making and non-decision making within a political system, as unmet *demos*' specifications remain visible with SE processes. The compliance to the SE processes by the contractors, political representatives can be assessed using the SEAM that was discussed above. SE minimises Lukes' second face of power through improved accountability to the *demos* as its processes are audited for failings in decision making (unmet specifications).

8 Systems Engineering Design Solution Is Evolutionary

To improve a system, the sources for change can be many; these changes can be split into technological ones and the end-user ones. SE is developed with an inbuilt continuous process improvement process. It maintains a system that it designs by identifying and incorporating the changes initiated from the people and those from technology (see Fig. 4). Note,

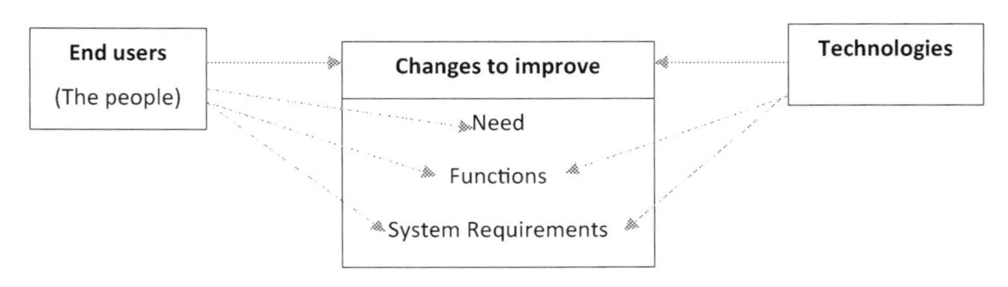

Fig. 4 Sources of change for system e-democracy ongoing improvement

only the people change the need as the system is built around their need as mentioned earlier.

When applied as mentioned in the previous chapter, technologies improve the performance of the system. SE develops technical systems, this means systems or objects that are designed to represent all types of human made activities. The 'technical system' is a collection of engineering activities that are performed in design where information is generated, retrieved, processed and transformed into products which can be tangible like an aircraft or intangible like services for example a logistics system (Blanchard and Fabrycky 1998: 13). The technical system also includes manufacturing activities, inclusive of the project management and cost estimation activities for both the product and society during the life of the system. An e-democracy designed by SE is a technical means to empower the people.

This technical system shapes the functions and the interface requirements of the system where communication takes place between components. Technology improves the efficiency of the system. The people can change the need for the system and for example decide collectively if the system e-democracy requires the people's rule. The people changes are the ones that impacts the need for the system. The people shape the form, for example an e-democracy that is inclined towards a liberal democracy or a direct democracy. Figure 4 highlights that it is not the technology which drives the ideological form but the people. A technical system is meant to make the lives of people better; the issue is to find a way to use technological progress to make adaption to use the technology less stressful on both the system (e-democracy) and the people. SE integrates technology to meet people's need; this SE link to technology was explained in the previous chapter.

To deliver people's rule, a SE process could for example have a paper-based system delivering SE's processes needs of storing and retrieving information. As mentioned in the previous chapter, the use of digital technology created its own changes to SE. The ICTs though improve the efficiency of the SE process with quicker information flow between the interfaces of the system (USAF 2013: 9). In a digital age, ignoring the capability of the ICTs is wasteful. The speed and capability to transfer, store, retrieve and transmit information using the Internet and wireless technologies like Wi-Fi creates new opportunities for developing better engagement between citizens and the governance system. Digital technology allows the tedium of managing vast data where records were lost and misplaced in a non-digital system to new higher level value added activities of analysing data to create good life information. Databases are inexpensive means to manage the information that society generates and uses. Distributed databases networked through now cloud computing capabilities open new options to release humans to innovate the good life, to people become information generators and users with the ICTs.

The system could be designed to generate alerts at the interfaces like exception reporting rather than routine activities of data retrieval and storage which could be automated. Such digital activities integration opens doors to significant efficiencies for routine tasks. When the tasks gets complex then what cybernetics used to call autonomous systems with complex programs like bots could be used as mentioned earlier. There are some who speculate that through the enhanced algorithms even an AI is in the realm of the feasible to innovate in these areas (Kaplan 2016). In Chapter 4, we mentioned the cybernetics scientists' speculation at the time about the potential role of information technology (Mead 1969). The digital technology like databases have an enabling role for the performance of the system by handling information but not in maintaining the existence of an ideology like people's rule. A SE system solution of people rule exists because of a need that is driven by the people, a need that SE meets but which is shaped by people's ideology.

SE can harness the technology and as new technology emerges, incorporate it to improve the product's performance to meet the system's ultimate need even when this could be technology like AI. The schematic also highlights the SE process maintains need purposiveness (democratisation) and the SE process dependency on people's decision making. In a SE process, the people are the centre of all changes. Figure 1 in Chapter 5 mentioned earlier depicts this centricity, and the SE process dependency on the people (as end user) ensures a system that listens and is people oriented. What it entails is that changes from technology can but more importantly should improve upon the existing system.

Through SE's techniques, a new or upgraded solution or system during its lifetime requires change or modifications to keep it relevant to meet the evolving needs of the end user, and this is normally achieved through approved modification of the system. An existing system is therefore modified to allow the system to persist through new functions that may be added from end user approved modifications. The analogy for this upgrade is to treat the existing system as a legacy system that requires change due to functional requirements no longer being met. Functional requirements changes are not efficiency measures they are new needs that may have surfaced as a result of changes from the end user or when the environment within which the system operates is changed. Such changes may have an adverse effect on the performance of the system and thus necessitate an upgrade. We have explained this change process as CM in the previous chapter.

9 From a Democratic Deficit by Design to an Enhanced E-Democracy

At the conceptual design stage, SE uses a top-down approach to derive requirements from the need which is the people's system e-democracy. This is the ultimate functional requirement at the early phases of system design. For the book, this is analogous to a literature review for an e-democracy improvement. From lessons learnt for engineering designs, the use of both top-down and bottom-up approaches augments the design process, and this is the SE's design approach (Blanchard and Fabrycky 1998: 24).

The bottom-up approach is used in the later phases of SE's preliminary design stage this is when actual design begins for when the components are manufactured and integrated into the specified system configuration. From the top-down and bottom-up approaches, SE addresses the need in a bi-directional manner. SE process is important as translating something abstract like a need into something more concrete like a lower-level component for manufacturing may result in a product that does not perform (Blanchard and Fabrycky 1998: 28). An approximate analogy is SE being applied to an actual context like a representative democracy, for example the Mauritius political system. A bottom-up approach is from the people as the end user this can only occur with a pilot study being run in the legacy system.

Traditional engineering designs are bottom up. The engineer starts from a known set of elements from research to develop the concept. This bottom-up approach according to Blanchard and Fabrycky (1998: 28) would have a problem translating to an applied form as they mention 'it is unlikely that the functional need will be met at the first attempt unless the system is simple'. The likelihood that a bottom-up design will meet the need for a complex system like e-democracy seems remote. E-democracy research has developed a significant scholarship (Scholl 2013; Päivärinta and Sæbø 2006; Grönlund and Horan 2004), yet it fails to implement.

An alternative is the SE process with its track record as a standard for solution development in Defence (Department of Defense 1993; Eisner 2011).

The SE design process is a directed method. It is stepwise process to deliver a structured solution with specific phases and specific decision-making points. The Waterfall model shown before in Chapter 5 discussed the steps in the process of design. The decision points in the model allow the system to be baselined such that it is a controlled means to deliver a solution. This allows a degree of control to the end user about the solution being proposed. Thus, an e-democracy is in this way a people's design where the people are the end user. This creates the e-democracy of the people by the people for the people. With SE, the people retain control for the system e-democracy for the life cycle of the system.

In the top-down approach, SE breaks the problem into distinct functions that can be aggregated to the system's function of an e-democracy, the FCI (Department of Defense 1993; Blanchard and Fabrycky 1998). In addition, no preliminary design is conducted until the need is firmly understood during conceptual design; this crucial step of need definition by SE is depicted in Fig. 5 as a top-down process.

A proven method in the military, the top-down SE approach, is consistent to deliver the required function (Blanchard and Fabrycky 1998). Audit of the requirements from the top down ensures that all functions derived from the top are met and the aggregates to create those functions (FCIs) can be linked to the component level (CI) (Eisner 2011; Blanchard and Fabrycky 1998). There are no orphaned requirements in

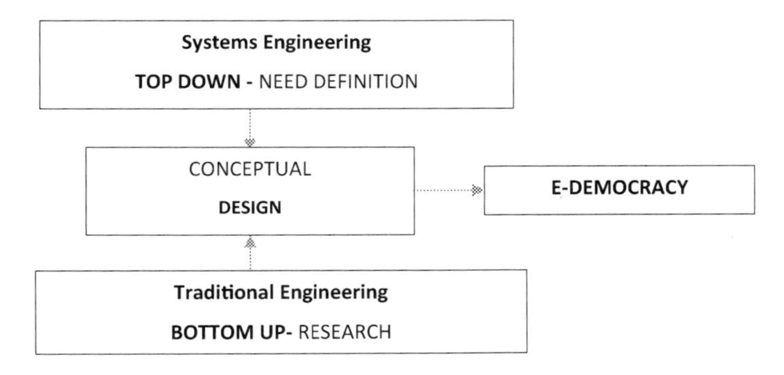

Fig. 5 The SE process to develop a conceptual design

the bi-directional manner from top and bottom, each functional and its component level requirements are linked. The databases mentioned earlier maintain these links throughout the design and operational stage, DOORS and Teamcenter. A requirement databases would allow the monitoring of the links when the system is developed and over time as the system evolved due to changes from upgrades to manage evolving end user needs. If we investigate each element SE designs recorded in these databases, each design is to meet the highest function the need or ultimate requirement to which it contributes through inputs and outputs and the way the element links with other parts. Blanchard and Fabrycky (1998: 28) mention the SE process at system level 'recognizes that general functions are involved in transforming inputs into outputs'. Therefore, at conceptual design stage it is about the functions that would be required to create the higher function (need) or ultimate requirement.

The lower-level interactions build the inputs and outputs until the highest function people's rule is created by the system. Easton at the system level investigated inputs being transformed into outputs (Easton 1971), but details were missing unlike with the SE structured approach, which applies inputs and outputs for creating identified functions in a directive manner. This directive SE design process is from a user's need (people need) to the solution cycle. Success for the system is from acceptance of the product by the end user (the people), this provides control to the end user and for an e-democracy it shifts control from the representative to the people. We expand how this deficit by design (Hindess 2002), can be removed and implemented below in every legacy system or nation-state if the political will exists.

10 Shaping the Ultimate E-Democracy Operational Requirement

In the context of this book, society is viewed as a purposive social system, which organises itself to create a better governance system to manage its affairs such that every citizen has 'equality in those things that make for a good life' (Estlund 2002: 39). It is with this conceptual goal in mind that we explore the SE's framework developed by the USAF (2010) and Department of Defense (1993). The context to which SE applies matters to deliver an operational e-democracy. The SE process ensures that the people's need is driving the operational e-democracy, as explained during the life of the e-democracy every function of the e-democracy is subordinated to the top function. This top function is why the system e-democracy exists, it is the operational function of the e-democracy.

The SE process came into being as 'entities can be enhanced by giving more attention to what they are to do' that is for SE 'simply stated, form should follow function' (Blanchard and Fabrycky 1998: 18). What this means is that for SE the purpose of an e-democracy is important. The system or form of the e-democracy is derived for a specific function during design, the ultimate need or primary function which is the top-level function or people's rule. SE designs a system whose functions meet a need, the people's needs in our case. For a successful system design, SE must identify if the need exists to upgrade the existing system, this need analysis is the first step or an analogy is a proof of concept for SE to proceed. This primary function which is people's rule is called Functional Configuration Item or primary FCI, it would be defined prior to developing a solution.

To design, SE creates a particular functional element that we call Functional Configuration Item or FCI at the system level and components of the system called Configuration Item or CI, refer Chapter 5. These CIs and FCIs apply the structured SE approach to design solutions which provides alternatives to the people (end user). CI and FCI are applicable to a whole class of systems, thus only a few FCIs and CIs are needed to conceptually represent the real system. This approach simplifies how systems are represented when the SE process designs to meet the need. There could be for example thousands of specific element for example like screws used to build a system but with SE only one CI uniquely represents all of them. One CI specification of the screw is used to design all such screws. The CI and FCI modules are the building blocks for creating the highest system's purpose, the solution to the need. We explain this link further below.

A need is defined as something that is required or a requirement at the highest level, this need is termed the ultimate requirement for an operational solution. From an operational need, SE is a structured way of thinking to solve a problem. The SE process integrates the efforts of a multidisciplinary team for this common purpose which is to satisfy a need (Blanchard and Fabrycky 1998: 105). This interdisciplinary team is called an integrated product team (Blanchard and Fabrycky 1998). The multidiscipline team brings the requisite expertise to find a solution to satisfy the ultimate requirement in a structured manner. However, an understanding of the product life cycle process is fundamental to SE as the design must encompass both support aspects and ongoing upgrades as technology changes so the system remains relevant in a changing context (Blanchard and Fabrycky 1998). This is done through the structured steps of the Waterfall model mentioned in Chapter 5.

SE is a process that designs a solution with an operational intent (USAF 2010; Lee and Miller 1998; Grady 2006; INCOSE 2007; Eisner 2011; Blanchard and Fabrycky 1998; Department of Defense 1993). SE as mentioned above is linked to technology and as technology advances it allows SE to upgrade systems that operationally remains relevant to the technological changes that ensue. Sources of change can be from technology which creates an opportunity for an e-democracy to improve its processes. Importantly though, it is from the people who through uses of the technology like ICTs are to decide if the representative democracy is inadequate in an era where better options could be available or simply put if the system must be upgraded.

An ultimate requirement may be shaped by new needs from the people or an upgrade to improve a latent need from the people like a democracy that democratises. This system level requirement is within the context of the state which supplies resources and that are its constraints. The state will require to supply the resources for the technology and infrastructure to allow SE to design the upgrade or new system we term e-democracy in our case. Figure 6 depicts the relationship that exists between the state's resource constraints which bounds the solution that SE can design, system e-democracy. We discuss this link as that is one of the strengths of the SE process.

A need is the 'ultimate requirement' (people's rule), and this need is the purpose of the design for the life of the system e-democracy when operationalised. This need using Requirements process, explained earlier, undergoes a refinement process and that is followed by a decomposition

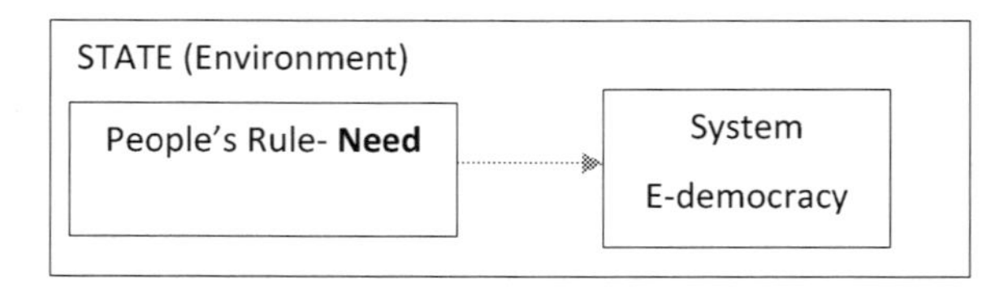

Fig. 6 The system level requirement (ultimate requirement or need) links to design (e-democracy)

process of the system's lower-level requirements as explained in Chapter 5 and above for an e-democracy (Grady 2006: 4). The SE process uses Requirement process to ensure that the need of the system e-democracy is correct and it uses CM so it is approved by the people. This is done prior to decomposing the lower-level requirements from which the final form system e-democracy is designed through the SE process, refer preliminary design and final designs SE flow process Figs. 2 and 8 in Chapter 5. The need is why the system is designed (purpose). In addition, when designed the SE structured solution meets the ultimate requirement. A need for SE is the highest function that the solution represents.

To meet this ultimate need in operation SE uses a design cycle, this is a process which encompasses the whole life cycle of the system, see the Waterfall model in Chapter 5, Fig. 8. When and if this ultimate requirement changes the SE design process will have lost its purpose (function), like a system e-democracy design is irrelevant or obsolete if democracy is no longer a people's rule or democracy is not what the people (end user) want. In both cases, that is either when transforming an existing system or building a new system, having a clearly defined need is crucial to the SE process success. This first step of involving key stakeholders is critical to the success of the design of the e-democracy.

11 Conclusion

This chapter explored how a thought experiment arising from SE can be applied in designing e-democracy. The Requirements and CM processes when applied by the state have the people at the centre of informed decision making using Configuration Item (CI) and Functional Configuration Item (FCI). The elicited needs from the people are iteratively transformed through SE's processes into the agreed requirements or Agenda to be delivered by the state as the primary FCI or ultimate requirement. The structured SE process, Requirements, shifts a degree of control which using CM puts the people in the decision-making seat. These two processes initiate an evolutionary democratisation process where using a staged development the SE design cycle implements the people's rule for a system e-democracy that is the people's. Application of the SE design life cycle like the Waterfall model then creates the function, an e-democracy as decided by the people. The decision-making function of the system e-democracy is from the requirements elicited from the people, and the e-democracy form is customised to suit the context to which the system applies and operates within. The elicited requirements when agreed reflect the people's will. The next chapter examines an existing representative democracy suffering from democratic deficit as a legacy system for a potential upgrade to a system e-democracy. This upgrade creates a system that reflects the people's will or Agenda for delivery by the state in a political sense. This SE upgrade of the legacy system, an existing political system, delivers control to the people. In the next chapter, we consider how the SE process might be adapted to an actual representative democracy.

References

Birch, A.H. 1993. *Concepts and Theories of Modern Democracy*, 2nd ed. London: Routledge.

Blanchard, B.S., and W.J. Fabrycky. 1998. *Systems Engineering and Analysis*, 3rd ed. Upper Saddle River, NJ: Prentice-Hall.

Briggs, C., and M. Sampson. 2006. *Tying Requirements to Design Artifacts*. Orlando: INCOSE. http://onlinelibrary.wiley.com/doi/10.1002/j.2334-5837.2006.tb02795.x/full. Consulted 28 January 2018.

Dunn, J. 2005. *Setting the People Free: The Story of Democracy*. London: Atlantic.

Dunn, J., and I. Harris (eds.). 1997. *Aristotle Volume II*. Great Political Thinkers. Cheltenham, UK and Lyme, NH: Edward Elgar.

Easton, D. 1965a. *A Systems Analysis of Political Life*. New York: Wiley.

Easton, D. 1965b. *A Framework for Political Analysis*. Prentice-Hall Contemporary Political Theory Series. Englewood Cliffs, NJ: Prentice-Hall.

Easton, D. 1966. *Varieties of Political Theory*. Prentice-Hall Contemporary Political Theory Series. Englewood Cliffs, NJ: Prentice-Hall.

Easton, D. 1971. *The Political System: An Inquiry into the State of Political Science*, 2nd ed. New York: Knopf.

Eisner, H. 2011. *Systems Engineering—Building Successful Systems*. Synthesis Lectures on Engineering, vol. 14. Washington, DC and San Rafael: George Washington University and Morgan & Claypool.

Estlund, D.M. 2002. *Democracy*. Blackwell Readings in Philosophy. Malden, MA: Blackwell.

Grady, J.O. 2006. *System Requirements Analysis*. London, UK: Elsevier.

Grönlund, Å., and T.A. Horan. 2004. Introducing e-Gov: History, Definitions, and Issues. *Communications of the Association for Information Systems* 15: 713–729.

Hague, B.N., and B. Loader. 1999. *Digital Democracy: Discourse and Decision Making in the Information Age*. London and New York: Routledge.

Held, D. 1983. *States and Societies*. New York: New York University Press.

Hilbert, M. 2009. The Maturing Concept of E-Democracy: From e-Voting and Online Consultations to Democratic Value Out of Jumbled Online Chatter. *Journal of Information Technology & Politics* 6 (2): 87–110.

Hindess, B. 2002. Deficit by Design. *Australian Journal of Public Administration* 61 (1): 30–38.

INCOSE, I.C.O.S.E. 2007. *Systems Engineering Vision 2020*. INCOSE Central Office, San Diego, CA, USA: International Council on Systems Engineering.

Kaplan, J. 2016. *Artificial Intelligence: What Everyone Needs to Know*. New York: Oxford University Press.

Kossiakoff, Alexander, William N. Sweet, Samuel J. Seymour, and Steven M. Biemer. 2011. *Systems Engineering Principles and Practice*, 2nd ed. Hoboken: Wiley. http://app.knovel.com/hotlink/toc/id:kpSEPPE006/systems-engineering-principles/systems-engineering-principles. Consulted 26 November 2017.

Lee, J.S., and L.E. Miller. 1998. *CDMA Systems Engineering Handbook*. Boston, MA: Artech House.

Lukes, S. 2002. *Power a Radical View*, 2nd ed. *New York: New York University Press*.

Mead, M. 1969. Cybernetics of Cybernetics. In *Purposive Systems: Proceedings of the First Annual Symposium of the American Society for Cybernetics*, ed. H. Von Foerster. New York: Spartan Books.

Meier, A. 2012. *EDemocracy & EGovernment: Stages of a Democratic Knowledge Society*. Berlin: Springer.

Nielsen, Claus Ballegaard, Peter Gorm Larsen, John Fitzgerald, Jim Woodcock, and Jan Peleska. 2015. Systems of Systems Engineering: Basic Concepts, Model-Based Techniques, and Research Directions. *ACM Computing Surveys* 48: 11–18.

Norris, P. 2004. *Electoral Engineering: Voting Rules and Political Behavior*. Cambridge, UK: Cambridge University Press.

OECD. 2003. *Promise and Problems of E-Democracy-Challenges of Online Citizen Engagement*. Paris, France: OECD.

Päivärinta, T., and Ø. Sæbø. 2006. Models of e-Democracy. *Communications of the Association for Information Systems* 17 (37): 1–42.

Papacharissi, Z. 2002. The Virtual Sphere: The Internet as a Public Sphere. *New Media & Society* 4: 9–27.

Parsons, T. 1971. *The System of Modern Societies*. Englewood Cliffs, NJ: Prentice-Hall.

Scholl, H.J. 2013. Electronic Government Research: Topical Directions and Preferences. In *Electronic Government*, ed. M. Wimmer, M. Janssen, and H. Scholl. Berlin, Heidelberg: Springer.

U. S. Department of Defense. 1993. *Military Standard Systems Engineering MIL-STD-499b*. EverySpec. http://everyspec.com/MIL-STD/MIL-STD-0300-0499/MIL-STD-499B_DRAFT_24AUG1993_21855/. Consulted 28 January 2018.

U. S. Department of Defense. 1997. *Military Handbook Configuration Management Guidance MIL-HDBK-61*. Falls Church, VA. http://everyspec.com/MIL-HDBK/MIL-HDBK-0001-0099/MIL-HDBK-61_11531/. Consulted 28 January 2018.

USAF. 2010. *Air Force Systems Engineering Assessment Model*. Air Force Center for Systems Engineering, Air Force Institute of Technology: Secretary of Air Force, Acquisition. http://www.afit.edu/cse/. Consulted 8 September 2013.

USAF. 2013. *SMC Systems Engineering Primer and Handbook*. Space and Missile System Center. EverySpec. http://everyspec.com/search_result.php?cx=partner-pub-0685247861072675%3A94rti-pv850&cof=FORID%3A10&ie=ISO-8859-1&q=SMC+Systems+Engineering+Primer+and+Handbook&sa=Search&siteurl=everyspec.com%2F&ref=&ss=20497j31595539j48. Consulted 8 February 2018.

Wilhelm, A.G. 2000. *Democracy in the Digital Age*. New York: Routledge.

Wright, E.O. 1994. Political Power, Democracy, and Coupon Socialism. *Politics & Society* 22: 535.

CHAPTER 7

A System-Engineered Approach to E-Democracy: A Small Island Mauritius

1 Introduction

As part of the analysis developed for the book, the question was of how might a Systems Engineering (SE) approach work in practice. As it happened, in doing background research on Mauritius and its political institutions, it emerged that Mauritius had a number of institutions that could lend themselves to implementing a SE approach. In particular, the SE feature of a Functional Configuration Item (FCI) had an analogue in the Mauritian political system, namely the National Development Unit (NDU). This presented an opportunity to use Mauritius as a type of proof of concept with which to test the possibility that a SE approach might work in Mauritius, or at least enable a rethinking of the architecture of Mauritian political institutions in a more democratic direction. This also involved conducting a series of interviews with key Mauritian politicians (across the political spectrum) and bureaucrats.

The purpose of the interviews was to gain an insight in how SE could be applied to the Mauritius system. The interviews were conducted during the course of three field trips to collect documentary data. The interviewees were chosen from a broad cross section of political actors and senior public servants involved in policy making on the basis of their publicly listed functions (see Appendices 1 and 2). An email request was sent for an invitation to participate in the interview process. In addition to the interview, I was able to observe the conduct of a national election which

S. Bungsraz, *Operationalising e-Democracy through a System Engineering Approach in Mauritius and Australia*,
https://doi.org/10.1007/978-981-15-1777-8_7

gave me the opportunity to observe how the political process occurred. I also used newspapers, policy documents and field notes of conversations with members of the public and political actors. The interviews were transcribed and data analysed using the interpretivist method of analysis as developed by Bhevir and Rhodes (Bevir 2006; Bevir and Rhodes 2002). These are described further in the appendices.

Blanchard and Fabrycky (1998: 24) pointed out that the SE 'process involves use of appropriate technologies and management principles in a synergistic manner' and when applied with a new 'thought process' (as explored in this book) can lead to a change in 'culture'. The prospect of a change in political culture was a vision shared by a number of those prominent political figures interviewed as part of the research underpinning this book (see Appendix 1). While every interviewee agreed that Information and Communication Technology (ICTs) had the potential to be a 'game changer', they perceived that this technological capacity had the potential to create a will to change. However, they also recognised that social media was becoming influential not just in terms of shaping of citizens' opinions but also with respect to its capacity to empower citizens. Consequently, they acknowledged that how this will to change might be implemented was a key question.

In this chapter, the particular characteristics of the Mauritian political system are discussed in relation to how they might lend themselves to the implementation of a SE-informed e-democracy in practice. The first part of the chapter describes the political context and political institutional arrangements in Mauritius. It then discusses several institutional features, in particular the NDU which could provide an effective starting place or interface for implementing SE technology. In the course of this discussion, the potential barriers to or readiness for an e-democracy as perceived by key stakeholders are also investigated.

2 An Overview of the Mauritius Political System

Mauritius is a small, middle-income island state that is multi-ethnic and multi-lingual, with English and French being mandatory languages taught in school and allowed in parliamentary debates (Government of Mauritius 2014). It is a land of migrants from every corner of the world. From its geographical location in the east coast of Africa, it naturally considers itself as part of the continent of Africa. As a representative democratic country, with Botswana, Mauritius is considered the African success story. It is one of the longest-lasting stable democracies in the region and, due to its economic performance, referred at one time as the little Tiger of the Indian Ocean. Mauritians call it a microcosm of the world.

Mauritius was named after the Dutch Maurice of Nassau in 1638 (Allen 1999: 9). When the Dutch left in 1710, the French claimed it in 1715. The English colonised it in 1815 by defeating the French who were using its port, Port Louis, as a strategic base to support their trade interests. The French were also supporting privateers who launched from Port Louis to prey on English shipping vessels from India. In 1936, financial distress caused significant social pressure in the English colony and Mauritius Labor Party (PTr) emerged as an organisation to represent the interest of agricultural workers, and led Mauritius to independence on 12 March 1968 (Allen 1999: 31). The island republic post-independence in 1968 consists of two main islands, a mainland Mauritius and a smaller autonomous island Rodrigues (Government of Mauritius 1968). The main island is 1860 square kilometres in area; its population is around 1.3 million with about 55,000 living in Rodrigues (Government of Mauritius 2014; Youngblood-Coleman 2014). Unlike the island country that it wants to emulate, Singapore, Mauritius is not close to any significant economies like those in Asia. Nonetheless, Mauritius aspires to become a services hub to Africa. To that end, it has therefore initiated Special Economic Zones agreements with Senegal, Ghana and Madagascar (Government of Mauritius 2017: 22). Some territorial disputes with UK remain post-independence.

Despite its lack of natural resources, Mauritius emerged to reach a middle-income economic status with strategic use of its human capital (Sandbrook 2005; Miles 1999; Meisenhelder 1997; Government of Mauritius 1968, 2014; Kearney 1991, 1990; Croucher and McIlroy 2013; Carroll and Carroll 1999; Bräutigam 1997). This economic improvement has had a stabilising effect on representative democracy, as the country moved away from its coloniser but in an amicable manner. Mauritius remains part of the Commonwealth nations.

The political system consists of a unicameral legislature broadly based on a Westminster system inherited from its British colonisers. Democratic experience in multi-ethnic and multi-lingual Mauritius is recent and is still evolving. In 1991, Mauritius amended its Constitution to call itself Republic of Mauritius, allowing Rodrigues to become an autonomous island to manage its own affairs yet stay within the republic system (Youngblood-Coleman 2014). This move was due to tensions bubbling to the surface from ethnicity issues that if left unchecked would have undermined political stability and undermined the basis for democratic rule. The role of governor which was authorised by UK under the post-independence arrangement was changed to a President decided by the Mauritian Parliament (Youngblood-Coleman 2016).

Mauritius has a democratic Constitution in which the President of the Republic is the 'Head of State' and the 'Commander in Chief' who ensures that 'the institutions of democracy and the rule of law are protected' and equality in terms of 'fundamental rights of all are respected' (Government of Mauritius 1968: CON 26). The President can also dissolve Parliament and call fresh elections. The Constitution ensures that there is a separation of power between the three key elements that provides good governance, the legislature, the executive and the judiciary. These three components are intended to be independent of each other.

On the recommendation of the Prime Minister, a President is elected by the General Assembly for a 5-year term which is renewable (Government of Mauritius 1968). The President formally appoints the Prime Minister who is the person with majority support in the assembly. The Prime Minister is the head of the governing body and is elected from popular vote in a First-Past-the-Post electoral system with a 5-year cycle. In consultation with the Prime Minister, the President appoints the Ministers forming the Cabinet. The President also appoints key functional heads like the Chief Justice, the Commissioner of Police and the various departmental heads in the Public sector as defined in the Constitution (Government of Mauritius 1968). The Prime Minister usually recommends these key appointments. These also include those who are principal representatives abroad like ambassadors and high commissioners. These constitutional appointments perform various state functions.

The Parliament creates the legislation aimed at ensuring good government, peace and order. Once passed by the Parliament legislation is then ratified by the President. The High Court is the highest court of the country and is independent. Constitutional matters are referred to the High Court for resolution. However, a citizen can also access the Privy Council in UK to contest a decision of the High Court (Government of Mauritius 1968, 2014). This colonial link seemingly provides an extra level of oversight over decision making of the legislative system. In 2017, the traditional selection of Prime Minister by the people underwent change. A father stepped down to allow his son to become Prime Minister in his place, this seemed like a dynastic power transfer without consultation from the people. It highlighted the extent of gaming that is possible in a representative system where people delegate power to political parties, and Mauritius learnt something new from this about a party agenda as opposed to an agenda by the people. The Mauritian Constitution allows such a power transition without consultation from the people.

The Mauritian Constitution clearly articulates that Mauritius must be a democratic state where fundamental citizens' rights are guaranteed. It is noted that freedom of the media is guaranteed in the Constitution, making it a pillar of democracy. A right to liberty is enshrined which even allows a convicted person with a 12 months jail sentence to be a member of Parliament (Government of Mauritius 1968). This modern *nomos* also mentions that every individual is protected by laws. The right to private property is similarly protected Under Section 47 of the Constitution; a 75% majority of Parliamentarians is required to change its provisions (Government of Mauritius 1968: CON 41). However, pressure groups ensure that such a majority cannot be formed. Most, if not all, of these groups have a vested interest in the status quo and carry significant economic clout. The party leaders in receipt of donations from these 'investor' groups are sensitive to their needs, and as a result, these groups gain more rights, or at least influence, within the system especially where the role of leaders of political parties is concerned.

The number of constituencies in the Republic of Mauritius is 21, and the number of representatives for each constituency of the 20 mainland constituencies is 3 each, while for Rodrigues it is 2. Over and above those electable three with the highest votes in each constituency, to ensure minority representation Mauritius developed a best loser system (implemented post-independence) to defuse ethnic-based violence that could disrupt the democracy. This best loser system aims to provide a voice to ethnic pluralities who might be under-represented following an election (Government of Mauritius 1968: CON 236). This ethnicity-based aspect of best loser system has been challenged by some pressure groups, other minor parties, as a human rights abuse for a secular democracy. However, the government still allows this system to exist. It removed the main contention that it was mandatory for candidates to declare ones' religion at an election event to seek a mandate. In the face of resistance from pressure groups, the best loser system continues with 8 additional representatives. Every representative draws a salary from the public purse, with the highest paid being the Prime Minister.

The team with a majority of elected members forms the government. The government consists of functions that make up the different government departments that as per the Constitution are managed by up to 24 senior Ministers excluding the Prime Minister. There are also 10 junior Ministers with the title Private Parliamentary Secretary (PPS) who report directly to the Prime Minister. The Parliament has a Speaker of the House who manages the Parliamentary Agenda presented to the Members of Assembly for debate when Parliament is in a session. The Speaker does not have to be an elected member. Similarly, the Attorney General who has oversight over the judiciary may not be an elected member.

The Prime Minister is the Chief Executive Officer in government and allocates the functions of government to his team (Government of Mauritius 2018). As noted, the President as Head of State ratifies this functional distribution. The following schematic, Fig. 1 shows these key roles.

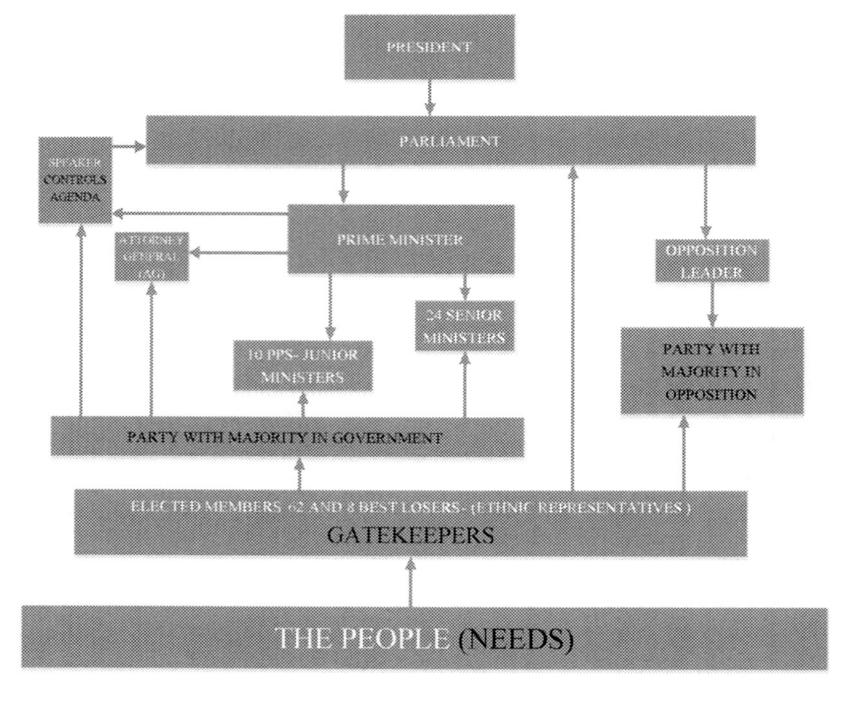

Fig. 1 The current system filters the people's voice

The Opposition Leader is a function for providing a monitoring role to constrain government excesses by raising questions in Parliament. However, the representational arrangements do not include any direct people's input to Parliament for policymaking (see Fig. 1). Once elected there is very little input requested from the people for decision making. It is the norm that all issues debated in Parliament are subject to filtering from the Speaker.

Figure 1 shows a top-down arrangement where the decision-making system of the Parliament is pushing downwards to the people without their voice being directly heard. In effect, the political system situates layers between the people and the Parliament where decisions are supposed to be made. This is typical of many pluralist arrangements in other states. Implicit is that Mauritius, like other similar states under a representative democracy system, is a tokenistic democracy in terms of its filtering. The representative system is efficient in filtering the citizen's voice even though the technological capacity may exist to communicate a citizen's needs. The communication is one way from the representative deciding whether to respond to any given request.

Mauritius has some unique features that reinforce its particular ethnic mix that has shaped the nature of Mauritian democracy through the development of the 'collage'. The meaning of collage is something like 'glued together', and it is a term used by Mauritians to describe temporary bloc formations between two supra-parties as an electoral winning combination. This is usually announced a few weeks before an election event. It is one of the key features that have stabilised the political system while at the same time undermining the more democratic of the features that a democracy should have, and in that way potentially distorted Mauritian democracy. Mauritius has a pluralist society drawn from a diverse number of ethnic groups. The country has a large Hindu population (50%), followed by other groups General Population (31%), Muslim (16%) and Sino-Mauritian (3%) (Statistics Mauritius 2017). The Prime Minister is normally from the Hindu group except for a short collage arrangement where a Franco-Mauritian the Mouvement Militant Mauricien (MMM) party leader emerged in 2003 (CountryWatch 2017). For political stability, this plurality provides some unique challenges to the country that it learnt from and adapted to. This pluralism makes it a unique melting pot of tensions from cultural issues. Because of these tensions and their impact at different times in Mauritius's history, a best loser system was introduced to ensure every ethnic group representation in Parliament.

This is allocated by the Electoral Commission following an election to ensure ethnic minorities have political representation in the parliament. The best loser process is part of the Constitution of the country.

Acknowledging this unique feature of ethnic tensions, to protect democracy, the party system and Parliament have been adapted to select candidates based on ethnicity. The seats for contesting an election are divided in terms of ethnic belonging by the parties. The main ethnic groups in each of the constituencies influence a candidate's choice by the parties. As the best loser system relies on ethnicity, post-independence the mechanism ensures that ethnic cleavages dominate Mauritian politics coming to the fore at every election cycle. This strategic ethnicity calculation from the parties and their leaders may have defused to some extent the tensions from the lack of representation that could emerge from such a sociocultural mix.

The country is also divided into constituencies where the people can elect more than one candidate per constituency. Mauritius is a federated system with 21 constituencies. This federated type of system allows a System of Systems (SoS) concept where each of the 21 constituencies can be treated as a sub-system (FCI) of the larger system comprising of the state, Republic of Mauritius. SE's elicited Requirements process makes the need of the constituency the focus rather than the need of the collage bloc which is vote maximisation strategy. Currently, the members elected to represent each constituency are drawn from their constituency's composition in terms of ethnicity and party affiliation.

On the mainland, the parties select three candidates to represent a mix that maximises their votes for that constituency. This arrangement seemingly gives a degree of choice for member representation as the competition for representation is relatively less than if only one member were to be selected per constituency. It also defuses tension as there is some consultation by candidates to work with the other ethnic groups to get their teammates elected. In each multi-ethnic constituency, this team of three seeks to represent the majority of voters in that sub-system. The representative system is reinforcing ethnicity at each election event. That is divisive and from these ongoing systemic cleaving, it may take longer to develop a secular mindset. This ethnicity has led to a collage behaviour as an effective strategy to get elected.

The parties within a collage preselect their candidates such that the constituency's ethnic representation in that electorate is a bloc of votes for securing a parliamentary majority. In developing this vote maximising strategy, well-financed teams are created by the coalition blocs from a voter's ethnicity representation for every constituency and the perceived popularity of the parties amongst the electorate. So, each constituency is further divided yet represented as a team which are aggregated to be formed into the two main contesting blocs. A super-party is formed from this collage to contest elections. This arrangement is based on lessons learnt by parties to reduce the influence of pressure groups given that they have to deal with multiple pressure groups with diverging demands. The blocs force teams that are pluralistic to come together to win.

2.1 The National Development Unit as the Legacy System Interface

One unique institution in Mauritius is the NDU. It 'was created in late 80 with a distinct purpose ... to bring development in every nook of ...[Mauritius] island' (Government of Mauritius 2016). In effect, it was set up to apply inclusive development in Mauritius with the specific function to provide people's inputs to government. The NDU aimed to assist people's inputs from every constituency to government by directly reporting to the Prime Minister's office. Any Member of Parliament (MP) can raise issues to be debated in Parliament; however, the junior Ministers as PPS have a role to elevate these citizens' voices to Parliament through the Citizens Advice Bureau (CAB). In effect, the MP acts as the interface to articulate people's needs to the Parliament. As part of this strategy, the NDU created the Citizens Advice Bureau (CAB) in 1989 based on a UK model (Government of Mauritius 2018). This process may not be transparent, and often, the layers are potentially unaccountable to the voters.

In Mauritius, the NDU is a dedicated body with an accountability to inform the public about government policy. Its key role is to facilitate requests from voters to the government for improvements to meet their needs. This is a channel from the government to engage with the people. As mentioned by ex-PPS Id No.: 5 in Appendix 1 during interview, the NDU is ideally placed to use ICTs for policy formulation directly from the people. In the pre-2014 government, the NDU reported to the Vice Prime Minister and Ministry for Infrastructure but now in 2018 it reports to the Office of the Prime Minister. Although it delivered on development issues, its focus was on infrastructure such as community centres, football grounds and the maintenance of existing social infrastructure.

The NDU is a collective institution that has CAB offices located in every constituency that makes up the state of Mauritius. The CABs interface with the 21 constituencies of the state, the role of the 10 Junior Ministers as Private Permanent Secretaries (PPS) was created to manage the CABs and their functions. The task of information gathering and informational empowerment is divided amongst the 10 PPS listed under constituency at Government of Mauritius (2016: 17). According to the functional description for the role, each PPS is given the responsibility to provide 'information and advice on existing services to empower citizens' (Government of Mauritius 2016: 3). The PPS process infrastructure issues pertinent to the people in their constituencies and represents them to the respective government department, and they may request a Minister from the appropriate functional department to assist with the people's needs.

The NDU web page provides further information to citizens about the CABs. There are 36 CABs located around the country (Government of Mauritius 2018). As an arm of the NDU, the CAB acts as 'an interface between Government and citizens' which according to the NDU's web page must be proactive in their approach to 'reach [out to] the people to reduce their hardships' (Government of Mauritius 2016). To deliver this outcome, the CAB mentions on its web page that Government of Mauritius (2018) aims to:

- Provide assistance/information to callers on issues of concern to them *[p]rovide assistance/ information to callers on issues of concern to them.*
- Enable people to express their needs and problems effectively.
- Act as intermediary between the people and government by transmitting the needs expressed by the people to government.
- Create a well-informed society that can contribute effectively to the development of the country.
- Integrate the local people in the process of development. *[i]ntegrate the local people in the process of development.*
- Bring about the collaboration of other agencies and institutions in order to offer assistance to those in need.
- Organise talks and seminars to inform people on subjects related to their welfare.
- Register suggestions from the citizens and NGOs for the improvement of their living environment.

- Adopt a proactive approach to serve as facilitator to come to the help of people particularly the needy and vulnerable group.

The words in italics indicate errors on these web pages. This indicates that the pages are not being monitored as the errors were still in evidence six months after they were accessed on 23 January 2017. Errors still exist on those web pages when last checked on 28 January 2018. There is provision to report such errors in another web page for such problems. This is through the ICT Minister's e-idea page, but if they have been reported then they have not been rectified over the twelve months the page was accessed (Government of Mauritius 2018c). It is possible that the web pages are just for show, that is to display processes or services that may not exist as Grönlund (2011) has pointed out in other contexts. The NDU web page also lists the location of the 36 CABs around the country which are delivering the above functions but there is no suggestion that the information being gathered is being analysed for policy making (Government of Mauritius 2018).

Given its strategic function and its long-standing presence within the Mauritian political system, the NDU presents a unique opportunity to apply a SE approach. The very structure of the NDU lends itself to a SE approach. The CAB functions of the NDU could be treated as a CI. Using digital technology like computers via the Internet, these CIs could be linked to each constituency and individual persons as a virtual network. People's needs could be allocated unique identifiers through a centralised relational database system, like the CoRE and DOORS database that are used to raise and manage requirements against CIs as discussed in the previous chapter. This would provide traceability to every issue raised in the state by the people.

Thus, the specific SE techniques of Requirement and CM could initiate the data gathering of people's issues that could then be systematised. The NDU framework would be able to adapt these two SE techniques to elicit people's requests within a SE framework. The SE embedded process is a pull strategy eliciting people's engagement through transforming their needs into requirements. It informs government about people's needs. SE's people need centricity augments democratisation in Mauritius. When successfully carried, this activity would germinate trust between the people and the governance system. It would also create a new and transparent process for eliciting the people's voice, or a listening governance system.

The member for Rodrigues Id No.: 1, who is a PPS, pointed in an interview the key role of the NDU and its potential for an e-democracy (IDNO 2016: 1; Government of Mauritius 2016). Similar views were also expressed by the five other PPS who were interviewed.

The following flow chart explains how the NDU may capture information through the CAB for SE, as a physical unit that is collecting and sharing information in the state. A SE proposed data flow that can be digitised is attached for this unit, Fig. 2. Citizen X has an issue and makes a request at a CAB office. The CAB officer discusses the citizen's needs and identifies which function is the appropriate Ministry to assist with the issue. The CAB officer contacts the appropriate Ministry, and the Ministry responds to the CAB officer. The citizen is informed and awaits a demand being raised by the Ministry for the request. If a demand is raised, then the Ministry delegates an officer to meet the citizen, and raise a requirement. The CAB officer schedules a meeting between the citizen and the delegate to discuss the requirement.

The information flow, as set out in Fig. 2, describes an input from citizens to the CAB office. The data that is generated by the CABs could be captured using technology like a database. Once collated in a central database system, the data could be analysed for synthesising requirements. These requirements would provide a basis for the government to understand the needs and wants of their citizens to reduce people's hardships. Over time, such a data set would provide a means to develop insights for long term and short planning for a good life agenda. The data elicited by the NDU would also facilitate better engagement between the government executive and the people in each electorate (CI). The NDU's engineering function is towards infrastructure projects; it responds to people's needs for repairs and new infrastructures. This is no doubt a good outcome for infrastructure development and maintenance, but its failure to engage with the public about policy formulation and policy inputs has diverted a potentially crucial public input to policy making. The intent and potential behind the formation of the NDU could have been policy making but in practice the institution fails to deliver.

A potential system for NDU's use to assist its function to support representatives and the people is shown in Fig. 3. The schematic below indicates how the NDU function could be redesigned with the SE techniques of Requirements and CM using databases to capture the information from the people.

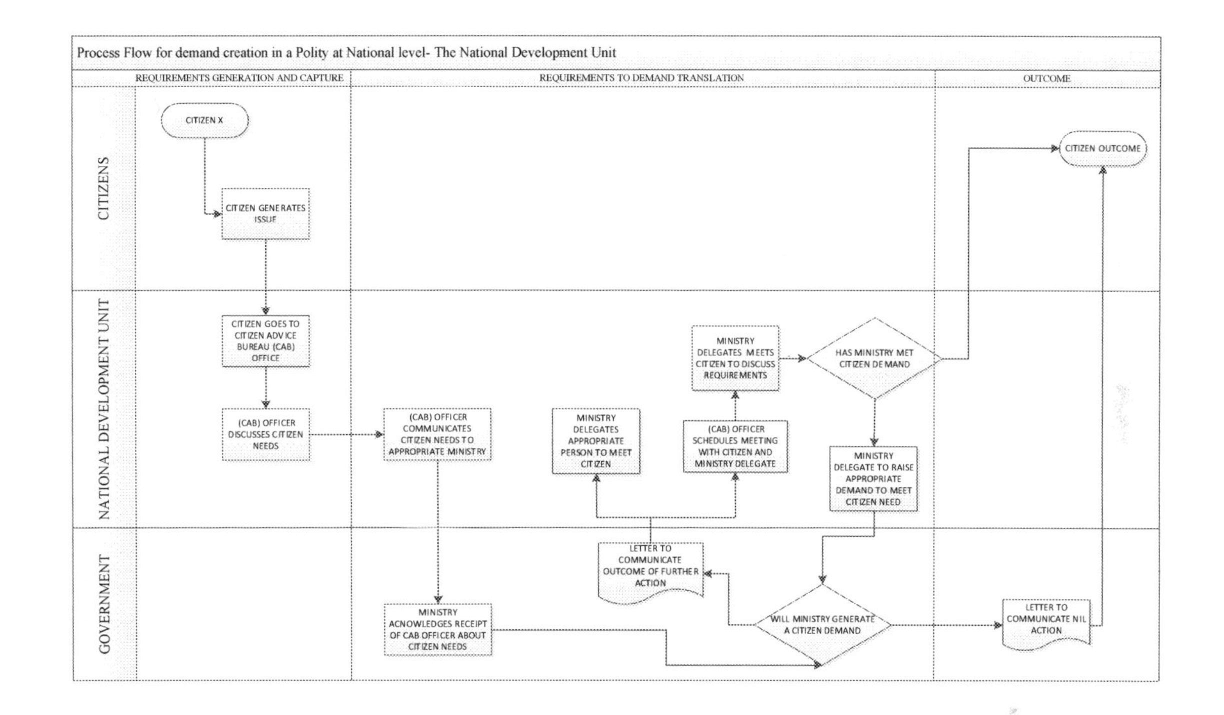

Fig. 2 SE proposed process flow for citizen demand

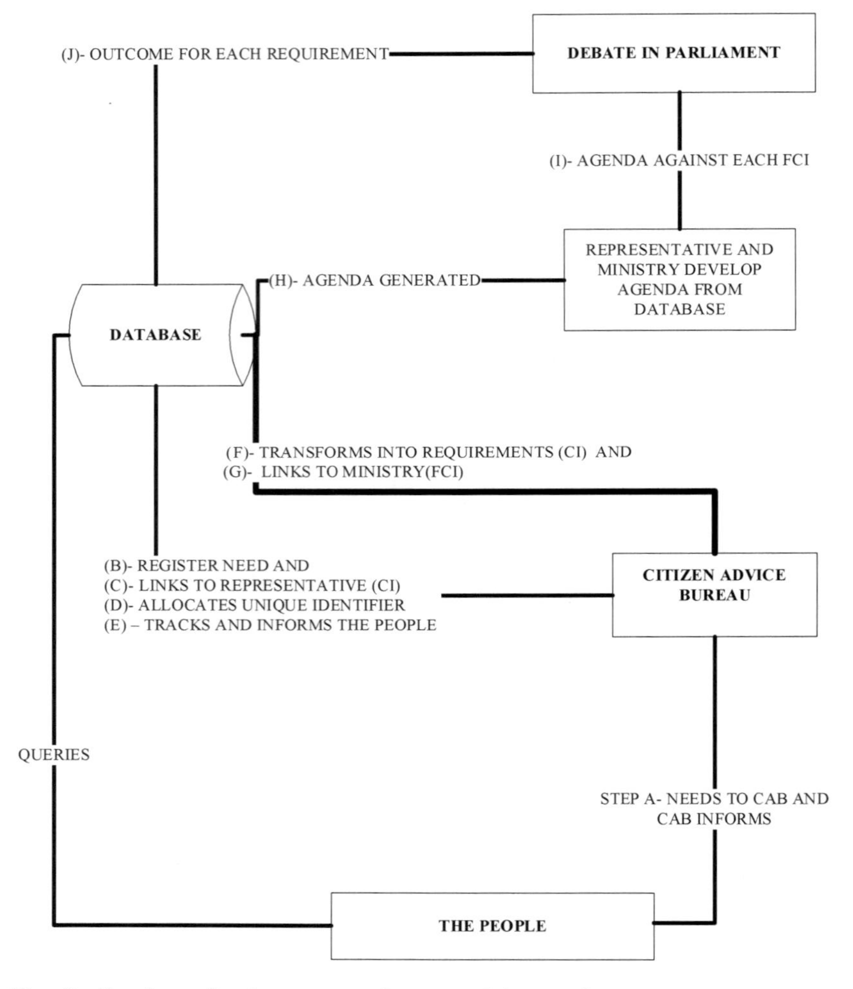

Fig. 3 Roadmap for data capture for a people's agenda

The process requires firstly to engage with the people at each CABs (CIs) using face-to-face and technology like the ICTs. People's needs can be translated into requirements as this interface initiates the key process for eliciting and managing requirements from the people to drive the decision making. This people and government engagement will elicit the raw data or primitive statements. From those raw statements that are generated at the CABs, citizen-specified requirements are created to shape the functions of the government (FCI). The CABs as CIs can enable the data collection for the whole state providing a basis for the state as the FCI to deliver what the NDU function was meant to deliver as explained above. However, it is noted as failing to deliver due to a lack of a structured process like SE being used by government. Here, Fig. 3 shows a potential flow for SE use by the NDU. Policy making in government is transformed from the two SE processes. Debates in Parliament will be from actual requirements from the people raised and recorded at each CAB office. We expand on the policy-making potential from this SE-transformed digital process in the context of Mauritius below.

Context is important; we investigate ICT use in Mauritius as an enabler to manage the data that NDU collects as described above. As has already been noted, technology is a core enabler for an e-democracy. There are lessons learnt from other states' ineffective digitisation experiences reported by the OECD to operationalise an e-democracy; a framework to create a roadmap is recommended (OECD 2003, 2005, 2010a, b, 2013, 2015, 2016). There are also, as mentioned earlier, risks from decision makers—like a political will for change, and technology where a framework for effective implementation of the technology whose capability is improving creates technological obsolescence that needs to be managed. Ineffective digitisation with vague policy implementation adds significant costs and risks like a digital divide from an information poor that exacerbates inequality.

The ICTs technology use in Mauritius is explored below to find if issues do arise and if the Mauritius ICTs policy address democratisation which an e-democracy promises. The next section of the book investigates the current perceptions of ICTs—their role, value and application for an e-democracy. These discussions are from interviews carried out with policy makers in Mauritius.

2.2 ICTs' Role as a New Pillar of Economy

ICTs have become a key pillar of the Mauritian economy and have been responsible for continuing the transformation from an agrarian economy to a service economy focused on call centres, education and tourism. ICT applications like cloud computing and disaster recovery systems for data management are being explored as potential areas for growth. The SAFE cable for Internet connectivity for Australia, Europe and Africa is via Mauritius. Big Data mining is another area to be targeted to facilitate business decision making. These are emerging areas in the field of ICTs for creating a knowledge-based economy. Facilitating government decision making through digitised SE seems feasible on the surface for Mauritius. Looking further below the surface from these ambitious promises reality is different.

Given the large aid input to the Mauritian economy, for 2012, 2013 and to 2014 it received annual aid of 185, 148 and 45 million US dollars, respectively (OECD 2016: 8); there may be a resources constraint concerning the sustainability of an overambitious digital system during its life cycle. In this respect, Mauritius will need to learn from the Spanish experience in which its Avanza plan resulted in significant costs of digitisation for which the Spanish economy is yet to reap the hoped-for benefits (OECD 2010b, 2013). This was primarily because there was a lack of alignment of all levels of government from local, state and national meaning that it was difficult for the overall objectives of the Spanish digital state to be met (OECD 2013: 8). In addition, there needed to be a proper focus on a system that would support better governance through better data provision. This would also require that the digitisation must ensure user uptake is consistent with investments in order to improve the overall demand. If user uptake is too low, this will significantly increase costs of the digital system turning it into a white elephant. Unsatisfactory user uptake was an issue for Spain (OECD 2010b). Another issue is the digital divide as explained by Norris (Norris 2001) that can create digital inequality. Lack of access reduces the ability of citizens to participate and that may result in tensions from digitisation. Barriers due to Internet costs may be a constraint. In a state where there already exists significant tensions from elitism, in an information era democracy may be at risk from a digital divide of have and have nots.

The use of the ICTs due to its mobility on smart phones was mentioned by the interviewees. Again, although they were referring mainly to its effective use by the Lepep collage party during the 2014 elections, they were well aware of that with the potential for novel ideas like virtual meetings. For example, the junior minister Id No.: 14 pointed out for the next election he would try to have virtual meetings using ICTs. With Mauritians becoming acquainted with parliamentary debates being screened using ICTs, electoral debates might do the same. With nearly a

2.5 ratio of mobile phones to citizens (a figure mentioned by both interviewees Id Nos.: 2 and 8), this seems feasible. In fact, it was noted by elected member for Constituency No. 2, the PPS Id No.: 3, that one of the areas where the mobile coverage was weak as in Constituency No 14 of Riviere Noire, the Lepep supra-party message seemed to have had less impact (at least according to her interpretation of her party's voting analysis). This mobility of ICTs via smartphones reduces the possibility of digital divide in Mauritius. This realisation about a technology link to a political system bodes well for an e-democracy in Mauritius provided the costs of such access are not prohibitive.

Another key development that bears on this discussion is that the island of Rodrigues is to be connected to the submarine cable with a third gateway making it part of a networked system globally. What is significant here is that it will link the island to the mainland as a System of Systems through the technology that is being developed. The government has issued an expression of interest for developing this third technological gateway to the rest of the world (Government of Mauritius 2018b). This integration will be significant for the application of SE in the context of a Rodrigues pilot case that we will discuss below.

2.3 Policy Making in Mauritius and the ICT Strategy

Mauritius has embarked on a significant digitisation project as it moves to diversify its economy by adding ICTs as a fifth pillar to its existing four pillars of economic activity: sugar, textile, tourism and financial services (Government of Mauritius 2014). This may potentially be good for e-democracy in Mauritius. The exploitation of ICTs is to create an information economy to diversify into an emergent sector for services. Mauritius aims to become a service-oriented economy, and thus, it views digitisation as a means to adapt to what is seen as a future area for growth. This policy is conducive to an e-democracy even when it is being forced due to economic imperatives. This ICTs policy move is due to the phasing out of the preferential guaranteed price of its sugar exports by the EU. Mauritius as a system is not immune to changes dictated by the EU system. This change from a guaranteed sugar price was anticipated to be disruptive to the economy. Mauritius has to adapt to absorb the economic impact from the anticipated transition.

The pricing shock also led the sugar barons to improve their sugar production processes through increased mechanisation. With a shrinking agrarian workforce, the government had to diversify to create new jobs for its displaced workers and growing populace. Faced with growing union-led radicalism and economic crisis, the EPZs policy allowed diversification into tourism and textiles which are both labour intensive and Mauritius maintained its direction towards a democratic developmental state (Sandbrook 2005; Meisenhelder 1997). This was timely according to Meisenhelder (1997: 282), adaptation of policies from the Newly Industrialised States (NICs) of East Asia like Taiwan allowed democracy to anchor itself in Mauritius.

The EPZs as alternatives brought comparatively higher-paying jobs in the textile and tourism sectors. Importantly, it also provided an outlet for women to enter the workforce. From the success and mistakes of the EPZs, Mauritian bureaucrats learnt the importance of diversification and the need to move to higher-valued activities. ICTs are claimed to create higher-income-generating jobs that Mauritius anticipates will shift the country from middle income to high income. Diversification through an 'economic revolution has transformed society from 80% agrarian to 8% agrarian in only three decades. Per capita income now approaches $10,000 as a result of the combined sugar and textile economies. The growing tech sector will likely drive this figure higher' according to Youngblood-Coleman (2014: 16). This projection into the tech sector is being experimented by the Mauritian government. Mauritius is aiming to develop a knowledge-based economy in the future. One of the architects of the ICT policy for Mauritius, Assistant Permanent Secretary Minister of Agriculture Id No.: 13, confirmed this strategy.

To embark on its diversification path, the government developed an ICT strategic document towards an information society supported by African Development Bank (Government of Mauritius 2014). This strategic document is an ambitious one, with 124 ICT projects identified of which it reported 24 were in progress in 2014 (Government of Mauritius 2014). It is noteworthy that the document has identified some key barriers to progress, for example aspects like Internet access and cost issues.

The report from IHS Global Inc. (2016: 3) mentions that 'The information and communication technology sector is pushing to take advantage of the country's high literacy rates and multi-lingual capacity, encouraging the development of call centers and business-process outsourcing'. Mauritius wants to develop high-end value-added services using the ICTs, and it is investing heavily in education to create an information economy.

The budget papers for 2016–2017 mention that at 'tertiary level, a dedicated Faculty of Digital Technology & ICT Engineering will be set up by the University of Mauritius' (Government of Mauritius 2017: 19). To develop into a knowledge hub using ICTs, Mauritius seems to aim to develop a whole supply chain of IT activities.

Given the prominence of ICTs, it makes the case study an interesting one for democratisation, as one of the objectives of the Ministry of ICTs is to democratise access to information as per Section 16 (a) of the ICT Act (Government of Mauritius 2002). We expand on this below as it complements the unique function of the NDU that has a role to operationalise this information democratisation for the people. ICTs role and support is discussed in this framework. The ICTs technology for future digitisation of any NDU processes that would require e-services to elicit and transform people's input into demands for an e-democracy is feasible given the government policy in that direction. It should be noted that ICT is not for democratisation, as it is to inform but not reform.

The current ICT thrust is an economic one. The aim is to progress to a high-income society where Mauritius intends to use ICTs to become an advanced technological society. Legislation has been passed to exploit ICTs technology in Mauritius (Government of Mauritius 2014). The policy documents for eGovernment explain the government strategy at Government of Mauritius (2018a). ICTs are to form another pillar of the Mauritian economy and allow creation of an information services sector, called i-Mauritius. The eGovernment document provides an insight into this area for the government's strategy to develop a knowledge economy (Government of Mauritius 2018a).

Digital services (eServices), though it empowers the consumer of those services to some extent for ease of access, these services at present are transactional (Government of Mauritius 2014, 2018a, c). Their digitisation as eServices does not necessarily democratise the political system. The imperative for ICT seems economic we expand on this below.

3 Mauritius's Readiness for an E-Democracy?

To augment democracy, the SE approach is bottom up from the literature research and top down from the interviews about the need for e-democracy. As mentioned in Chapter 5, an analysis of the e-democracy idea must be conducted to ensure that the need for democratisation exists in Mauritius and thereby a system can be built to meet that need. For a successful system, SE's reality check from users of the system is essential (USAF 2010; Eisner 2011; Blanchard and Fabrycky 1998). Prior to this step, the book conducts an analysis of key stakeholder's use of ICTs in politics and this shapes their understanding for an e-democracy. Once the need is understood, then the second step is to analyse how SE through an existing or new function within the state will be facilitated to develop an e-democracy.

Now, this functional requirement for implementing SE in Mauritius is through the NDU interface as we discussed above. As an existing legacy system, that NDU interface was identified and was explained from a top-down view of the Mauritian political system functional analysis. The following interviews provide the insights for key stakeholders' readiness towards developing an e-democracy. This is to be considered as a preparatory stage for ensuring that the key ingredients like technology and a will to democratise by the political class exists. In a legacy system, it is anticipated that even though the interface like the NDU exists, it is the policy makers who will facilitate an e-democracy implementation. So the decision to change the system must come from the political class. The interviews conducted were to gauge the willingness of the political class in Mauritius to embrace change whereby SE could be applied to augment the system to an e-democracy.

3.1 Insights for a Modern Participatory Democratic System

Semi-structured interviews were conducted to gain insights into the policy makers' thinking about the Mauritian political system where technology could be used to enhance democracy. In terms of establishing an e-democracy, it is essential to determine the potential efficacy of a SE-based design. If the political will and appropriate ICTs do not exist, then SE cannot support the development of e-democracy in Mauritius. Such an analysis aligns with the pre-concept stage of the standard SE design process, as was discussed in some depth in Chapter 5. It is the first key step underpinning the conceptual design stage (Grady 2006; U.S. Department of Defense 1993, 2008; Blanchard and Fabrycky 1998). This step is to be undertaken through the pilot study suggested by junior minister for Rodrigues (Id No.: 1); however through a series of interviews and discussions with key decision makers in Mauritius, it shows some positive signs (See Appendices 1 and 2). The interviewees are involved in a range of decision-making processes in what may be termed a pragmatic welfare state constrained by its limited resources. As a result of these discussions, it was possible to identify some of the constraints that Mauritius's e-democracy would face. Mauritius is a welfare state by intent with some unique features like pluralistic representation through multi-member seats, a degree of pragmatic liberalism and consensus generating processes like arbitration and tripartite negotiations to reduce strikes. These aspects may create particular SE-relevant interfaces that would initially shape the requirements needed for an e-democracy in Mauritius.

There are other general characteristics that may also assist the creation of a SE-driven e-democracy. The size of the island nation is close to the area of what Athens in Greece would have occupied when democracy prevailed in the fifth century BCE (Dunn and Harris 1997; Dunn 2005; Dahl and Tufte 1973). The population is relatively large but with its choice of three representatives per constituency the system reduces the effect of the tyranny of size mentioned by Dahl (Dahl and Tufte 1973). As per the Mauritian *nomos*, these representatives are authorised to run the affairs of government, which as a welfare state Mauritius's focus is on delivery of the good life to its citizens. Mauritius already has a dedicated team of 10 junior Ministers (PPS) overseeing a National Development Unit function (FCI) to achieve this outcome (Government of Mauritius 2016). Using the specific techniques of SE makes it possible to understand how an innovative approach to the use of ICTs might be developed to facilitate a successful e-democracy.

The NDU as mentioned above has the institutional capacity to enable an e-democracy to be designed using the SE techniques of Requirements and Configuration Management mentioned earlier by the book. Its engineering unit could facilitate the data for people's requirements to be elicited in each constituency. There was interest expressed by the member for Rodrigues (Id No.: 1) Honourable Jean Francisco Francois, a surveyor by training, and also the member for Constituency No. 5 Id No.: 11, a civil engineer about the SE techniques. Now both of them are PPS in charge of managing some constituencies. The member for Rodrigues suggested during the interview that Rodrigues could be a pilot site given its natural advantage of a fisherman monoculture and being an autonomous political system within the larger system of Republic of Mauritius which was more complex. Developing a SE capability that integrated and allowed the people to elevate their needs and wants was welcomed.

The PPS Id No.: 1 even suggested that he would be engaging with the local University of Mauritius where he was doing a Master's in Public

Administration to find out what the academy was doing in this sphere. As a politician, he was very supportive of the innovation that this book could herald for the country as a whole and for autonomous Rodrigues in particular. This is also indicative of a reflection of the mood for change from the people due to the increased capability from technology. A view shared by other interviewees that in a technological environment which is increasingly more capable one must change or perish as a politician.

The political context has changed from the introduction of ICT which we discuss below. The people have shown in no uncertain terms that they want a better system, and the politicians reflecting on the 2014 elections outcome understand that the collage system which used to work may no longer be appropriate for changes that ICTs herald. A more informed public is more critical of the systemic scandals that they have been subjected to. The changing context is discussed by these interviewees and mentioned below. Politicians feel that the opportunity exists for a better form of participatory democracy using technology like the ICTs to improve democracy in Mauritius and we expand on these views below. SE could integrate this technology to assist Mauritius to develop a better system given the need exists, and this technology integration is mentioned earlier in the book, Chapter 5, as one of the strengths of the SE design framework. A need for the ICTs has been identified, and we discuss this next.

3.2 Mauritius's Need for an E-Democracy

Key actors, listed in Appendix 2, were approached to see if there is a willingness by those key actors, current and past representatives in Mauritius, who act or acted as gatekeepers to policy making, to embrace ICTs as part of a transformation of existing governance arrangements to create an e-democracy. However, as a result of scheduling circumstances some were not interviewed during the two-week window allocated for the task. However, 15 policy makers (listed in Appendix 1) were interviewed about e-democracy for Mauritius. The interviews were conducted to explore the degree of willingness on the part of policy makers to embrace a digitised e-democracy. This issue of political will is widely acknowledged as a key problem for this government system (Moss and Coleman 2014); these collated views are discussed further below.

The fact that Mauritius already had a dedicated Ministry of ICTs meant these; the political candidates, who were interviewed, were well aware of the importance and potential of ICTs, especially in the development of economic well-being, but they also felt that the ICTs application for politics needed to be analysed further. For some of the candidates, ICTs offered an opportunity to develop better rapport with the electorate to provide information about what the government is achieving in Parliament by providing those debates live using ICTs, an e-Parliament concept. The rationale, as explained by the junior minister for Rodrigues (Id No.: 1), is that it would allow people to see and understand what their representatives were debating about and show how hard they were working. This would seem like a form of performance-based reporting. It assumes that the people are interested to watch and listen to some debates to which they are just a spectator.

Nonetheless, a common theme emerged that it was necessary to find a means to improve the effective use of the ICTs in political activities, government and the economy. Interviewees from both the winning and losing parties emphasised that the key element was the capacity for the strategic use of ICTs to make electoral campaigns effective. The winning side achieved this communicative edge. The ineffective use of the technology for the other side was reportedly reflected in their electoral loss. A concern was that there seemed to be a sense of disconnect between the youth of Mauritius and the politicians in terms of the needs of the youth not being understood and met by the current system observed by two of the interviewees, Id Nos.: 2 and 7 in Appendix 1. This had flow on effects in campaigning when using the traditional media; the messages were not connecting with that audience.

3.2.1 Views About Technology and Risks

There were some cautions expressed from the experienced politicians like ex-Minister of Foreign Affairs, Id No.: 2. He thought a need to manage the integration of the technology into society was essential as it may be disruptive in a politically negative sense. His example of disruption included the Arab Spring events where ICTs facilitated a revolution. He argued for a wait–and-see approach for multi-ethnic Mauritius. He pointed out that it mattered how the ICTs were implemented by other countries and Mauritius could use their model and lessons learnt for a Mauritian unique e-democracy, Id No.: 2.

Politicians and policy makers view ICTs as a tool that can be a game changer due to its information conveying capacity. From those interviewed and also from the literature, the book found that ICTs use is not currently used to reform to empower the people in direct decision making. The proponents in Mauritius mentioned that they were keen to engage with the citizens using the ICTs but it was for ICTs use as a tool to communicate. The candidates interviewed mentioned that ICTs were used to win elections. They could see ICTs importance in the 2014 election where its communicative power was used to engage with citizens to push their agenda. While some did not perceive an empowerment of citizens in direct decision making, however they thought this role was not ruled out in the future when the right process was in place. A process that SE initiates for Mauritius is through the book. It confirms and creates the awareness of this need to decision makers at the conceptual level through the needs analysis process of SE mentioned earlier, Chapter 5. The view of candidates was collated to explore progress towards an e-democracy given the technology ICT is being exploited by Mauritius. The form mattered and this is discussed below.

The argument against direct citizen's rule was from a concern expressed about people having a myopic view by the CEO (Id No.: 8). He illustrated this by using an analogy of passengers driving the cab rather than a skilled driver. This view is for cautioning about potentially chaotic situations emerging, and risks were stressed by some of the other candidates like the junior Minister (Id No.: 3) who was wondering about ICTs impact on certain cultural norms amongst groups like women. As a lawyer, she was seeing many divorce application due to ICTs misuse in a culture that was still learning about its capabilities. This learning curve was mentioned by ex-PPS Id No.: 5 and detailed below.

This concern about ICTs risk in politics is mitigated by SE's need analysis explained in the book. Though the input from the citizens were welcomed, this was more for understanding what had to be delivered or the notion of an Agenda in future. The interviewees could not see how decisions made by citizens would be for the benefit of the country or democratisation as some of their demands would be short term around their individual needs. There was need for some form of aggregating process to transform these people's inputs into policies that the book explains, refer Chapter 5. To improve democracy, there was a role for the politicians in that transformation for managing the transition to a system that used technology like the ICTs. A role that the NDU unit can meet as shown by the book through its current function to explain government policy to the people.

The Ex-Vice Prime Minister and Minister of Infrastructure, Id No.: 7 could see the benefit of the ICTs use to 'sense the pulse' of the people prior to making a decision. His view was that those in power could use the ICTs to feel what the people's mood was. He explains this decision-making input using an example of when as a Minister he was introducing speed cameras. He could see the consequences of not sharing decisions with the public through this example about road safety. To lower accidents due to speeding, he was introducing speed cameras for safety and to change driver behaviours. ICTs assisted the opposition to distort his message and changed the public opinion for the project from a safety measure to one about an imposition on the motorists, with hints that he was getting some benefits from the project. He mentioned ICTs could destroy the real message of motorists' safety by his opponents to disrupt and misinform. He countered that the ICTs could also have helped the government if he had used it to engage in public discourse through this medium.

The ex-Minister (Id No.: 7) was supportive of the concept of a form of direct democracy where the people participated in decision making using the ICTs for opinion polling. He mentions online polling with ICTs as being instantaneous, and this was a significant improvement in decision making as

> with the conservative system our public opinion poll that could be done once at a time, at intervals, but with the new systems that we have like ICTs we can do it almost every day and on each issue. On each issue, we can know the pulse of the nation and that can help those who are at the helm of power. We can feel the pulse of our people on every issue and policy decisions can only be voted according to what people feel about it. (IDNO 2016: 7)

This decision de-risking with polling is a form of input that he feels the ICTs can provide to inform him for how to support a policy in Parliament. With Requirements process to elicit information for good decision making, the role of the politicians is changed, as they can actually make policies that are relevant to the people, based on needs—needs that can be traced to requirements like road safety in the above case and the government intent (solution) to bring changes to driver behaviour to reduce road fatalities.

Furthermore, he adds that the ICTs were more efficient as he could link to the people directly through the technology:

> Say we are moving for a deal, instead of going for opinion poll and the traditional method of circulating papers, we can easily get it through the ICTs, through the system. You can easily get it and for every issue you can get it now. So if you want to become a populist government go through the ICT. This is what you got I call that in modern terminology Direct Democracy. See what you got is direct democracy.

> Ruling directly, there is no, there won't be any problem in that 'couma to pe prend ene decision par example' [every time you take a decision for example] as far as controlling the road accidents you want to impose certain things, some decisions you want to take, you then immediately can get the response. Within a day you can get the response and the following morning you take the decisions. (IDNO 2016: 7)

> That is, that is the Participative Type of democracy where everybody participates through this system. That means there is a first time that we have such a system. There is a positive element that we have and secondly we are a small nation, small nation where everybody participates. (IDNO 2016: 7)

The ex-Minister Id No.: 7 reflected that if the leader had implemented such a transparent system that was participative, then his party (Labor Party) would not have lost the election. Now, for the CEO (Id No.: 8) there is a concern expressed about the massive inputs and whether it would be possible to make the country move in the right direction from this massive flow of demands, while for Id No.: 7 keeping the public in the decision-making process through polling the technology added value. It created social capital allowing for a more inclusive decision making where the people are part of the process through technology. There is tension here between these two positions as one is about reflexivity which causes stress to strain the system and the other plurality in decision making.

Unlike the CEO (Id No.: 8), the ex-Vice Prime Minister (Id No.: 7) was anticipating that ICTs could develop the solution. He assumed it allowed people to participate in direct decision making through polling. This participation for the ex-Vice Prime Minister reduced negative feelings towards government decisions as the people directly through the ICTs were allowed to inform the government (parliamentary debate).

This would create for the politician, Id No.: 7, better support for the government and its rulers.

While the response of the CEO (Id No.: 8) seems reflective of a current paradigm of a representative model, the politician Id No.: 7 is open to exploring new ways of implementing the participative and direct democracy paradigm. The CEO is a professional managing professional bodies as head of the Medical and Dental guilds while the ex-Vice Prime Minister Id No.: 7 a seasoned politician is analysing the defeat at the election and reflecting on a new means to create a populist government. The ex-Minister (Id No.: 7) is perhaps after a government that listens to the people. It becomes democratising as it is participative and indirectly allows the people to shape decisions in Parliament. The SE technique, Requirements improves such a participative intent. A flow in Fig. 2 shows a potential process analysed by SE to support the ex-Vice Prime Minister's intent while also mindful of practical issues like demand stress from too many inputs. This flow in Fig. 3 is built on to suggest changes to the NDU function, as mentioned it provides the support to address both concerns.

In the previous government, the NDU was under the ministry of the ex-Vice Prime Minister (Id No.: 7). If the NDU had used the SE process, its functionality could have developed an effective decision-making process of the kind explained earlier in the book. SE could accommodate the CEO's concerns for a reflective response as analysts at the NDU are able to effectively communicate with citizens to explain the merits of improving safety through speed cameras. The NDU's engagement would have provided the agreed requirements from the people to make the project a success. SE's needs analysis of accidents would have provided the justification for speed camera's requirements. However, such a participative process like the one suggested by the book in Fig. 3, and alluded by both the interviewees CEO Id Nos.: 8 and 5 an ex-PPS, is absent in Mauritius.

The ex-Vice Prime Minister as a seasoned politician was aware of what went wrong to a collage that was perceived as the strongest. The collage had leaders who stopped listening to the people, and even their own members in Parliament. The people were kept out of the decision-making loop, and the outcome was an erosion of support that was swift and decisive. Due to the new public appetite for change that the ICTs could generate, the ex-Vice Prime Minister one of the most popular politicians undefeated for decades lost his seat. A change targeting the leaders and their collage system but which also reflected on the members of the party like the ex-Ministers (Id Nos.: 2 and 7), who shared the same fate of loss of public support that flowed from a loss of public confidence in their leader. There is an apparent confusion about what must be done to develop an effective system that can respond to the new capability from the ICTs to communicate.

For some, the government must remain in the driver's seat for if there are too many drivers the state does not progress. So even though both interviewees, an ex-Minister and a CEO (Id Nos.: 2 and 8), felt that the inputs from the people were necessary using ICTs, these were only for providing a supporting role to decision making. ICTs were mentioned by the CEO for eliciting an agenda to which the government can work on within the resources it has at its disposal. He brought in the sense of prioritisation for the people which could be done in a transparent manner. However, a means to gather information that allows prioritisation for Parliament as explained by the book in Fig. 3 is yet to be developed. SE techniques were not mentioned during the interviews.

The interviewees acknowledged that there are times that the government has to do things which would benefit the country but not the people in the short term. These pragmatic observations are crucial to understand that the notion of every person making decisions in an e-democracy may be impractical and too many inputs lead to chaos in the absence of a structured approach. If the demands of the people are not met like the ex-Vice Prime Minister (Id No.: 7) mentions, then their next mandate could be rejected by the collective. There is a tension between these two positions, a tension that the SE framework can defuse through system thinking by creating an Agenda from people's inputs. Such a system for creating an Agenda is yet to be developed in Mauritius but the design process to develop it is explained by the book.

Applying SE, the scoping through an Agenda would constrain power abuse to shift some control to the people and reduce corrupt behaviour from scope creep given a transparent Agenda making process. The problem is that the means to aggregate the collective so that it is both informative and legitimate is missing. An alternative being suggested by the book is that inputs must be collected from the people and synthesised into agreed requirements by entities like the NDU. Once aggregated, the politicians need to discuss with the people about the prioritisation of these inputs and that can form the basis of policy making. It raises a key issue of a SE-generated public Agenda and the quality of information that politicians work from in their decision makings. This is a core area that is flagged below as a need that ICTs are currently failing to meet without a structured approach like SE to provide the appropriate knowledge for politicians' good decision making.

One final point about the risks and constraints of pursuing an ICT strategy concerns the nature of Mauritius's state. For all intents and purposes, Mauritius is understood by its citizens to be a welfare state. This generates expectations that will need to be managed in an effective manner. Mauritius's welfare system has created significant challenges for governing. Despite its lack of natural resources, Mauritius has significant cost outlays to meet its welfare programmes. This is a key constraint for applying SE as the commitment to its welfare system requires significant budget outlays (Government of Mauritius 2018a, b). Its welfare system subsidises key food items, provides free public transport to students and pensioners and has a universal pension independent of income. In many respects, some scholars see Mauritius's approach to welfare provision as instructive for other states (Willmore 2003). Nonetheless, the point to be made here is that investment in expanding ICT infrastructure and increasing easy access to ICT-based facilities may require a rethink of budgetary priorities. At this stage, it is not clear whether Mauritians or their government will be prepared to trade off welfare provision to fund their ICT strategies. And this raises the key issue of political will.

3.2.2 *The Political Will in Mauritius: Policy Makers' Views*

There seems to be a will to embrace ICTs in politics as mentioned above. In what follows we explore why and how some of the politicians feel it should be applied. The ex-Minister of Foreign Affairs (Id No.: 2) mentioned above-expressed caution about being a leader in the implementation of an e-democracy using ICTs in Mauritius. This caution though is in terms of Mauritius being a leader and then not having the appropriate controls and resources to mitigate the risk some which he acknowledges would be unknown and arising from this new technology itself and while others may be due to the way it is applied. Implementation mistakes without the luxury of prior experience he claims could be foolhardy. This concern is also in the context of a fragile equilibrium within which Mauritius operates as what he calls a secular state where the rule of law applies. A question that seems to emerge is that whether Mauritius is ready for an e-democracy. However, at the same time the ex-Minister of Foreign Affairs agrees that the context has changed and one could not be e-stupid.

In general, there seems a realisation that the context has changed and that the politicians in Mauritius have realised that ICTs had a role to play and they must be pragmatic and change the status quo. The politicians understand that there is constant erosion of trust from the ongoing scandals of the political class which was mentioned by the ex-Minister of Culture and Arts (Id No.: 6). This erosion is such that the collage of the two traditionally strongest parties can no longer actually guarantee electoral success as the people have spoken very succinctly at the last election for change. This is an important motive to introduce ICTs. One of the problems identified by Moss and Coleman (2014) in the UK is a lack of political will to embrace ICTs in governance. From lessons learnt for better government, this support for adoption of the ICTs was also identified by the OECD as a need for the top leadership to ensure successful implementation of ICTs (OECD 2003). In Mauritius, ICTs potentiality is recognised. The changing context from ICTs and some emerging issues is discussed below.

The issue of information quality and overload was raised by the Minister of Agro Industry and Food Security Id No.: 12 who mentioned that there was no dearth of information on the web but he needed to know which one he could rely on. He wanted to be able to analyse and synthesise the information, and accessing correct information through ICTs was crucial. He mentions

> The idea also is how we are going to make use of those technologies in promoting or enhancing a democracy in a society and for me in an e-democracy you are supposed to allow people to voice out their views allow people to be aware of what's happening, Information flow should be very fluid, there should be more transparency. I say there should be more participation and in all decision making processes. So for me through e-democracy we can use those platforms to reach out and get those to facilitate the processes in a better way. Again, as I said here we are a small island we don't probably feel it but if we look at big countries where people are in very remote places where sometimes getting information to those places is very difficult. Now, with these technologies information availability is much more easily accessible, so it is better, definitely it enhances democracy. It removes the tyranny of distance sure. Distance should no more be a barrier and probably one of the reasons that has hampered probably good practices and good democratic development in some societies is because when you are in a big country where information is not readily

> available people are left out. Then, you can find discrepancies in the way some people are treated. (IDNO 2016: 12)

The informative capability equalises but the Minister wants to improve the decision-making process in the system as well. He is open to the idea of having the people's input. He is aware that representation also brings about a performance-based assessment for the elected for quality decisions. The Minister's views were that

> My concerns is that sometimes when you are overflowing with information, how do you digest all that information? How do you assess whether this information is right is wrong? It is sometimes, because if you just take it as it is. What I am getting at is this has to be correct information else you could be in trouble. You have to be sure, for me whatever information you get you have to analyse. Analyse them first then you make your decision. We are overburdened nowadays with information. It is a question of how you assess it to make sure that you use the right one in any decision making process, so that you don't go out of the target that you have set yourself. So that's one thing that I see as I keep saying to myself don't take whatever you get from all the mass of information that is available. Always try to synthesise analyse it first, and take what is the right one. (IDNO 2016: 12)

The reflexivity from information is creating new problems from ICTs. Information overload aggravates the ability to use the quality of information to make good decisions in a timely manner. To make informed decision, a means to synthesise and analyse is felt by the Minister. As explained in the previous chapter, this need for quality information that is synthesised and analysed in the form of requirements can be met by the NDU adopting the SE process.

The NDU with the Requirements technique could assist the Minister in reducing the problem of both overload and accuracy. As SE is a means to synthesise and analyse information, it would allow the needs for decision making to be met. The flow in Fig. 3 could capture the relevant information to support decision making. The Minister is also mentioning the scope of his decision making; he mentions targeting of the right information. Creating an agenda for his functional requirements as a FCI allows him to identify how he can target the right information. A functional agenda transformed into specifications by SE analysts in the NDU will provide him a means to work on actual needs rather than raw and imprecise data. Specification as developed by SE is used to develop complex products (U.S. Department of Defense 1985, 2003).

In the need for precise and accurate information, he is joined by the Minister of Business, Enterprise and Cooperatives, Id No.: 9, who brings a new element the speed of propagation of information. He mentions that ICTs as a tool in his view have the potential to innovate.

> My understanding is that it is a very strong tool in disseminating information. It has characteristics inherent in ICT first it is precise. It is precise. The information that it disseminates is very precise, secondly it is very rapid, swift and rapid and also it offers a lot of opportunities especially to our young people to get jobs and you know it is something which is ever-changing. It brings around a lot of innovation. I mean it facilitates life for one and all. Yes of course, ICT is inherent in all walks of life including, including politics, including a political system. I just take the example of the last general election which was won I can say on the nose of social media and which disseminates you know striking information to people and which creates awareness and if it is done, its prepared with a lot of efficiency and effectiveness, it has a reach that any political party can expect. I can say that the winners of last election I mean the government, it owes much of its success to ICT. (IDNO 2016: 9)

Though the Minister is explaining the capability of the technology, he is also alluding to new uses for innovation that the capability may bring. SE integrates technology as explained earlier in the book it looks beyond the obvious which is the transmission of information for getting elected to creation of a transparent and effective political system. SE's Requirements process ensures that elected representatives deliver what the people need as the Requirements process elicits the need information to accurately articulate it in the form of requirements sets called specification as mentioned in Chapter 5. There were similar views to the above that were mentioned by both sides of politics. However, there are also some new areas that are emerging, which is the impact of information conveyed by ICTs and the behavioural risks it creates in a fragile democracy.

The Minister of Agro Industry and Food Security mentions below, for example, that he is also expressing concerns about the impact of information through ICTs use on society, there is potential of misuses and abuse.

> The other thing which sometimes you have to be careful about is how kids, young ones can learn on the basis of this information. We all know what if we don't set some kind of control on some people young ones especially those who are not mature enough they could be influenced by certain things that they get access to and that could have detrimental effect on their lives their future and it affects family the whole society itself. These are the kind of things that we have to be very careful about. But it's a question of how we sensitise our kids and our people that we care for so that they don't get trapped into that kind of problem. (IDNO 2016: 12)

A concern is about the new behaviours that are emerging from the informational deluge that is affecting the young and immature, those who are gullible and tend to be manipulated by what is on the web. Some extreme forms are they may be trapped in virtual reality, others is the potential for radicalisation that the ex-Minister of Foreign Affairs alluded to above.

There seems to be an emerging role for the government to educate and create a means to filter the true from the false. The dangers from misleading information have social consequences which then reflects on the political through the family and society. In a volatile environment like pluralistic Mauritius, the existing ethnic cleavages tension such behaviours and destabilises democracy. SE transparent processes through NDU create trust through channelling the tensions to mitigate this informational risk. Providing people access to the database repository mentioned in Fig. 3 will inform based on facts. Issues can then be taken up with the elected representatives if further clarification is needed. The advent of the ICTs creates a risk to misinform that government should be concerned about and address.

This ICTs danger is further explained by Id No..: 4, in Appendix 1, a Member of Parliament (MP) when interviewed but was promoted to a Junior Minister (PPS) attached to the NDU after the 24 January 2017 reshuffle of cabinet. He narrated the case of a youngster who committed murders as being disconnected from the real world. Having a social worker background, he is concerned about young people shutting themselves in the virtual world and disconnecting with reality. The following is his narrative

> Sure we are concerned with all these incidence and we have been talking as I mentioned lastly with these persons, these staff at the centres, like the community centres, the social centres. They have their own schedule

> of work and program for each and every age group of people. We have also discussed amongst our colleagues at the various departments to find ways and means to assist. The incident I was talking earlier on was about a crime a young guy committed. A 17 years old guy killed a 54 years old woman. After that crime, he waited in the house for hours, for another girl to come in and he also killed that girl who was 14 years of age. So what's happening to this society? When you ask people living in the vicinity they say that 'we have never seen that guy come out of his house'. We never saw that guy roaming around how can he do such a thing, but we have this question which brings us to reflect on that, here is a youngster of 17 years, he should be involved in certain activities, in social activities, and if so then perhaps these things would not happen. (IDNO 2016: 4)

The information era brings a new paradigm for new thinking to accommodate the technology. The M.P Id No.: 4 is not against progress that technology brings but rather his concern is about the new behaviours that need to be addressed. The political environment has changed, and new behaviours are emerging that the politicians need to deal in their electorates. The impact of these new behaviours to disrupt democracy which is fragile is unknown.

The ex-Foreign Affairs Minister Id No.: 2 noted that an 'Arab Spring' type dynamic driven by ICTs might destabilise a political system and be destructive to orderly change in society. A means to appropriately inform and increase trust in society emerges, and this may be a challenge given the nature of the Internet. The data set created by SE, Fig. 3 refers, is a start for developing a better political environment from quality information that allows better insights between electorates and their representatives. If the process is made transparent as proposed in the book, Fig. 3, the SE process potentially improves the system to mitigate the emerging risks by creating trust. This could be areas for research in future when a SE process is adopted by Mauritius to develop system e-democracy.

The brief Athenian experiment with direct democracy was comparatively small compared to modern democratic states (Dahl and Tufte 1973). As mentioned in the book, direct democracy experiment on a large scale has not existed nor did the French Revolution allow such a model to persist where the many were involved in decision making. This makes digitisation of a model like direct democracy hypothetical though not impossible. There is no precedence in recent times of a decentralised system where every citizen decides except in the form of voting during a plebiscite. In a representative system, such popular votes are not very successful mechanism to convince as the experience of Australia found where out of 44 only eight referendums were successful (Williams and Hume 2010). To create engagement as a routine activity, a paradigm shift may be required; there is a learning stage to integrate change from a representative model. To operationalise an e-democracy, the member of Constituency No 9 ex-Junior Minister Id No.: 5 takes this thinking further:

> I think e-democracy should be a *sine qua non* you name system, you name it any name. It is a step further, you name it as a need in democracy. If we really want to go a step further, we have to adopt e-democracy because it will widen our scope of thinking. It will make the simplest citizen proactive. It will give the common man a say. The common man will be more participative in the decision process of his country. The common man in turn will be in a better position to explain to his younger son or daughter what is happening in his country and the younger generation will be able to learn how his country is progressing in a democratic way. So we can say that e-democracy in itself will be very educative especially for the younger generation. I take the example, how it will, it will unfold. It took 30 years for the young to learn. It took 30 years for people in Switzerland to learn that they shouldn't litter around in the streets and for 30 years people have been explaining to the younger generation and these youths who as the adults later in turn started telling their younger ones that they shouldn't litter around, So you have as a country today Switzerland is considered as the number 1 country in terms of cleanliness.
>
> So in the same pattern e-democracy will be educative for the younger generation. So that one day, we will have a very solid democracy in all the countries which will adapt and will adopt e-democracy. And I am sure, that with so many conferences around the world, with so many working sessions, with so many interviews, with so many forums, it will gather pace and it will come gradually, you name it, Africa, South America, you name it. Asian countries will have to go through it and the very simple example of we are not talking about e-democracy. We are talking about democracy. What happened in Burma, because people kept fighting for years, and some people who said that it will take generations and generations, but ultimately democracy came so in the same way if we keep on talking and sharing views, implementing an e-democracy, it will come. (IDNO 2016: 5)

As e-democracy is linked to democratisation, it is but a question of time that it will emerge due to people's demand for systemic improvements. This positive view for e-democracy due to anticipated popular demand bodes well for Mauritius. It is to be noted that this ex-PPS, Id No.: 5, is an experienced IT practitioner with a long career in the ICTs. His views are indicative of the malaise that exists in many Western and non-Western democracy where a system designed for an industrial era is under pressure from an information-driven era.

Importantly from his example on littering, he flags a period of learning about the new system. SE can design a new system e-democracy that would be relevant to the new environment and keep upgrading as the context changed to remain relevant; this design life cycle was explained in the book in Chapter 5 and its centricity to the people in Chapter 6. SE processes used by both politicians and the people in a new partnering arrangement that system e-democracy creates is feasible for Mauritius. The NDU has a critical role to play in the transmitting of the learning of the SE process to every stakeholder. This democratisation function to educate will need to be created as it will create a new form of partnering between the elected and their mandates. New culture and new behaviours emerge given a system which is appropriate for an information era is used.

From the interviews of both collage blocs, there was no evidence of the participation of citizens directly in decision making; however, the interviews indicate that in Mauritius it seems that the political will exists for an expanded role for the ICTs. However, in adopting the proposed SE framework to create the data repository in Fig. 3 the role of the defender of democracy, the Mauritian President could be augmented. We discuss this potentiality below as areas of future research to develop the support for the changes to an e-democracy in Mauritius. This was suggested by Honourable Jean Francisco Francois that a system e-democracy that improves democracy must be entrenched within a modern Constitution so that it is protected. Given the elitist culture in Mauritius, collage for example, this recommendation may be crucial to change the entrenched political class and the vested interest of the many pressure groups. Reforms cannot survive if the modern *nomos* does not reflect the people's aspiration. The defender of democracy as per the Mauritian *nomos* is the function of the President which SE suggests is a role that will be augmented to defend democracy using technology. However, a key stakeholder the people's views towards the current system needs to be discussed as well. The Mauritian public seems apathetic and tolerant to significant scandals from their lack of active resistance to a leadership which is reported as tainted, see (Youngblood-Coleman 2016, 2014). These country watch reports about scandals are in line of those reported in the newspapers of the country and the Transparency International Corruption Index (Transparency Mauritius 2013, 2015, 2017). The interviewees agree that a culture change is necessary and that SE could bring about the transparency and accountability to improve the current system's operation.

3.2.3 *The People's Will—Mauritians Mood for Change*

The garnering of people's will in liberal democracies is normally exercised through voting at an election. This construct has been conflated to authorise the delegation of power by the people to the elected; the member who gathers the maximum support during a voting event then decides on behalf of the rest. Scholars have noted that electors do not show 'interest in or [have] knowledge of the political issues involved in the election' rather they voted based on their political preferences and that preference was culturally driven (Birch 1993: 83). Selecting representatives is meant as expressing of the 'people's will' and the team managing to cobble together a majority of support from representatives elected by the people forms a government to rule. As a means to aggregate the people's will, in practice voting allows the elected representatives freedom to follow their own agenda. Election makes the act of voting a tokenistic one for decision making by the *demos*. The elected member is free to implement the promises made to the electorate but in practice the politically powerful tends to ensure that their interests are maintained over those of the *demos*. The elected member is free to implement the promises made to the electorate but in practice the politically powerful tends to ensure that their interests are maintained over those of the demos. The electoral system is used with the economic clout of a few (the politically powerful) to effectively undermine any change that threatens those powerful few interests. Thus, the people's will even with voting rarely emerge under these circumstances. An election cannot overcome the systemic control of the politically powerful. From the empirical evidence, there is voters apathy in most representative democracy, it has led some scholars to question whether democracy incorrectly assumed 'the average citizen' was interested in political participation (Birch 1972, 1993). Voting is the accepted norm; it is a mechanism to adjudicate when consensus could not be sought; it was not meant to replace a democratic means to seek consensus from the *demos* or self-government as a process.

The views from the policy makers seemed positively inclined towards democratisation in Mauritius. However, we also analyse some observations from the bottom up about the people's will. As mentioned by the ex-Minister of Arts and Culture, Id No.: 6, the system in place is far from fair and open which makes the quality of democracy from an inherited Westminster system questionable in its actual application to Mauritius. The Mauritian political system exists but it is not democratic, and though it is creating a ICTs technology capability, this is mainly for economic reasons. There is an awareness by the politicians in the interviews that the status quo can no longer be the case. The trigger for change was the 2014 election event where the realisation has set in that the public mood is for improvement to the political system in place and political parties' behaviour. The public wants a say and they want better government, the collage strategy failed as the public made it fail.

The public rejection of the collage system at an election indicates the public mood is for change. This was observed during the 10 December 2014 election event where a lowering of tolerance amongst the public for scandalous behaviour was rejected. Manipulating the public which is becoming more aware and informed through the ICTs is no longer an option; the silos of power through information manipulation is open to disruption as observed by the interviewees Id Nos.: 6, 7 and 8. This has brought an awareness in the political class that as the public becomes more informed the politician must follow being forced to adapt and innovate by a demanding public informed through the ICTs. According to the interviewees, an e-democracy through ICTs is but a question of time for the people demand such a change, and it is *sine qua non* for ex-PPS Id No.: 5. Also, as some of the politicians put it, a few family dynasties cannot stop the progress for a more democratic society.

The interviews show a new consciousness amongst the policy makers that public support is changing. This shifting public mood is driving the need for them to change and adapt to the potential from digital communicative technology capabilities like the ICTs. However, as mentioned above there is an implicit need expressed here to find a way to improve the democratisation by implementing change. This has resulted in interest expressed by many of the candidates about the potential for SE to give them the means to address the people's voice in government. It appears new thinking is needed and SE is of interest to the Mauritian political class.

There was a suggestion by the junior Minister Id No.: 1 to use Rodrigues as a pilot case concerning SE for an e-democracy and then implement it in Mauritius as it is a more complex system than Rodrigues. Such a strategy allows the bugs to be ironed out prior to full scale roll-out or in effect adopting a System of Systems approach. Rodrigues is autonomous yet part of the larger system, the Republic of Mauritius. The politicians seem to understand the problems they confront but they lack a means to fix the problem. A request was also made by member for Constituency No. 5 the PPS Id No.: 11, a civil engineer by training, amongst others for assistance once this book is complete. It is anticipated that some entrenched representatives will resist but given the public mood for change, these will not deter those willing to embrace a SE approach to democratise. Politicians understand that trust from the public is on short supply; hence, they need to do something to regain support. ICTs have been used for some changes mainly to inform, to communicate during the elections but more change is required.

ICTs have introduced a mood for change from the public to the politician. There is an eroding electorate trust for the collage system. The strategic system of collage that political parties had previously modelled their electoral success seems under pressure in this new technological environment. The ICTs were effective allowing first mover advantage to the Lepep supra-party to cause collage system disruption. This informational edge may not last as the other parties adapt to get on board as indicated during the interviews. The ICTs use in the election event of 2014 does not show any application to the development of a decision-making process to democratise yet as suggested by the book. Direct and participative democracy was mentioned by the ex-Vice Prime Minister, Id No.: 7. There was no mention of the direct empowerment of the people in decision making by those interviewed using a virtual type *demos*. At this point in time, it was deemed too early and unfeasible as the system must be developed first and the people and politicians must learn and improve from use of the new system. Under the previous team, Mauritius did invest in eVoting.

A theme that emerged during the interview with the candidates was that the ICTs have the potential to engage citizens in the political process. This makes ICTs a force to be reckoned with in electoral campaigning. For many politicians in Mauritius, ICTs appear to be indispensable for the future of the political system in Mauritius. As several politicians noted, mobile phones use is widespread with the number of phones estimated to be more than double the number of voters in Mauritius. In their view, this offers a good means both to disseminate information and for voters to access it. However, underlying this change is the fact that they also understand that the people are changing the way they use information and that is shaping their voting behaviour which is becoming unpredictable from previous systems like the collage. This uncertainty is contributing to a willingness to try innovative solutions like a SE-designed e-democracy.

In fact, one of the candidates, the elected member Id No.: 3, pointed to a bounce effect whereby the information the party wanted to disseminate went overseas and then the overseas recipient contacted people in Mauritius to discuss the political event. With many Mauritians having smartphones and friends and relatives overseas with smartphones, and various communicative applications that were free for examples Skype and Facebook, local political events were debated using the virtual sphere. The technology seemingly increased interest in the unfolding political events around the elections in 10 December 2014 and this was a new form of e-engagement. There was a heightened awareness and increased engagement amongst the voting public about the Alliance Lepep winning team's discourse on ICTs. The ICTs allowed media files to be shared with other interested stakeholders creating a domino effect. Both local and overseas Mauritians shared and debated local issues using ICTs. Distance was no longer an issue. The opposition bloc, the Alliance Lepep in the 2014 election, used ICTs, namely YouTube and Facebook, to create a need for change by depicting the incumbents in government as offering nothing better than a sense of *déjà vu*. The other team were using the local TV as their main propaganda channel secure in the knowledge that they had the best collage ever, and they did not engage using this media. My conversation with the chief strategist during one of the field trips confirmed this ICTs engagement strategy for the Lepep supra-party prior to the election event. And my brief conversation with the outgoing Prime Minister and interviews with some of the candidates who were defeated confirmed this oversight from the other team. Many of those interviewed were seemingly surprised by ICTs effectiveness though and the Vire Mam strategy was also observed by the local newspapers.

This new participative mode aligns with the Norris findings mentioned earlier that despite democratic deficit people were using alternative means to engage rather than giving up (Norris 1997). The ICTs' newness brought a sense of freshness and excitement to the unfolding events. A new form of discourse converged to create a general feeling of expectancy in the public's mind for creating a change of government. This mood for change was very palpable during the scrutiny of the vote counting process. The whole island was infected by a new wave of expectation of good economic times to come if change happened. These promises by the opposition to win support via social media were popular amongst most young Mauritians and with potential swing voters where unemployment was high. This may not seem right as hinted by some interviewed; these promises are unfeasible. Coupled with rumour mongering in a small island culture, the ICTs helped create through negative news a strong sense of expectant sensationalism. This people's mood was observed by the researcher onsite during 4 weeks of the 2014 election campaign.

Newfound public appetite for experimenting to create change could spell disaster for those leaders who fail along with their party to understand this new public mood. Such a pattern of behaviour of misleading the public creates cynicism. Cynicism is from learning and exposure to previous manipulative outcomes of these entrenched representatives who promise big yet deliver little if any changes to the people's lives. This reduces trust in the representatives' system of governance. It is hard to predict how this erosion of social capital will shape the new behaviour of a cynical electorate in future elections. As even with a choice of other parties and individuals that competed no independents or minor parties emerged. The two major parties bloc (collage) were seemingly still effective to attract the attention of the main electorate. However, as highlighted by the interviewees who belonged to both blocs, government required a new system that listens to the people. This is necessary to maintain support they argue, and this could be a SE-designed e-democracy they suggest. The uncertainty created by the people reacting to being taken for granted from collage may be prompting the pragmatic politician to look for improvements like that which SE can bring to improve the existing system.

4 Conclusion

Technology through SE extends an individual's ability to act autonomously as it empowers them to progress their requirements in a transparent and accountable manner with SE processes, Requirements that assures transparency and accountability that is through CM. In Mauritius, applying the Requirements process will ensure that a system e-democracy allows people's say in decision making through the NDU interface. Mauritius, using the FCI and CI concepts, points to the feasibility of SE's adaptability in upgrading a legacy system. A pilot project was proposed by one of the policy makers, the member for Rodrigues. He suggested that the autonomous Constituency (and island of) Rodrigues would make a good pilot site for developing an e-democracy as a System of Systems concept for Mauritius.

The importance of technology and its emerging capability was understood by every interviewee. They anticipated that SE can provide a means to improve the current system to harness the digital technology which they understand cannot be ignored. Even though it is for economic reasons that ICT technology is a government priority for Mauritius, there was no evidence (IDNO 2016), with the exception of being used for elections to promote a party's message, that it was used in a structured manner to empower the people.

As we have seen, the NDU's functions align with the goals for a system e-democracy where peoples' input allows SE techniques to transform people's needs into requirements. However, the NDU's engagement capability can be improved upon as shown above. For Mauritius, NDU can be transformed into a SE interface for an evolutionary system e-democracy that implements people's say. In fact, some of the Permanent Parliamentary Secretaries (PPS) along with experienced politicians are keen to implement change through the capability that technology brings but the lack of a design framework like SE allows a less than democratic representative democracy system to persist. This positive outlook for Mauritius for a system e-democracy bodes well for its people as implementing the SE process augments the current system to allow accountable and transparent policy making where the people becomes central through the Requirements eliciting process of a transformed NDU.

SE requires a database to capture the peoples' inputs like those used by the US military in its design projects (as discussed in Chapter 5). The database is to capture transformed requirements which are recorded to meet the NDU's existing mission to anticipate and progress the people's need for resolution. This transformation of people's inputs by NDU provides a novel process for people's inputs in policy making. In Mauritius, the NDU, upon adopting the SE processes, becomes the interface for an upgrade to a system e-democracy design described in Chapter 6. Through implementing such an arrangement, it is anticipated that some control would shift to the people. A suggested input to decisions workflow software is mentioned in Appendix 3. The interviewees' suggestions are supplemented from the site visits' observations and through engaging in conversations with the people as well as observations made during a key event like an election, these along with the book findings from the literature provide some political insights which are discussed in the next chapter.

References

Allen, R.B. 1999. *African Studies: Slaves, Freedmen and Indentured Laborers in Colonial Mauritius*. African Studies Series 99. Cambridge, GB: Cambridge University Press.

Bevir, M. (ed.). 2006. *Interpretation and Method: Empirical Research Methods and the Interpretive Turn*. Armonk, NY: M. E. Sharpe.

Bevir, M., and R.A. Rhodes. 2002. Interpretive Theory. *Theory and Methods in Political Science* 1: 1.

Birch, A.H. 1972. *Representation*. Key Concepts in Political Science. London: Macmillan Press.

Birch, A.H. 1993. *Concepts and Theories of Modern Democracy*, 2nd ed. London: Routledge.

Blanchard, B.S., and W.J. Fabrycky. 1998. *Systems Engineering and Analysis*, 3rd ed. Upper Saddle River, NJ: Prentice Hall.

Bräutigam, D. 1997. Institutions, Economic Reform, and Democratic Consolidation in Mauritius. *Comparative Politics* 30 (1): 45–62.

Carroll, B.W., and T. Carroll. 1999. The Consolidation of Democracy in Mauritius. *Democratization* 6 (1): 179–197.

CountryWatch, I. 2017. *Mauritius Country Review*. Houston, TX: CountryWatch Incorporated. http://web.b.ebscohost.com.ezproxy.newcastle.edu.au/ehost/command/detail?vid=0&sid=83858dab-466d-43a1-8559-41e43a834c04%40sessionmgr101&bdata=JnNpdGU9ZWhvc3QtbGl2ZSZzY29wZT1zaXRl#jid=DWE&db=bsu. Consulted 28 January 2018.

Croucher, R., and J. McIlroy. 2013. Mauritius 1938: The Origins of a Milestone in Colonial Trade Union Legislation. *Labor History* 54 (3): 223–239.

Dahl, R.A., and R.E. Tufte. 1973. *Size and Democracy*. The Politics of the Smaller European Democracies. Stanford, CA: Stanford University Press.

Dunn, J. 2005. *Setting the People Free: The Story of Democracy*. London: Atlantic.

Dunn, J., and I. Harris (eds.). 1997. *Aristotle Volume II*. Great Political Thinkers, 2. Cheltenham, UK and Lyme, NH: Edward Elgar.

Eisner, H. (2011). Systems Engineering: Building Successful Systems. *Synthesis Lectures on Engineering* 6 (2): 1–139.

Government of Mauritius. 1968. Part 1—The Constitution GN 54 of 1968, March 12. Mauritius: Government Gazette. http://attorneygeneral.govmu.org/English/Documents/A-Z%20Acts/T/Page%201/THE%20CONSTITUTION,%20GN%2054%20of%201968.pdf. Consulted 28 January 2018.

Government of Mauritius. 2002. *Information and Communication Technologies Act—Act 44 of 2001*, 11 February. Mauritius: Government of Mauritius. http://mtci.govmu.org/English/Rules-Regulations-Policies/Acts/INFORMATION_AND_COMMUNICATION_TECHNOLOGIES.pdf. Consulted 23 September 2016.

Government of Mauritius. 2014. *Government of Mauritius—Consultation Paper on Electoral Reform—Modernising the Electoral System*. Mauritius: Government of Mauritius. http://www.gov.mu/English/Pages/default.aspx. Consulted 14 April 2014.

Government of Mauritius, M.o.F.a.E.D. 2017. Budget Speech 2016/2017. Government of Mauritius. http://budget.mof.govmu.org/budget2017/budgetspeech2016-17.pdf. Consulted 29 January 2017.

Government of Mauritius, M.o.F.a.E.D. 2018a. Summary of Expenditure by Votes 2016–2020. Government of Mauritius. http://budget.mof.govmu.org/budget2017-18/V_00_112017_18ExpbyVotes.pdf. Consulted 6 January 2018.

Government of Mauritius, M.o.F.a.E.D. 2018b. Statement of Government Operations 2016–2020. Government of Mauritius. http://budget.mof.govmu.org/budget2017-18/V_00_102017_18SGovOperations.pdf. Consulted 6 January 2018.

Government of Mauritius, M.o.T., Communication and Innovation. 2014. *National Information & Communication Technology Strategic Plan (NICTSP) 2011–2014: Towards i-Mauritius—Copy Requested Through Minister of Arts and Culture*. Mauritius: Government of Mauritius—Ministry of Technology, Communication and Innovation.

Government of Mauritius, M.o.T., Communication and Innovation. 2018a. Republic of Mauritius e-Government Strategy 2013–2017 l *Empowering Citizens l Collaborating with Business l Networked Government l*. Mauritius Government of Mauritius. http://mtci.govmu.org/English/Documents/eGovernment%20Strategy%20finalv1.pdf. Consulted 28 January 2018.

Government of Mauritius, M.o.T., Communication and Innovation. 2018b. *Expression of Interest for a Market Sounding Exercise For the Implementation of a Third International Gateway Through the Installation of a New Submarine Cable for Both Mauritius and Rodrigues*. Mauritius: Ministry of Technology, Communication and Innovation. http://www1.govmu.org/portal/sites/mfamission/pretoria/documents/bids/ict/Third_Submarine_Cable_EOI.pdf. Consulted 7 January 2018.

Government of Mauritius, M.o.T., Communication and Innovation. 2018c. *E-Ideas Online Service*. Mauritius: Ministry of Technology, Communication and Innovation. http://mtci.govmu.org/English/Pages/e-Ideas-Online-Service.aspx. Consulted 28 January 2018.

Government of Mauritius, N.D.U. 2016. *National Development Unit, Customer Charter—Obtained from Honourable Jean Francisco Francois, Private Parliamentary Secretary (Personal Copy)*. Mauritius: Republic of Mauritius, Prime Minister's Office.

Government of Mauritius, N.D.U. 2018. Citizens Advice Bureau, Web Page Accessed on 23 January 2017. National Development Unit, Prime Minister's Office: National Development Unit, Prime Minister's Office. http://ndu.govmu.org/English/Citizens%20Advice%20Bureau/Pages/default.aspx. Consulted 28 January 2018.

Government of Mauritius, P.M.O. 2018. List of Ministers. http://pmo.govmu.org/English/Documents/LIST%20OF%20MINISTERS%20as%20at%2024%20January%202017.pdf. Consulted 28 January 2018.

Grady, J.O. 2006. *System Requirements Analysis*. London, UK: Elsevier.

Grönlund, Å. 2011. Connecting e-Government to Real Government—The Failure of the UN Eparticipation Index. *Electronic Government* 6846: 26–37.

IDNO. 2016. *Policy Makers in Mauritius, List of Interviewees—Interviews from 6–19 March 2016, Appendix A of Book*. Nvivo.

IHS Global Inc. 2016. *Country Reports—Mauritius*. Lexington, MA: IHS Global Inc.

Kearney, R.C. 1990. Mauritius and the NIC Model Redux: Or, How Many Cases Make a Model? *The Journal of Developing Areas* 24 (2): 195–216.

Kearney, R.C. 1991. Mauritius: Managing Success by World Bank. *African Studies Review* 34 (3): 136–137.

Meisenhelder, T. 1997. The Developmental State in Mauritius. *The Journal of Modern African Studies* 35: 279–297.

Miles, W.F. 1999. The Mauritius Enigma. *Journal of Democracy* 10: 91–104.

Moss, G., and S. Coleman. 2014. Deliberative Manoeuvres in the Digital Darkness: E-Democracy Policy in the UK. *The British Journal of Politics & International Relations* 16 (3): 410–427.

Norris, P. 1997. Representation and the Democratic Deficit. *European Journal of Political Research* 32: 273–282.

Norris, P. 2001. *Digital Divide: Civic Engagement, Information Poverty, and the Internet Worldwide*. Port Melbourne, VIC, Australia: Cambridge University Press.

OECD. 2003. *Promise and Problems of E-Democracy-Challenges of Online Citizen Engagement*. Paris, France: OECD.

OECD. 2005. *Report from OECD Forum 2005 to the OECD Ministerial Council Meeting*. Paris: OECD Publishing. http://www.oecd-ilibrary.org/

docserver/download/0105101e.pdf?expires=1517024994&id=id&accname=guest&checksum=6E744C2E44301A5B640ADBF534EFC209. Consulted 27 January 2018.

OECD. 2010a. *The Development Dimension—ICTs for Development—Improving Policy Coherence.* Washington, DC: Organization for Economic Cooperation & Development, OECD iLibrary. http://www.oecd-ilibrary.org.ezproxy.newcastle.edu.au/docserver/download/0309091e.pdf?expires=1517024328&id=id&accname=ocid194270&checksum=61AB79A0E079DDDBBFE3B6E4568A2B0F. Consulted 27 January 2018.

OECD. 2010b. *Good Governance for Digital Policies—How to Get the Most Out of ICT—The Case Of Spain's Plan Avanza.* OECD Publishing. www.oecd.org/gov/egov/isstrategies. Consulted 27 January 2018.

OECD. 2013. *Reaping the Benefits of ICTS in Spain Strategic Study on Communication Infrastructure and Paperless Administration.* Paris: Organisation for Economic Co-operation and Development, OECD iLibrary. http://www.oecd-ilibrary.org.ezproxy.newcastle.edu.au/docserver/download/4212081e.pdf?expires=1517023432&id=id&accname=ocid194270&checksum=28A93F4721E71AADB84F3539BCB992DE. Consulted 27 January 2018.

OECD. 2015. *States of Fragility 2015 Meeting Post-2015 Ambitions.* Paris: OECD Publishing. http://dx.doi.org/10.1787/9789264227699-en. Consulted 27 January 2018.

OECD. 2016. *Development Aid at a Glance Statistic at a Glance 2.0 Africa.* OECD Publishing. http://www.oecd.org/dac/stats/documentupload/2%20Africa%20-%20Development%20Aid%20at%20a%20Glance%202016.pdf. Consulted 27 January 2018.

Sandbrook, R. 2005. Origins of the Democratic Developmental State: Interrogating Mauritius. *Canadian Journal of African Studies/La Revue Canadienne Des études Africaines* 39 (3): 549–581.

Statistics Mauritius. 2017. Table 4—Population Enumerated at Each Census by Community and Sex, 1962–1972. Statistics Mauritius. http://statsmauritius.govmu.org/English/Pages/POPULATION–And-VITAL-STATISTICS.aspx. Consulted 16 November 2017.

Transparency Mauritius. 2013. *Corruption Perception Index 2013.* Port Louis, Mauritius: Transparency Mauritius. http://www.transparencymauritius.org/corruption-perception-index/corruption-perception-index-2013/. Consulted 25 December 2016.

Transparency Mauritius. 2015. *Corruption Perception Index 2015.* Port Louis, Mauritius: Transparency Mauritius Organisation. https://www.transparencymauritius.org/wp-content/uploads/2016/01/CPI-2015.pdf. Consulted 25 December 2016.

Transparency Mauritius. 2017. *Corruption Perception Index 2016*. Port Louis, Mauritius: Transparency Mauritius Organisation. https://www.transparencymauritius.org/wp-content/uploads/2017/06/CPI-2016-25-01-2017.pdf. Consulted 30 October 2017.

USAF. 2010. *Air Force Systems Engineering Assessment Model*. Air Force Center for Systems Engineering, Air Force Institute of Technology: Secretary of Air Force, Acquisition. http://www.afit.edu/cse/. Consulted 8 September 2013.

U.S. Department of Defense. 1985. *Military Standard Specification Practices MIL-STD-490a*. Andrews Air Force Base, Washington, DC. http://everyspec.com/MIL-STD/MIL-STD-0300-0499/MIL-STD-490A_10378/. Consulted 28 January 2018.

U.S. Department of Defense. 1993. *Military Standard Systems Engineering MIL-STD-499b*. EverySpec. http://everyspec.com/MIL-STD/MIL-STD-0300-0499/MIL-STD-499B_DRAFT_24AUG1993_21855/. Consulted 28 January 2018.

U.S. Department of Defense. 2003. *Department of Defense Standard Practice Defense and Program-Unique Specifications and Format MIL-STD-961e*. Fort Belvoir, VA: Defense Standardization Program Office. http://everyspec.com/MIL-STD/MIL-STD-0900-1099/MIL-STD-961E_11343/. Consulted 28 January 2018.

U.S. Department of Defense. 2008. *Systems Engineering Guide for Systems of Systems*. Office of the Deputy Under Secretary of Defense for Acquisition and Technology, Systems and Software Engineering. Systems Engineering Guide for Systems of Systems, Version 1.0. Washington, DC: ODUSD(A&T). http://www.everyspec.com/. Consulted 28 January 2018.

Williams, G., and D. Hume. 2010. *People Power: The History and the Referendums in Australia*. Sydney: University of New South Wales Press.

Willmore, L. (2003). Universal pensions in Mauritius: Lessons for the Rest of Us, ST/ESA/2003/DP. 32. DESA Discussion Paper No. 32. United Nations DESA Discussion Paper Series. Division for Public Economics and Public Administration, Room DC2-1446, United Nations, New York, NY, 10017, United Nations.

Youngblood-Coleman, D. 2014. *Mauritius: 2014 Country Review*. Houston, TX: CountryWatch Incorporated. http://ezproxy.newcastle.edu.au/login?url=http://search.ebscohost.com/login.aspx?direct=true&db=bth&AN=99017947&site=eds-live. Consulted 8 January 2018.

Youngblood-Coleman, D. 2016. *Mauritius: 2016 Country Review*. Houston, TX: CountryWatch Incorporated. http://web.b.ebscohost.com.ezproxy.newcastle.edu.au/ehost/pdfviewer/pdfviewer?vid=1&sid=ddbc9060-af26-4b18-8af7-09fd6888144c%40sessionmgr120. Consulted 8 January 2018.

CHAPTER 8

Upgrading Mauritius a Legacy Political System

1 The Proposal

The issue of ongoing corruption in Mauritius brought to my attention that as an engineer from my experience the political system could be improved using Systems Engineering (SE) and technology. With the advent of the Information and Communication Technologies (ICTs), a significant scholarship on e-democracy emerged promising to fix representative democracy with more direct input from citizens (Grönlund 2001). However, despite the potential for ICTs to engage citizens and allow bottom-up participation in policy making (OECD 2003), their successfully operationalised e-democracies have not been able to be sustained (Moss and Coleman 2014). What we might call the e-democracy problem sits inside a larger problem which is the problem of how people can rule their lives or, or to put it in a slightly different terminology, of how the *demos* can be empowered. In many respects, this is a democratic deficit design issue within representative democracy in the sense that representative democracy has this problem built into it from the very beginning (Hindess 2002). This book aimed to show how SE might contribute to addressing this problem as well as enabling the democratic operationalising of e-democracy.

The book aimed to answer the question of *how might a Systems Engineering approach address currently perceived problems with e-democracy*? To

S. Bungsraz, *Operationalising e-Democracy through a System Engineering Approach in Mauritius and Australia*,
https://doi.org/10.1007/978-981-15-1777-8_8

do this, it set the stage by exploring the general understanding of democracy and different attempts to establish e-democracies that might improve on the limitations of representative democracy. To the core driving question the book engaged in a thought experiment that outlined and justified an approach to e-democracy that made innovative use of ICTs through a Systems Engineering (SE) approach. The aim was to develop a model of a democracy that democratises. Essential to this thought experiment was the view that SE offered a new and innovative way to design a solution to establish a people's e-democracy.

As the book demonstrated, SE is a systems approach derived from the field of engineering. This was discussed in some depth in Chapters 5 and 6. As was demonstrated, SE starts from agreed requirements that are used to develop a technical system as a solution, an e-democracy in the book (Blanchard and Fabrycky 1998). To contextualise SE in political theory, the book discussed Easton's systems theory first and then it explained how SE resonated with it in a political sense though not necessarily in a methodological sense. The book introduced some of the terms to be adapted to the e-democracy solution and then applied them to show how they might create a democratic e-democracy. This e-democracy model, as developed in the book, was then overlayed on an actual representative democracy, Mauritius, which also suffers from democratic deficit. In this context, the case of Mauritius served as something like a proof of concept. It was treated as a legacy system for upgrading to a democracy that democratises. Using Mauritius as a proof of concept to show how SE developed e-democracy could make it possible for citizens to have direct input to important channels of decision making in practice. Through an augmented version of Mauritius's National Development Unit (NDU), a SE-enabled democracy could shift meaningful control to citizens. Key features of the e-democracy solution that make up the book findings are discussed below.

2 Findings

The scholarship on e-democracy and its use of digital technology, namely the ICTs, suggests that government can be improved to allow people's direct participation in decision making (Grönlund 2001). However, in practice, in each state with a representative democracy, the degree of democracy varies. Now, variance is not something new in government systems. Aristotle had observed this variance earlier when he studied forms of government and had suggested alternatives, at least six forms with democracy as people's rule being called a bad form. After 2500 years of evolution, the modern representative democracy is a compromise it is designed for governing but without people's rule (Hindess 2002). Contemporary governments still reflect the dialectic between the neoliberal form which protects and is protected by the market, which in turn undermines the perception and feasibility of forms of direct democracy, forms that might prove inimicable for the wider context of the capitalist system (Schumpeter 2010). In most states, representative democracy as a compromise is a regime in equilibrium that exists from its own unique evolutionary path in that state. A term called democratic deficit emerged to describe and quantify the lack of democracy in a representative democracy (Norris 1997; Hindess 2002). In fact in forms of representative democracy, the democracy problem is within a larger problem which is that of a deficient design of democracy as people's rule in representative democracy itself (Hindess 2002).

The scholarship suggested e-democracy is the solution to this idea of a democratic deficit (Rios Insua and French 2010; Dahlberg and Siapera 2007). However, as discussed in the book, an e-democracy model has failed to appear in practice, at best what has emerged have been piecemeal digital attempts or fragmented experiments (Päivärinta and Sæbø 2006). The variance in government systems complicates matters, as digitisation derived from broad policies to develop an e-democracy to empower the people is complex in practice. The degree of democraticness (Wright 1994) affects e-democracy attempts to digitise a process capable of empowering the people in decision making. This is despite the technological capability of ICTs potentially to democratise through e-engagement and e-participation (OECD 2003).

ICTs allow improved participation and deliberation through its communicative capability; however, the actual attempts to digitise have been failing to translate broad policies into democratic outcomes. The OECD's study of those twelve countries that attempted to implement e-democracy recommended a digitisation framework that was supportive of people's participation in decision making (OECD 2003). Some technological software to digitise existing electoral processes like Smartvote uses digital technology but that is to improve the electoral process by providing information to voters to select their best candidate (Meier 2012); this Swiss customised software is yet to allow an e-democracy to emerge. Rather,

Swiss e-democracy remains a promise. Use of the software traps Switzerland in an existing representative democracy paradigm. It assumes that improved use of technology leads to a knowledge society capability which then creates an e-democracy (Meier 2012). The focus is on the technology rather than direct decision making to empower the *demos*. Without transformation, digitisation of existing representative processes will at the most show e-democracy failures to empower the people as the current processes do not fix the design deficit (Hindess 2002).

The lack of an e-democracy to operationalise in the manner envisaged by its proponents led to a different framework being considered by this book, namely Systems Engineering. As has been discussed in some detail, SE is an engineering governance system applied to manage a multidisciplinary team to create an agreed solution. To solve a problem, SE purposively elicits and integrates each individual requirement to develop and maintain a technical system for the life of the system (Blanchard and Fabrycky 1998). The SE technique is applied in the book to a state which it treated as a legacy system made up of Configuration Items (CIs) and Functional Configuration Items (FCIs) to create a technological political system, or a SE-enabled system e-democracy. Using SE's approach, an e-democracy that democratises is a design to enable people's rule through a NDU type interface.

SE's central positioning of the end user shifts control to the people. Basically, SE builds into an e-democracy the idea of democracy or people's rule as an integral part of the system of e-democracy. The book explained SE's novel techniques, namely the Requirements and Configuration Management processes that when adapted meet the need for decision making from the people's elicited input. The book showed how these techniques allow the identified weakness of the Easton systems theory to be overcome for representative democracy. A critique of Easton's systems theory was that people generate inputs but this was incorrect from a normative perspective (Birch 1993). For Birch, in an actual representative democratic system, the representatives and senior public servants are in charge of policy making not the people. SE's process as applied by the book showed that the direction of inputs is from the people where eliciting requirements become deliverables through a parliamentary system. This SE flow if adapted to policy making by the people is for the people. This shifts control to the people in a meaningful way. It enables an e-democracy to emerge that democratises. As was noted in the beginning of the book, such a development sounds all very good in theory but how will it work in practice? The discussion of Mauritius and its specific characteristics, especially its NDU and the willingness of key political actors to embrace the ICT technology, provided what the book termed as a proof of concept. That is, it provided some indication that translating the theory of SE into a practical application would have some real-world purchase.

2.1 Systems Engineering's E-Democracy from a Deficit to a Democratic Pre-requisite

In the book, SE explains the development process of successful systems that informs the politicians about the people's needs using technology. In this sense, SE meets the identified need for a technological framework as outlined in the OECD's e-democracy study about engaging the people in policy making (OECD 2003). This OECD recommendation was confirmed by those interviewed in Mauritius. There is a new thinking emerging to fit politics to the digital era, but importantly for some, also to reform what is a deficient system through innovation. Information is making the flaws of a representative democratic system apparent. In synthesising those interviews, representative democracy limits what is feasible with technology like novel forms of participatory and direct democracy or an e-democracy.

The common goal for democratic forms of governance is assumed as the good of the people (Dahl 1995). This good for the people is what differentiates democracy from other forms of government. This is a shared view amongst those Mauritian policy makers who were interviewed. The politicians and public servants profess that they want a system that works for the good of the people. Using technology is treated as a good thing as it could be harnessed to develop policy.

The book argued that SE creates the transparent and accountable processes that could formulate policy for Mauritius. The NDU as explained in Chapter 7 could initiate the transformation process to a system e-democracy as described in Chapters 5 and 6. Importantly, in the government processes of Mauritius, the SE processes described by the book do not conflict with the existing role of the NDU; in fact, the NDU's function is augmented by SE. SE's system e-democracy ensures that technology when applied is relevant for the needs of the NDU both now and in the future. SE's change process allows value for money investment in technology.

A SE-transformed NDU type institution meets the need of honest politicians to transparently deliver the good life outcome for their mandates in existing democracies. The Mauritian NDU institution's existing engineering unit is the interface identified by the book for how SE upgrades and augments the legacy political system to an e-democracy. The unit's engineering team with SE processes allows a multidisciplinary group to emerge to articulate the various needs of the people. This unit identifies ongoing changes to the e-democracy system from emergent changes identified by the institution's existing function to educate and conduct policy knowledge transfer to the community. Costings of needs can also be carried out to ensure that the resources exist and can be allocated to meet needs, while scope creep is reduced as policy arises from actual rather than perceived needs. SE is the governance aspect to this new NDU process for Mauritius to augment it to a democracy that democratises.

Most political systems already exist so the question for SE is how it upgrades these legacy systems to first one that is democratic using technology to assist its transformation and then improve the transformed system during the life of the system. As technology becomes more capable, the system e-democracy with the SE's life cycle approach allows continuous upgrades. Configuration Management (CM) allows the system to adopt more capable technology and the notion of autopoiesis is feasible through technological upgrades as technology develops to improve systemic self-sustenance through automated and accountable processes. This makes a SE-enabled e-democracy a system that persists to maintain people's rule. The concept of a staged upgrade for Rodrigues provides the beginnings of a System of Systems (SoS) construct, which will then pave the way to allow SE to upgrade an existing system in a structured manner.

Now, this upgrade will require a NDU type function (FCI) and this may vary from one state to another based on the analysis of its existing functions (FCIs). In the case of Mauritius, analysis of its institutions shows the NDU FCI aligns with those for the SE Requirements eliciting process of people's inputs in decision making. In other states, this NDU type interface may have to be designed. So, if the political will exist and the NDU type function to facilitate SE can be created, then an e-democracy can come on its own. So policy formulation would be facilitated through e-democracy using the NDU (FCI) to drive and deliver the performance of the e-democracy for policy making.

SE designs a system that informs politicians about the people's needs. Thus, in terms of risk with SE, it is one of performance. With the SE process, the corrupt political actor becomes constrained to act honestly as the system is transparent, recorded in networked databases. Scope creep would require justification in open parliamentary debate.

The book has demonstrated that the e-democracy form is capable of being customised by SE as the e-democracy design is centred around the people and by the requirements from the people. Now, in some states new, more capable technology may automate the data capture. The automated data capture may require a different interface to a NDU type data flow described in this book so that it can accommodate both face-to-face and direct inputs from citizens. However, no matter the level of technology that is adopted, the proposed SE system is upgradable by CM to create an augmented e-democracy that can use the more capable technology. As data capture becomes mainstream, the data could be accessed by the people to pluralistically provide further ideas to improve policy making for their country. The centralised databases mentioned, through further research in the data mining for example, provide a means for the NDU type function to anticipate people's needs with a higher degree of accuracy. This potentiality of involving every citizen seemed of interest to most of those interviewed as it unleashes a potential capability from technology that builds a stronger community where decisions taken are understood by all. Transparent decision making reduces the potential for negative news and improves trust in the government.

3 Recommendations

In analysing an actual democracy, the context for digitisation mattered. Mauritius, from the evidence gathered in the book, showed the potential for applying SE for improving the political system through the National Development Unit's existing function. Each nation-state as a system would be confronted with its own unique constraints for implementing ICTs. SE upgrades such systems by treating them as a legacy system. SE in this sense is independent of the system to upgrade but dependent on the will to upgrade. The will to apply SE must exist in the leadership for the system e-democracy to allow people to become actively involved in decision making (OECD 2003).

Leadership issue was one outcome of the study from the OECD and the recommendation to ensure that the senior leadership promoted a citizen empowerment remains for the Mauritian political context and it is identified as a risk for an e-democracy by the book (OECD 2003: 10). Note that the OECD assumed that perhaps in representative democracy there was no dynastic leadership entrenched as highlighted for Mauritius by the ex-Minister for Arts and Culture, ID No: 6. In each political system, the context would vary from state to state as the democratic and institutional arrangement for checks and balances would be different in each case. With the constitutional inclusion of an e-democracy that democratises, it will also be opportunistic to explore a role change of the President's function or similar role in other states to defend democracy through SE's life cycle approach. Thus, further research is recommended to fix some of the identified issues like constitutional barriers that must be removed to protect a democracy that democratises. SE opens some areas for further research like policy data mining from the data repository built by a transformed NDU which when opened to the public allows new forms of participation to policy making.

4 Conclusion

There is no best model of democracy. In practice, as Mauritians learn through new generations adapted to the concept of a SE-embedded e-democracy, the customised form of better government will emerge in a continuous cycle of improvement driven from the people and the capability from technology that SE incorporates to augment the existing system in a purposive manner. There is no one standard, or one best democracy, as there was no model or processes that were delivered by the Greeks just an idea that anyone could rule and be ruled. Implementing this ambiguous idea requires a framework that can translate the idea into outcomes, like a SE framework as shown in the book.

SE has an iterative design with people at the centre of the design solution for the life cycle of the system. With its application in different context, a general theory for good government system or real democracy may become the norm. Thus, SE creates through design a technological system we call system e-democracy that may finally shed the bad reputation of democracy as chaotic and a tyranny of the majority.

However, SE requires a new way of thinking and a new framework. It also requires a Constitution that supports the effort towards this customised form of governance by the people which is appropriate to the context of each state as the FCI and its constituencies as CIs. A Constitution that SE Requirements elicits for an e-democracy design from the people and of the people is through a NDU type function. SE Requirements process transforms the people's needs into a database repository used to inform policy making.

This systems-based e-democracy is a new vision of modern democratisation that is made feasible by adapted SE processes and technology that continuously improves so that e-democracy is upgraded and remains relevant to the needs of the people in each era. With SE, a democracy design to create the *demos* is feasible as this book has argued. The opportunity to initiate an e-democracy in a pilot site like Rodrigues creates a prototype to upgrade existing government systems for others to adapt the SE's system e-democracy concept. The SE process as shown in this book addressed the fundamental design deficit identified by Hindess in representative democracy (Hindess 2002). The member of Rodrigues in his budget speech for 2019–2020 on 19 June 2019 flagged the importance for a vision of change. The member's commitment to an e-democracy is now part of the Mauritian Hansard; in his view, this is a guarantee that a Rodrigues' e-democracy solution is not for a too distant future. The book initiates the journey for change to a better government system designed by the people and for the people. SE eliminates this deficit in representative systems through a systems-based process that democratises.

The above was for a small nation-state, so in the next chapter SE explores a *what if* for an Australian context as a large nation-state. In the nation-state of Mauritius, politicians are not averse to democratise despite the challenges they confront as they cannot be e-stupid (ignore technology and its benefit); Australia a comparatively large nation-state is explored next with a notable reticence for change identified from strategies adopted at the 18 March 2019 election. Reticence by politicians for a *status quo* is still a significant stumbling block. Thus, a different strategy is proposed for this continental-sized nation-state. From the author's observations, it could be said that the member for Rodrigues is more enthusiastic for change to empower the people for Rodrigues than a larger and wealthier democracy where there exist two major parties that apply old tactics to win elections rather than improve the political system using technology.

References

Birch, A.H. 1993. *Concepts and Theories of Modern Democracy*, 2nd ed. London: Routledge.

Blanchard, B.S., and W.J. Fabrycky. 1998. *Systems Engineering and Analysis*, 3rd ed. Upper Saddle River, NJ: Prentice Hall.

Dahl, R.A. 1995. Justifying Democracy. *Society* 32 (3): 386–392.

Dahlberg, L., and E. Siapera (eds.). 2007. *Radical Democracy and the Internet: Interrogating Theory and Practice*. Basingstoke: Palgrave Macmillan.

Grönlund, Å. 2001. Democracy in an It-Framed Society. *Communications of the ACM* 44 (1): 22–26.

Hindess, B. 2002. Deficit by Design. *Australian Journal of Public Administration* 61 (1): 30–38.

Meier, A. 2012. *EDemocracy & EGovernment: Stages of a Democratic Knowledge Society*. Berlin: Springer.

Moss, G., and S. Coleman. 2014. Deliberative Manoeuvres in the Digital Darkness: E-Democracy Policy in the UK. *The British Journal of Politics & International Relations* 16 (3): 410–427.

Norris, P. 1997. Representation and the Democratic Deficit. *European Journal of Political Research* 32: 273–282.

OECD. 2003. *Promise and Problems of E-Democracy-Challenges of Online Citizen Engagement*. Paris, France: OECD.

Päivärinta, T., and Ø. Sæbø. 2006. Models of E-Democracy. *Communications of the Association for Information Systems* 17 (37): 1–42.

Rios Insua, D., and S. French. 2010. *E-Democracy*, vol. 5. Dordrecht, Heidelberg, London, and New York: Springer.

Schumpeter, J.A. 2010. *Capitalism, Socialism and Democracy*, 1st ed. Florence: Taylor & Francis.

Wright, E.O. 1994. Political Power, Democracy, and Coupon Socialism. *Politics & Society* 22: 535.

CHAPTER 9

A System Engineered Approach to E-Democracy: A *What If* for Australia

1 Introduction

Australia aims to be a progressive country, and representative democracy development in the Australian context creates its own paradigm about democracy, its politicians' rhetoric is that the political system must be fair. However, democracy's application by the different actors who rule remains subject to public emotions as democracy is an essentially contested concept. In the Australian context due to factional power struggles to rule democracy's evolution has led to instability within the government. Democracy *a la Oz* is certainly not a foregone conclusion, as each government 'enforce their own quaint meaning' lacking the luxury of universities to explore and define these, they and their factions tend to justify some of the misunderstanding associated with this term (Birch 1972: 7). The meaning of democracy used by these power entrepreneurs is imposed through the Cabinet and machinery of government, and through both action and inaction influences, they define in terms of a government serving the public, the Australian political way of life—a way of life where the public has no direct say.

Australian factions in political parties are detrimental to democracy, and factions in their original fights which pitted monarchist against non-monarchist led to the polyarchy model. These factions may have outlived their usefulness to citizens in an era where technology can look at 'a level of democratization beyond polyarchal democracy' (Dahl 2005: 197). In

S. Bungsraz, *Operationalising e-Democracy through a System Engineering Approach in Mauritius and Australia*,
https://doi.org/10.1007/978-981-15-1777-8_9

Australia, most processes in both government and the private sector are conducted online, and so these transactional processes are digitised. Email and online forms are commonly used between the representatives and their constituents at all levels of government. This digital communication is also supplemented by written media, yet despite extensive digitisation to communicate the governing in Australia remains chaotic. Electoral re-engineering to improve had some unanticipated outcomes for Australia where three systems of voting are used to reform the electoral system. However, a key function the Prime Minister's role has become a political musical chair from perceived populism of the incumbent and pressures by opportunistic personal and factional agendas. Using perceptions from polling, it is the factions aspiring to rule that actually decide instead of the Australian people. Australian political parties' behaviours are creating a volatile system for Australia. This is despite an electoral system that has attempted some reforms as discussed below. It appears that Australia's representative democracy has a system design problem that SE can fix through an upgrade of its political system.

This chapter considers the representative democracy of Australia, which is a large island state and an ex-British colony, as a political system. Australia is deemed to be an exemplar of democracy, this is despite representative democracy is a design failure of democracy (Hindess 2002). Systems Engineering (SE) explores Australia as a top down, desktop or preliminary analysis of its political system in a *what if* experiment described below. With its varying size and distance of communities (village, town, cities) in this continent type island, Australia's population agglomeration is concentrated in the large cities which except for Canberra are coastal. The large size and democratic evolution of Australia as a political system differs from others for a SE application. Despite being geographically part of the Asia Pacific, Australia seems desperate to link itself to its coloniser UK as some in Australia regards this nation-state as an European outpost and resist changes to delink itself from its colonial past.

The isolation of the island continent from Europe appealed to its coloniser, but in a different way. It was primarily used to penalise and this may have shaped to some extent the deep-rooted distrust in Australians about politics and politicians. Australia was sent convicts as part of a British empire penal colony and it also had some emancipists (former convicts free to hold land), this settlement was established while the island continent boasted an indigenous population. A system with feudal representatives, the governors, as local rulers was empowered to manage the colonies (states of Australia) in the name of their monarch. This ruling system was not unlike those for the USA and Canada as colonies. It was a representative system that was organised in similar fashion to many other parts of the world that Europeans colonised at the time. For Australia, this penal history and its migrant policy created an increased complexity that modernising Australia as a political system must acknowledge to progress to a traditional democracy. While still a colony, starting with parliamentary discussions to aim for a 'social democracy' it seemingly took 6 decades of 'spasmodic official effort' to create the Commonwealth of Australia (Crisp 1983: 1). Progress to the idea of an Australia was at times tenuous and resisted, and a democracy in the true sense is still a work in progress.

Australian ruling parties have tried to adapt to its unique challenges by using a combination of federalist and Westminsterian system of governance (Smith et al. 2012). One may argue that the resultant political system of the Commonwealth of Australia follows a federal democracy which is modelled from the USA and Canada at the time of its inception. There are also some uniquely Australian creations emerging from its colonial system legacy. It led to the current two main party systems where strong union links are associated to the Australian Labor Party (ALP) for creating a socialist inclination while its main political adversary is from a Coalition group that opposes unionism tending to conservatism. These two main party systems contest for their claims to rule. Some scholars suggest that Australian socialism is without doctrines while its conservatism is dismissed as being without traditions (Hughes 1968: 1). Commentators regularly report the deep infighting from the factions in these main parties to seek power on behalf of their followers, factionalist tendencies that stress the Australian political system which is already quite complex.

Now broadly, in both main political parties, the elected members seem to follow a liberalist type paradigm at times moving to a centre and then out to either left or right based on perceived populist demands (from polls) of the electorate. So *what if* SE is applied as a framework that upgrades this political system to increase the democraticness of the system. An exploration of how a SE approach may result in creating the people's inputs for those who wish to represent them is considered in this chapter for Australia. Firstly to contextualise, the below covers some of the elements of Australian political system and institutions, before investigating its upgrade to an e-democracy that fixes democratic deficit. Democratic deficit is a representative democracy system design issue that we highlighted before (Hindess 2002). Corruption claims by politicians have also impacted Australian standards with calls from the Opposition (ALP) for a body to address corrupt practices at the national level (Transparency International 2018; ALP 2019).

The current system suffers, like other political systems, from a lack of trust of its politicians who tend to abuse their privileges in parliament. Now despite reform leading to a Parliamentary Privileges Act of 1987 (Craig 1993: 66–68), there continues to be media reports of breaches by elected members. For a democracy, the behaviours of elected representatives from its systemic turmoils appal the general public. Modern Australia, as any other representative democracy model discussed earlier, was never designed from its inception to be a model of democracy as meant by the traditional use of the term (Held 1983, 1993, 1995, 2006). Australia's political system though has both the resources and the technological infrastructure to improve and implement an e-democracy but its evolution is subject to the actors, those aspiring for representing the people, if they are willing to do so. Some experimentation towards an e-democracy has been tried in Australia as reported by the OECD (2003), though without the reported use of SE framework. Australia can upgrade to an e-democracy system using the SE framework, we explain further below.

2 As a Lead Democratic Political System Context

Australia's ambition is to be a lead role model of democracy for itself and in the Asia Pacific region. It is a country that gives people a fair go so this bodes well for democracy. Australia is also deemed to be a developed economy and it is economically amongst the top nations of the world with a relatively small population and it has a significant amount of natural resources. As a nation-state, Australia tries to maintain a strong commitment to fairness which is part of its national anthem. It has a global presence, and Australia is building strong links in the region where it is situated, and depending on the government of the day as a global citizen it tries to do its fair share. However, its political system of representation is not without challenges and more recently with five Prime Ministers over a 10 year period it has been deemed in 2018 to be the capital city of coups from comments of its ex-Foreign Affairs Minister from the ruling Coalition party. A former Victorian Premier of the same party alludes that 'part of the problem is that Australia is now a country with no direction' (Brissenden and Andersen 2018). Some scholars see Australian politicians as 'pragmatists par excellence' (Hughes 1968: 1). However, commentators often blame these pragmatic politicians with a lack of vision for the country, accusing them fairly or unfairly of being self-absorbed rather than serving the greater good for the country. In a democracy, ruling is for the greater good, a service to the community, a privilege not a job.

Australia in the region where it is located attempts though to lead as an example of a good government with a parliamentary system where the people vote for their representatives. This does not mean that the system always gets things right as similar to other political systems, the media reports various instances of corrupt practices and bad behaviours of elected members, with some being sued in court. Australia is in this sense like many other government bodies confronted from time to time to behaviours that are unbecoming for its elected members. This has resulted in calls from some elected members for an independent body being established to investigate corrupt practices at the federal level (Doran 2018). Mauritius is a young democracy, recently decolonised from Britain, with an independent body that is institutionalised to fight corruption but it still has corrupt practices being reported in the media. In fact one of the Mauritian Ministers, not listed, while being interviewed during the interviews mentioned in Appendices 1 and 2, had to resign due to allegations of corruption. Furthermore, two of the listed interviewees who were promoted to junior Minister and senior Minister were forced to resign after allegations made in the local media. Even in Australia, local media have revealed inappropriate behaviours resulting in demotion from Ministerial positions. Systemic weaknesses can always be exploited by the powerful as political parties are entrepreneurial organisation, legislation or institutions are not enough given the Mauritius case. We expand on this below. A weakness is that parties competing to rule do so without clearly defining by systemically seeking what the people need or want, which is a people's agenda developed by the people. Thus, despite being amongst the top 10 most industrialised countries in the world, Australia suffers a democratic deficit.

Australia also forms part of large donor countries that assist other countries, developing nations, with development aid and poverty reduction of their populace. Australia can aspire to lead as it has a strong push towards technology with very strong academic and research institutions to assist these countries. To overcome the digital divide the Australian government has invested heavily in the National Broadband Network (NBN) so that despite its large size as a country (continent) and a relatively small population spread over large distances, the NBN provides connectivity to every Australian. These investments of Australia to develop other nations in the world and its commitment to technology may bode well for an e-democracy if the political will exists. However, the Australian system is static, and its *staticism* ails the current political system as despite its potential the system regresses in today's technological environment to dysfunctionalism as its political paradigm. A paradigm that appears fixed for a lower *techne* environment, in that it has a Constitution with a colonial mindset, which in a sense forces it to revert to a status quo (remaining static rather than progressive). Australia still operates within the framework of its colonial vestiges, and it is a political system that is inclined towards maintaining equilibrium rather than facilitating change that improves it in a systemic manner. Despite the existence of significant resources, in terms of technology, the Australian political landscape is one of keeping the people's voice out of parliament, one of democratic *staticism*.

Political *staticism* is the state of a political system being so stable that it no longer meets its core objective of allowing dynamic change to improve even in a fast-changing technological environment. This systemic inertia arguably is because as an obsolete political system it does not have the means to alter itself, like with a system that has a SE approach. As mentioned SE allows a system to upgrade and remain effective when technology enables improvements, SE counters obsolescence, SE is an evolutionary framework it is dynamic. Now this Australian *staticism* contradicts the earlier statement of the system being unstable to govern for the people, the current dysfunctionalism though is from party infighting. Intraparty conflict has led Australia being described as the coup capital in the world by a deputy leader and senior Minister of a deposed Turnbull government. An upgrade to improve the political system is through SE involving Australians, as the sovereigns for Australia in an e-democracy. Now the representative democracy upgrade to an e-democracy that democratises is not a one-step process, it needs an evolutionary or phased approach which SE enables. A key component of the political system that of voting in Australia has been experimented upon, yet it still produces dysfunctional government. This is because the election of senate members can be with only as little as 19 citizens directly voting for that person (Mckenna 2017). A 19 vote person who then as a senator can become a national embarrassment for delivering his maiden speech in parliament by making use of a quote to Nazi final solution and for promoting a white Australia European policy (Conifer 2018). It would appear this senator as per Mckenna (2017) may not even know the 19 people who voted directly for him. Despite changes to the system to allow minor parties to emerge from changes to the voting system, from the voting reforms the quality of the parliamentary debate is to be desired so far.

The establishment of the Australian representative system was with a Constitution for federation in 1901, it was in an agricultural context and as a penal colony. System change has been actively thwarted since. Recently, for example to Australia becoming a republic, the static political system maintains a status quo. The political *staticism* one assumes profits a small group with vested interest in a colonialist past which allows them a degree of control and relevance. In the interim, the silent majority watches its agents or elected representatives at work destroying what could have been so that what has been is retained and maintained, basically a political system that regresses. SE therefore explores how it can be applied to upgrade the current system so as to shift power to the people through political agenda control. In so doing SE improves the democraticness of the system moving towards *astaticism* to facilitate ongoing change from the systemic dynamism inherent from a SE approach. This journey from *staticism* to dynamism with SE allows through use of technology like software, new ideas to surface and the institutions of knowledge like universities have a role in its facilitation. These institutions as interface add value to validation and verification of ideas (requirements) that are transformed to policy (specifications), and as new technology is created and operationalised, it allows even the ongoing improvement to the SE methodology itself. SE as mentioned earlier counters obsolescence for its approach implements technological upgrades, upgrades that are ongoing as new technology emerges.

As we shall discuss below, the Australian federation was not constituted because politicians wanted to forge a greater country; rather, it was the need to have a common defence for all the colonies that led to a federal government system. A federal government concept that at many times was threatened by discords leading towards its derailment as each premier from the other states disputed the goal of the Premier of New South Wales (NSW), a leader for a federation. The other colonies timidly approached any change from the status quo. Victoria openly deferred to vote for a common government that could command the Australian military which was the intent for federating to a system that was to become the Commonwealth of Australia. Finally, it was the leadership of the people of NSW who on 3 June 1898 voted yes to the concept of a federal system following the Canadian model, and importantly it was without the bloodshed of civil war that the US federation had to go through (Wise 1913). So the journey to modern Australia was not smooth or visionary, it happened due to the perceived threat to its defence while it was still a colony; however, fortunately its creation was a transition that did not lead to a revolution and loss of life.

Now according to Smith et al. (2012: 21), Australian policies tend to be refinements rather than ground breaking 'as the decision-making forums are fragmented' from its institutional architecture. An argument is that in these messy arrangements good ideas are forced out of politics. In 2018, many elected representatives (by the people) had to resign as they had dual citizenship while the head of the Australian nation-state is a foreign monarch residing in United Kingdom (UK). This may seem incongruous and its constitutional arrangement illogical as one may reflect about the need for continuation of the tacit status quo from a colonial heritage. This situation is despite Australia trying to get the best political arrangement based on its copying facets of the republican model of the USA and the UK parliamentarian system along with division of powers for a mixed representational system. Also, to provide checks and controls, it has an independent judiciary and free press. Yet, the resulting system is seemingly still inadequate for good decision making in parliament as Prime Ministers are chosen by the party's factions rather than elected by the people for a fixed tenure.

Australian version of democracy has issues. With its outgoing Prime Minister in 2018 calling the latest leadership change an act of madness on the national television platform of ABC Q&A programme (*The Guardian* 2018), one may pose the question of how a thriving and exemplar system of representative democracy aspiring to lead in the region has succumbed to such a state of chaotic governing. This is not a corrupt leader that is being expulsed or deposed, and this sacking of the elected leader is in a democracy, where the norm one aspires for in representative democracy is that it is the people that chose their leaders at an election which is held for every cycle, or term as defined in its constitutional arrangement. At the federal level which has overarching responsibility for the federated system, Senate elections are every 6 years with half the Senate renewed every three years, while the representative or lower house members are elected every 3 years. Yet, such chaotic changes to leadership roles are not new as since its federation in 1901 leadership turmoil is a feature of the Australian political landscape. With even one governor, Kerr being involved to initiate a seed of leadership chaos in 1975 by allowing a minority team to govern (Lovell et al. 1998: 113–120), and systemic chaos has grown to become ingrained in the Australian political system. It is the norm in representative democracy that the party with a majority is to lead as approved by the governor, a norm that Kerr rejected creating a new precedent for governing. Now through a party system, it is factions that rule.

The aim of the ruling party having a majority is to retain power or acquire the power to rule. Political parties' conflicts for power are the means to that end despite the elected members claiming to represent the people's will. It would appear that in this immigrant nation in general the 'elected politicians were anxious to sustain and expand the control of their parliaments and to mobilize the majority population, if necessary, against minority groups and opinions' (Jupp 2018: 8). Jousting for power is the norm in the Australian political system which as a consequence is that it leads to public aversion about politics and politicians, especially when a group of 19 votes can secure representation that encourages extreme behaviours.

In this political system, it would appear that the people as a key stakeholder to provide inputs seem to have been lost from political consideration. People inputs may reduce the internal warring for power in these factionalised parties. With people's inputs through a SE application, the focus turns, in this highly industrialised nation, to outcomes mandated by the people for the people and importantly from the people. After an election event, the people whose inputs are the one that should matter the most are a silent impotent mass subjected to chaotic governing. Equally, the rhetoric of the warring parties vying to rule adds to the existing chaos through ideological debates for left, centre or right orientations, a positioning to suit the actor and their cohorts wanting power. Below we explore the political system of Australia by providing a brief overview of a functional system in disarray. A system that seems to have lost its way according to observers (Brissenden and Andersen 2018).

Even though the new person in the Prime Minister's function taking over in 2018 is seeking to retain its party's grasp on power, some commentators are questioning his decision making in recompensing those actors instigating instability (Coorey 2018). A similar argument was presented to the electorate by the Labor party to justify replacing an elected Prime Minister in 24 June 2010 and again in 26 June 2013. The key issue apparently here is that it is not about what the people want but how the Coalition keeps the main contender Labor out of power and vice versa. Each main party would go to great lengths to stake and retain their claim to rule with the factions adjudicating. Given the system in place, the minor parties those with 19 votes representation aim for extremes, so they can exist. Extreme views are even seeping into the mainstream parties from enterprising opportunists. Australian political system requires an upgrade to allow the people's inputs so extremist views are curbed through better government.

The rhetoric for both main parties is that they are the best choice for the people yet some of them also understand that they do not seem to know what the electorate wants and needs. However, in spite of the party rhetoric it would appear the electorate's input is not the party members' or elected member's priority. An example is the recent national referendum about marriage equality which showed in the current representative system how out of touch many of these politicians are in terms of their electorates' wishes. Simply put there is a growing disconnect between what the electorate wants and what the elected member believes from party inputs the public wants and needs. This is resulting in a lack of trust of the public in its representatives and thus a wasted opportunity for improving the communities making Australia and its political system. The e-democracy system improves on the Australian representative model as using SE it facilitates the public's inputs to emerge with digital technology.

Now an e-democracy experiment is not unknown to Australia as reported in the OECD report. The reported use of technology for policy making in 2000 is discussed below for one of the government departments, Defence, as it involved the public for policy making (OECD 2003: 116). There was however no mention of use of SE framework, and more remains to be done for an e-democracy to emerge. Governing in Australia is tending towards instability from both main parties' behaviours and like in other nation-states public trust in the main parties to govern for the people is eroding (Levi and Stoker 2000). E-democracy that democratises is unfeasible when those who are in control the guardians or powerful gatekeepers wish to protect themselves by maintaining a status quo (Qvortrup 2007).

A challenge for Australia is there is a lack of capacity by political parties, as the agents of the people, for new thinking. In Australia, for example, people are captives from free market ideology mantras. This is an old paradigm that promises a lot through freeing the market but it is yet to deliver on Australia's initial social democracy aim at its inception to a federated model. One may need to question this free market paradigm given the many failures that are now becoming apparent. Systemic change defined by the people, and for the people as a new paradigm for people's contributing to their own rule is lacking even though the technological capability exists. As reiterated, a change of thinking to one with a SE framework using technology has potential to upgrade the Australian political system.

Change from the people can be facilitated by SE given the existence in Australia of the communication network from technology like the NBN. NBN provides the digital means to connect every Australian. A common claim by elected members in Australia like elsewhere is that they represent their constituents' views, but if the recent referendum in 2018 on marriage equality is an indicator, it would appear that these claims are unfounded. The electorates voted for marriage equality, and yet as shown post-referendum during the policy debate in parliament, elected members were strongly defending their own dubious views, not their electorate's. A disconnect exists between elected members views in parliament and what people actually want. A means to support what the people of Australia want from their elected representatives like the people's will is feasible from the SE process using technology. Exploring this idea leads to political system improvement, and it does seem imperative given the increasing systemic volatility from factions that disrupts good government. But first, we discuss some of the events that shaped and still shape the current complex political landscape in Australia.

3 An Overview of the Australian Political System

Australia is an island continent settled by Aborigines for around 40,000 years and the first British to visit its east coast was in 1770 with more coming to settle permanently in January 1788 (Whitlock and Carter 1992). It would seem that Australia was sighted by the Portuguese in the sixteenth century and the Dutch in the seventeenth century. A brief visit to the north-west by the English in 1688 was not a positive assessment of the country, but in 1770 the east coast with its lush green changed those negative views of Australia with more Europeans coming to stay from January 1788 onwards (Whitlock and Carter 1992: 73–74). As part of the British empire expansion, Australia became a supplier of wool to

UK in return for imported British manufactured goods (Gare and Ritter 2008), a captive market for the penal colonies. As a penal colony of Britain, Australia post a century of settlement in 1888 was starting to feel the brunt of the convicts forcibly transported to the colonies. In 1833 in a population of around 60,000 in NSW, one of the colonies, there were 24,543 convicts and 26,064 free adults (Smith 2008: 138). Free settlers supported the emancipist's (freed convict's) civil rights and 90% of the children that were born had at least one convict or ex-convict parentage. To clear this shame the politics of convict transportation underwent significant transformation (Smith 2008). The convicts though helped shape the society which was to emerge to its *laissez-faire* form and egalitarianism where humour and the ability to laugh at oneself create some unique synergies. Building on its transportation history, the modern Australia is a multicultural society and its immigration programme continues to boast in 2007 a population of around 21,015,000 that is projected to increase to around 27,000,000 by 2026 (ABS 2018). This mainly migrant originated population is concentrated mostly in the coastal regions and all of its main cities, except Canberra, are on the coast.

Australia was attractive due to its isolation as an island prison (Hirst 2014). In each of the Australian state, as penal colonies of Britain, a representative system of government with governors as representatives was initially established to reduce political tensions in the empire. Each state was a colony of Britain with its own bureaucracy, and the migrants of Ireland and Britain were brought to Australia through a state-assisted system which lasted till the 1970s. Migration was opened to other European countries through a white Australia policy, land was granted to settlers as free franchise but this grant was not extended to Aborigines or Asians (Jupp 2018: 33). Australian federated system of colonies post-1901 followed this white policy in a pursuit perhaps to replace the black indigenous population who were declining in numbers through this massive white subsumption. Some authors allude to a public policy designed to maintain this decline with support from a Catholic and Protestant system derived from a clientele from Britain and Ireland that assisted the process (Jupp 2018: 8). This type of policy influence to subsume indigenous populations is not uncommon in other countries which have been colonised, for example USA and Canada.

British loyalty gave access to land and allowed franchisees to a number of small farmers but the wealthy settlers were in power and remained influential; they still maintain large landholdings (Jupp 2018: 33–34). A class society was created and the fear of Asians perceived as inferior or dangerous instilled and maintained while those who were wealthy Asians and invested in businesses were welcomed (Jupp 2018: 34). Despite Australia having a parliamentary representative system born in 1901 change to this white migration policy was only in the 1970s. Australia signed the international treaties for human rights, rights which were denied to its early convicts and the indigenous population who did not have a right to vote until the 1960s, but also according to some views (Jupp 2018), are denied to refugees. Refugee policy is an emotive issue that polarises the electorate, and when effectively used at elections, it perpetuates fear and diverts attention from other more important policies like the economy, jobs, etc. Political parties are adept in diverting attention or a divide and rule strategy.

Australia has a distinctive legacy of British colonial subsumption that still reflects in changes attempts to its political system. It is reflected in that it arises as an issue time and again as those in power want to sustain and increase parliamentary control by if necessary pitching a majority against its minorities (Jupp 2018: 8). There is a strong resistance from some quarters for Australia to become a republic and delink itself from the UK monarch as the head of the Australian state. This resistance is despite the legacy of Britain to use Australia as a penal colony. There is also pursuit in the courts of elected federal members who as an Australian citizen have dual nationality, but the Queen of a foreign nation remains Australia's head of state. In 2018, for those who have or inherited dual nationalities from their migrant status or lineage, their eligibility as elected citizens to represent other Australians was made invalid by the High Court. This is despite the head of the Australian state being a foreign monarch as mentioned earlier; none of those elected in the main parties challenged this issue though. There are unique tensions in the society from the displaced indigenous populace, convicts and migrant ancestral heritages that are yet to resolve themselves out. These tensions create unique complexities in a parliamentary system where a three-tier government local, state and federal exists to manage the affairs of the country.

Federation in 1901 was through the states giving up some of their powers to create at the federal level a government system for the whole country through a written Constitution that sets out the division of powers. The process to a federation was slow and it was brought about due to the need for defending this part of the British empire (Wise 1913). An Australian army needed to have a central body to command it. This need for defence was due to 'the passing excitement of German interest in New Guinea in 1882' and in 1885 to protect Australia a Federal Council was formed from an Act in the British parliament (Crisp 1983: 5). New Zealand and Fiji seem to have influenced the formation of this Federal Council but along with South Australia both failed to join the Federal Council. There was significant debate about the colonial states' relinquishing of their powers to the gestating federal entity. If it were not for the leadership of NSW against the arguments from the premiers of the other states wanting a status quo, and also without the tacit support of the governors as representative of the Crown in each of the state, this crucial step to federalism could have been messy.

In Australia, if a different path like in the USA was adopted, then civil war may have been the outcome for federalism (Wise 1913). The British had learnt from their US colonial experiment where they were uprooted and therefore to defuse simmering tensions in Australia like for NSW they allowed reform that culminated into federalism. It was a calculated risk so that in the colony the influence for England's control remained from its predominantly British settlers who were used to maintain and sustain it (Jupp 2018: 33). In the 1880s, the idea of Australians as sovereigns may have been a joke of its ruler, the British monarchy. The British manipulation of the system was so it retained its colony or at least a colonial link through an established class system, those with land franchise and those without. A colony which otherwise would have been lost or worse competed against its interests if the tensions in the system had built up like for the USA which underwent political change through a revolution. As mentioned earlier, manipulation of agenda by those in power is hard to detect in a representative system (Lukes 2002). The Act to reconstitute the colonies as one called Australia was passed on 9 July 1900 in UK, and its content ensures that the colony remained subservient to the British monarchy and its successors (Lovell et al. 1998: 865). This is to continue unless a new system is developed to supersede the existing one. Systemic change is still thwarted by the politically powerful few who like in other nation-states are well resourced and set upon maintaining their inherited privileges—privileges that a static Constitution helps them maintain.

The Constitution that was drawn in the years preceding 1901 provided for unification of the colonies making up the island continent. It authorised a bicameral system with a Senate and the House of Representatives. The Constitution was meant to establish a sovereign people, Australians. This is a thought-provoking idea given the head of state for Australians was and remains the monarch of a foreign country more than a century after its formation, we discuss this further below. Subsequently in 1977 through a referendum, the Australian Constitution was amended and a governor-general's role was appointed to represent the Queen and given the powers to prorogue parliaments. This governor-general's constitutional function includes appointing the Federal Executive Ministers, and also judges and ambassadors technically on behalf of the Queen of England. The function also approves legislation from parliament and issues the writs of election by deciding the sessions of parliament along with the dissolution of the House of Representatives. Now lower house representatives have 3 year terms while the Senate is divided into two classes those who are Class 1 have three years terms to vacate their seats and Class 2 are those with six years terms (Craig 1993: 12–13). The number of lower house representative is based on population, and the number of representatives is 150 for the House of Representatives membership; while senators are 12 per states with the Northern Territory and the Australian Capital Territory each currently represented by two senators, thus there are 76 senators (Parliament of Australia 2019). The Senate has the same powers as the House of Representatives. Each constituency elects one House of Representative member to make up the current number of voters represented for Australia. The number of representatives is not fixed, and they are based on a ratio for population bounded in constituencies; these constituencies electoral redistributions are managed by the Electoral Commission (AEC 2019). An electorate has an average of 100,000 voters to elect 1 lower house representative.

In the Constitution, a list for disqualifications of members of parliament is at Sections 43–47. This has an interesting contradiction which needs to be highlighted as even though the British queen is the head of state of a foreign power and Australia, yet an Australian with dual citizenship is not eligible to be a parliamentarian (Parliament of Australia 2010; Lovell et al. 1998: 865–888). This dual citizenship for the Australian citizen may in some cases involve a second citizenship of UK. Two classes of citizenships were formed insidiously with this contradiction that remains. In 2018, both the government and those in opposition used these sections of the Constitution for questioning the eligibility of those citizens elected to represent the people by the people. It would seem even in exemplar democracies some privileges are made to outlast generations of underprivileged Australian citizens. To note that instead of addressing the class discrepancy in a colonial inherited document, the resources of the warring main parties are spent on decimating the ranks of their elected opponents through the High Court. A High Court which was interpreting the appropriate sections of the Constitution that the warring parties claimed disqualified the elected opponents so that each party could disrupt parliamentary sessions. Parliament is meant to work to deliver outcomes for the people when in session, but it is yet another zone for addressing interparty conflicts. These sessions which are short become ineffective to deliver outcomes for the people as conflict is the *modus operandi* of the representative system of democracy.

This citizens' class issue used by the warring political parties came to the fore in 2018 with several elected members effectively being forced to resign. Some elected members despite being born and raised in Australia as citizens of this country lost their seats. Some Australians therefore had less rights than others, while others like the Queen of England and her family, present and future, are privileged in Australia through a document drawn in the 1900s which maintains this status quo. Contradictions for inequality though are not surprising given the white only migration policy was only removed in 1978 by the Labor government, and the Australian indigenous population whose ancestors preceded migrants by 40,000 years were allowed to vote only in the 1960s. Some questions are raised amongst scholars about what it means to be the people in Australia (Smith et al. 2012) and have equal rights. Those issues need to be answered given contradictions remain in the amendments to improve the Constitution. A change to the Constitution was in 1967 when Section 127 of the act was repealed to count 'aboriginal natives' as people of Australia (Lovell et al. 1998: 887; Parliament of Australia 2010). As observed by Groom in 1919, 'the Constitution, was never meant to be a hard and fast piece of machinery incapable of alteration' but it would appear that any change becomes baffling as Menzies in 1951 remarked from 'the amount of muddled thinking and speaking that can proceed' from an educated elite Crisp (1983: 40). Now Section 128 allows the Constitution to be altered but yet it does not get altered as the elites resist, we explain below.

Change is undermined with the strong debates in the media where a few powerful actors tend to prevail to shape and distort key tensions embedded in the Constitution. In the media balanced debate seem rare, opinionated views are the showcase. As had been noted by Norris and mentioned earlier in the book in politics, there is effective use of negative news, and in Australia, it shapes politics like in other nations with democratic deficit. Rather than a factual discourse of the merits of continuing with illogical elements still embedded in an important document like the Constitution which does not reflect modern Australia or its aspiration, these debates are deferred. In Australia, constitutional debates that can be contentious are rarely encouraged by each main parties bent on winning the maximum votes to rule. An example is the demise of the debate for a republic through the effective prosecution of a no campaign using the media; it is anticipated that contradictions will remain until such time as a better political system emerges to allow reason rather than emotion to prevail (Kant 2000). Reason that when reflected in an Australian Constitution is to drive a better political system. However for reason to prevail, importantly a leadership to allow such debates to emerge is crucial. If it is to be a fair and equal society, Australian political reform benefits every citizen. Reforms started in the late 1890s, it was towards a social democracy but its intent is yet to be fully actualised as in practice the leadership seems missing.

This leadership need given the timidity of the main parties is anticipated to come from the people with SE facilitating the people's inputs as described below. A Constitution is meant to be a living document not a dead weight that constrains progress. Even though constitutional reviews are carried out, implementing change is difficult. It would seem the reviews rarely get to improve the Constitution; they are defeated by a conservative mindset that profits a few opposing change in a system developed for a 1900 paradigm (Craig 1993: 51–54). An obsolete systemic paradigm that may persist given for some scholars the 'mediocrity in Cabinet' exists despite 'the present stage of spectacular national development and unparalleled scientific, technical and educational advances' that is not being taken advantage of (Crisp 1983: 88). Thus, unless the political system is upgraded; real progress will remain unfeasible even though the technology can improve the political system. Technology for communication is used by elected members but not in a way to harness the people's will that SE explains below.

The powers of the parliament are listed under Sections 51 and 62 of the Constitution, and it created the executive government as an entity to advise the governor-general in ruling the country. The governor-general appoints those who administer the Ministries and is the commander-in-chief of the military forces. The governor-general also appoints the justices of the High Court. As per the Constitution, money is to be drawn from the Treasury through appropriation bills. States could borrow from the Commonwealth, and the trade between states was made free from duties. The States' public debt could be taken over by the federal parliament. The idea was to homogenise and the law of the Commonwealth (federal level) overruled that of the State where and if there were inconsistencies.

In this federal system of government, there have been calls for a removal of the State level tier given the complexity of policy making. It appears that State legislatures 'have a diminishing and diminishingly-valued role' (Crisp 1983: 86). An argument for change is that seeking agreement from all the states can be problematic given the rivalries between party lines as there could be a Coalition government at federal level and Labor government at state level or vice versa. To the critics it may be that there is a demand for unfettered power by the main parties at federal level; as from a management perspective for a political system, it is not incongruous to have a strategic entity at federal level and then a tactical one at state level and finally an operational arm at the local level. The problem is that of alignment between the three tiers of government, strategic, tactical and operational. Alignment would require common goals, a common vision for the whole of Australia shared at all levels of government. Something is unfeasible with the current political system.

Australia adopted what was a hybrid system with some facets of the British parliamentary system and some aspects of the US federated system. It has however developed a feature of election where every citizen has a mandatory requirement to cast their vote or compulsory voting. This compulsory voting was introduced through an Act in 1924 (Craig 1993: 176); this makes Australia different to the US federated system where one has a choice to vote or not to vote at an election. This compulsion to vote has resulted in informal voting or *donkey voting* at all three tiers of government but in Australia whether this conflicts with the democratic rights of citizens has apparently not been an issue (Craig 1993: 177). So Australia does experiment and attempts to improve albeit sometime with some unanticipated consequences from those changes, like donkey votes, and it has senators from minor parties who are selected with as little as 19 direct votes with the power to shape parliamentary decisions.

The appeal of the mandatory voting system perhaps is that its intent is to more accurately reflect which team the Australian population wants in terms of a majority to rule. In elections of 1903–1917, the First-Past-the-Post (FPP) was used and then preferential voting (PV) was introduced in 1919; however, Aborigines who were descendants of those who first settled Australia were not allowed to vote until the state legislation was corrected in 1962 (Craig 1993: 167). In the federal Constitution, Section 127 of the Australian Constitution mentioned that the 'aboriginal natives' were not 'people of the Commonwealth', and it was only in 1967 through an amendment that this section was repealed (Lovell et al. 1998: 887). There is also a tendency for the political parties to incentivise by enticing the populace through promises which are unfeasible, as sometimes these promises are not well debated or costed in terms of affordability. This has attracted terms like pork barrelling from commentators. There also seems to be a large disconnect about what the populace actually want and what is actually being proposed in parliament. Benefits delivered by the government to some segments of the populace are seen with cynicism as a vote buying activity. At federal level, political parties are focused on winning elections every three years and their decision making seems reflective of a voting market behaviour, with marginal seats being their focus while safe seats are neglected.

Opposition rather than collaboration becomes the norms as the idea is to beat one's opponents through promising with a largesse that may not be sustainable. Tactics to win also involve an undermining of progress through use of fear with rhetoric like for example economic mismanagement to balance the budget. This strategy makes for speculative type policy making as the intent is to beat the other party's promises, and between a two main party system where it is either Labor's turn or the Coalition's the voice of the people is rarely if ever emergent. The system allows and caters for the loudest and noisiest of lobbies and pressure groups which with their significant resources advertise strongly to promote their agenda in the media forums to even defeat government's policy. The most vocal group tends to sway the politicians' agenda, like the mining tax which was undermined by the mining lobby (Marsh et al. 2014). The lobbying is to the detriment of the silent majority. The system defeats under-resourced good ideas as the lobbyist's negative publicity are effective in manipulating the outcome in favour of the lobbyist. Negative publicity threatens those wishing to maintain their power to rule. In the current system, a weakness is the people only gets to choose who to elect rather than what those elected need to deliver that is relevant to their life. The system does not cater for the means to have an agenda set by the people or the systemic process for people's inputs for agenda-setting which is the goal of SE's e-democracy approach. However, before we look into this we explore further some institutions that shape modern-day Australia as a political system.

3.1 Some Key Institutional Arrangements

Federalism emerged under the vestiges of the tensions from a set of colonies coming together to make laws in the constitutional Parliament 'comprising the Queen, a House of Representatives and a Senate' (Parliament of Australia 2010: v). In the early 1900s, the idea to establish through compulsory voting that every citizen is to make a choice rather than voting as a civic duty allowed for a different political system. However, if the citizens must attend a polling event and cast a ballot, they can still decide not to choose to elect a member. Non-voting could be a form of protest against the policies being proposed by every candidate. Here, one needs to understand that in Australia the act of voting is enforced and there is a fine for not voting. However, the citizen is still left with the choice as to give their vote to any of the candidates on the ballot to represent them or have a *donkey vote*. Normally, political parties must convince voters that they must come out and vote, and secondly, give the voters a reason to vote for them. In Australia, voting benefits political parties as every citizen must vote or get fined. Donkey vote as mentioned above is an issue for those who feel disenchanted with every side of politics as there is a general perception amongst the Australian public that the politicians are out of touch with the people's needs and wants. This behaviour is exacerbated from negative news in that it also affects the genuine politician as it taints the whole political system.

There has been accusation of inappropriate behaviour for even the governor-general's function, for example when the Fraser government came in power with what was perceived to be a coup in 1975. Normally in a representative democracy, the majority in parliament is the team allowed to rule. That governor's act of dismissing a majority team in 1975 led to significant unrest, and that unrest was despite Fraser who took over with a minority declaring that he was to govern for the common good for the average citizen. One may dispute the evidence for what this claim was based on, and there is still debate whether Kerr, the governor, at the time behaved appropriately (Lovell et al. 1998: 113–126). Despite the unrest organised by the losing side, the ousted government, which was in this case a majority team in parliament, at the referendum that followed then would tend to show that the Fraser government, because they won a significant majority in the elections that followed, may have done the right thing to step in to remove the majority team. One can speculate if the preferential system had some effect to provide the Fraser government with a significant number of seats. When people are compelled to vote without a clear idea of deliverables, then claims and counterclaims are meaningless. Voting becomes irrelevant as a choice of candidates for a set of deliverable like an agenda defined by the people from the people is missing. Thus, people take a punt or change for the sake of change given their perception of party behaviours at the time.

In a mandatory voting system, the entrenched political parties profit from voters apathy as at times the voter's ideology like freedom of choice or individual's rights are at odds with this compulsion to vote. Also, some entrenched political actors are less equal than others due to an inherited following (traditional voters), thus these actors may not be attentive to what voters actually want. Using polling they are able to predict and thereby manipulate to some degree voting behaviour by incentivising some segments of the populace, their voting market. Also, with the formation of two major political blocks technically the voters have been restricted in terms of who can govern in Australia. Support of compulsion to vote by the major parties is due to a fear perhaps that people may not turn up at the election event. Also, the public provides funding from the public purse for political parties due to the number of votes they acquire at the polls; in 2019, it is 275.642 cents per vote (Australian Electoral Commission 2019). This is a major source of funding for the political parties to reimburse some of their significant costs, namely advertisements that they incur during an electoral contest. Mainly costs are for advertising on the media channels and for supporting their campaign activities like letterbox drops of party's agenda or promises. There are also now web pages and social media used for promotion. This chase for votes makes organised parties into vote maximising enterprises that are in the business of seeking power for their own party's agenda. This party agenda may not be aligned to the people's agenda making even Australia in practice fall short of in terms of democratic criteria as a large scale democracy (Dahl 2005). Dhal questions how people can have effective control on the agenda so they can have an input in government decision making in democracies; this agenda control cannot be addressed by everyone just voting for an individual or a party. In Australia, voting is limiting democracy's evolution without a people's agenda.

Australia has tried various institutional voting arrangements to try to reflect voters' intent more effectively, and it has a preferential and proportional system for voting. An argument is the FPP system provides stable government but it profits large parties as in a multiparty system a member with a territorial base can be elected with 35% of the votes despite being rejected by 65% of the voters; this percentage can be even smaller if there are more parties (Craig 1993). In the preferential system, voters are made to also choose in terms of an order of preferences all the candidates on the list. These preferences are then tallied and the candidate with the requisite 50% and one vote is elected, this can produce very different results from the FPP system. Australia has experimented with proportional representation which is based on the preferential system variants to more closely reflect voters' choice of a candidate after preferences are distributed to meet the seat quota for that state. In Australia, at federal level, there is a combination that is used, with PV for single-member representation like for the House of Representative and proportional representation (PR) for the Senate seats quota.

At State level, state-wide, a proportional representation is for the legislative committee while a PV is used for the legislative assembly. The electoral reform was to reduce the impact of FPP on minor parties at state and federal level. However, at the local government level though, the FPP is still used. Despite such arrangements getting changes through parliament can be challenging. Changes require both the Senate and lower House to approve the change being proposed. With interparty rivalries and intraparty factionalism, the good of the country gets side-tracked. This is despite a conservatism ethos in the Coalition ranks which further complicates and similarly the leftist one in Labor ranks. The rivalries of individuals in political parties have led to claims from commentators of a lack of vision for the country from the political class. It has also led to an erosion of trust of the public for the major parties and politicians in general. This is despite significant communication infrastructure in place like the NBN and an ability for the government of the day to implement strategies like a more participative form of government. Thus, even though voting has been reengineered as Norris recommends (Norris 2004), this has not been sufficient in itself. With new technology that is now available, further improvements are necessary to get the people's inputs to parliament. Representatives must be informed about what their electorate needs and wants are, and there is technology which with the SE process can assist to provide them with a common Australian agenda for debate in parliament. A change with SE allows for common goals (Requirements) at local, state and federal level thus with a Requirements process a potential alignment for the three tiers of government.

In terms of representatives, the Australian federal parliament elects at the federal level 76 senators for the house of review and 150 lower house members, or it is a 226 members institution for decision making (Parliament of Australia 2019). These 226 elected members are for a projected population of around 24 million which as per the Australian Bureau of Statistics (ABS) estimate will grow to 27.237 million by 2026 (ABS 2018). In 2018 based on voters, NSW elects 59 members, while Victoria has 49, Queensland has 42, with 28 in Western Australia, while South Australia it is 23, and Tasmania has 17, with 4 in Northern Territory and 4 in Australian Capital Territory (Parliament of Australia 2019). Australia given the large distances is yet to use electronic voting even though through the NBN the capability exists to implement electronic voting.

Some forms of electronic voting as per some studies conducted by scholars are not trusted as mentioned earlier but as mentioned it is anticipated if people understand what they are voting for that is a people's agenda, this resistance to technology may shift. The focus of the voting becomes the communal agenda, an Australian agenda that one contributes to and it is the agenda that incentivises voting. Technology is not for a channelling of votes only or just e-voting for selecting a candidate with a party agenda that may or may not be delivered as is often the case. Voting at an election should not be about party politics. Rather voting is to reflect the needs and wants of the community which must be transparent and its delivery accountable to every member of the public. This reform may require a consultative approach that technology like the ICTs enables through a SE framework. First, we discuss a consultative approach that was experimented in Australia using ICTs and is reported by the OECD.

4 E-Democracy Report of an Australian Consultative Approach

Technology use for public consultation in policy development is not new to Australia. Australia was amongst the twelve countries where the OECD examined the use of the ICTs towards an e-democracy experiment. Australia as an OECD member provided expert support to the team compiling a report on ICTs use in twelve e-democracy case studies, and OECD

(2003, 73) describes a public participation activity in policy making where Australia made use of technology. The report identified that one of the challenges was effective integration of the information throughout the policy decision-making process. The Australian case study described the public input for the 2000 Defence Policy. This review was being conducted by the Australian government. In 2000, for three months July to September, consultation was carried with the Australian public. The consultation requested Australian views about the Defence Forces future direction; the review teams had both public meetings and invited online submissions. The team received 1100 submissions with 80% being from individuals, and the technology used was email.

There was no dedicated (customised) technology created for the task except for an email address, and information about the submission was provided on a defence web page. It is interesting to note that the report mentions that the vast majority of those who participated were keen to participate and build upon that initiative for a policy-making process. The positive outcome is encouraging for a SE process that designs the technology to develop policy making using inputs from Australians in a systematic manner. The report notes that there were no obstacles and the engagement developed a high volume of traffic. There was no mention of the SE framework being used in this policy-making experiment (OECD 2003). In politics, where there is a need for transparent agenda-setting by elected members, this event showed use of technology is a feasible process as a policy-making experiment. An Australian leadership or institution for growing this process further is necessary though to deliver this agenda-setting function, we explain further.

Australia has a governor-general function albeit one for constitutional reporting to a foreign entity like the Queen of England, currently a vestige of its colonial inheritance given a lack of support in parliament for change. This role's monarchic link seems to be a bone of contention between the Coalition and some Labor members (ex-Prime Minister like Paul Keating) in whose view Australia is a country that is mature enough to sever this link and go its own way. However, the timid approach of the Labor members and the antagonism of many members in the Coalition and other fringe parties in parliament have resulted in a stalemate. The issue is perceived as being an emotive one rather than one for change to a better system, and it therefore remains as unresolved. The role of the governor-general with a SE enabled system could be enhanced though as a Functional Configuration Item (FCI). This FCI with SE embedded processes is an upgraded function of this role to ensure that democracy prevails through the enhanced political system. When transformed, this FCI function uses technology to monitor if the executive and parliament are on track to deliver a people's agenda that could be generated through a SE approach. In so doing the governor-general's office becomes the overall functional head to drive effective outcomes mandated by the people at every election or the primary FCI. This upgrade would use support from those that reports to the function, lower FCIs—the executive or Ministers and CIs—as elected members from each electorate making up the political system. Basically, a monarchically delinked governor-general function with additional resources becomes the protector of democracy or people's rule; this political system means changes to the Constitution. It could start as a pilot project for a consultative approach.

The existence of a significant military presence in the form of infrastructures across Australia would supplement the reach of a pilot project for a consultative approach to develop and enhance the agenda-setting process building on an initial experiment in policy development for the Defence FCI (OECD 2003). The development of the primary agenda (key FCI) in itself would make Australia as a democratic oriented political system a systemic leader in the region; it is a role it aspires to and for through development aid. Email technology is a means that political members (CIs) commonly use to both respond to issues from constituents or as a channel to disseminate information to the members of the public, and thus, the technical infrastructure (226 CIs at federal level) exists for elected members to seek inputs from citizens using digital means. However, the data capture of the inputs is seemingly sporadic as it is yet to be homogenised throughout the country, a potential tool for the data capture and processing is suggested at Appendix 3. Technology has improved significantly since the experiment conducted by defence for policy making in 2000. Also, the costs of those digital technology have plummeted, see Appendix 3 link. A degree of automation using AI for data collection and processing is feasible now.

Data is the key aspect of an evidence-based approach that a governor-general's office may seek for understanding the needs and wants of the people and then ensuring as the primary FCI that as the oversight arm of a government for the people, it approves for whoever is elected by the people the outcomes that relate to the delivery of the people's agenda. The office of the governor-general's function which has oversight on every activity of parliament thereby becomes reinforced through the availability and access of the agenda (specifications set or FCI) generated by the people at every election or a people's mandate. The governor's office with an electoral mandate of the people generated through technology would ensure that the parliament, and Ministers each sub-FCI, are operating along the outcomes defined by the people or within its electoral scope.

This FCI would effectively under those circumstances defend the interests of the people as a collective improving democracy. Transparency is from its reports on the delivery of outcomes at the end of each mandate or during the mandate at specific intervals. The report from the governor-general as FCI can be debated by each member in parliament during their tenure, and when seeking to be voted to represent the people at an election event with their electorate for their mandate's approval and renewal. Other people competing for election can also debate their version of a people's agenda. This new FCI function that SE enables would shift a degree of control back to the people and each constituent. It would improve transparency and accountability in decision making at parliament. Scope creep in parliament will have to be justified and bad decisions re-examined for improvements. We now explore the costs of doing nothing, in the context of Australia and its current arrangements, where decision making that is meant to benefit the people rather than people's generated agenda is through a party's agenda. A party system bent on conflict as its core activity. Conflict that is wasteful for delivery of people's outcomes given at times non-decision making from conflict can assist ongoing bad behaviours. The SE process upgrades the current political system's decision making that is inadequate.

5 Cost of Inadequate Decision-Making System

Government mostly regulates to control, and regulation is developed with an intent to change behaviours as issues emerge over time that requires society to be protected. Laws constrain as there is a threat of litigation. Parliament normally legislates to ensure that the laws are effective in providing for individuals and businesses a framework so that they can go about their affairs in a secure manner. It provides room for both individual and business to grow. In a defective design which is the representative democracy system, there is a limit to what laws can do in practice as litigations are costly. Also, at times bad legislation can cause more problems than it addresses. It conflicts with other laws for example. The only winner is the litigation industry growth, which the well resourced can avail of. A better outcome wherever feasible is to design out the problem without overburdening individual citizens. Now individuals have rights and one of the ways government encroaches over these individual rights is through legislation. Thus, in the current system bad governments can persist by legislating to protect a system which allows them to rule unfettered and with impunity for their bad decision making—bad decision making which each parliamentarian is linked to given the system in place.

Constitutions provide a certain degree of protection to individuals but as technology improves or new knowledge allows new thinking, further improvements are feasible yet cannot occur due to legislation which protects a few (well-resourced entities) against the interests of the many. Also, these initial protections perhaps even with well-intended checks and control in for example a less technological environment, paper-based compared to digital, these become burdensome to progress or even obsolete at times. It may seem an incongruity in that many political parties call themselves progressive yet they are regressive when in government due to their increasing government coercion through more legislation. Many of the imposed legislation or coerciveness is due to political parties' lack of a holistic or systemic view of the impact of the legislation on people's lives or simply put a lack of alternatives to solve the problem differently based on science rather than rhetoric. Government and elected members through non-decision making can also exploit a representative system's weakness to perpetuate blame rather than actions meaningful to solve what the people need. A solution that designs out the problem is more effective. SE is a design methodology to create solutions for a defined need.

SE provides alternative solutions that can be discussed and a preferred solution from the set of solutions is then selected as mentioned earlier. Though academia has for long been a rich source of ideas, it has not been sufficiently engaged in the realm of politics due to nature of the system in place, namely technology that can become an enabler through SE for delivering outcomes for the people. Few academic institutions investigate the design of political systems and its systemic upgrades using technology with a SE framework. System design is normally a realm of engineering like computer systems or technological systems. Over time enterprising teams or political parties have hijacked the people's authority through the weaknesses of the political system in place leading to inadequate decision making; it is an abuse of power delegated by the people. This pattern of behaviour may have resulted in significant waste of the countries' resources, as the solutions are not the optimal ones from a set of SE generated solutions that get implemented.

The bad decisions termed as inadequate are like a thought bubble. If these thought bubbles are untested rather than linked to what the people want or need, they are risky for the country as some problems could be perceived rather than linked to real needs and wants. Need and wants that are not being effectively captured and progressed in a systemic manner described by the SE processes, and it increases the risk of bad decision making in parliament which can be working on perceived issues. The parliamentary decisions given its current design which is from monarchic roots in its operation are coercive, it excludes the people's inputs, it is imposed and not suited for people-centred decision making like proposed by SE. The relevance of some of the decisions reached in this manner becomes questionable when technology can improve decision making.

The current system does not allow participation or consultation in government decision making by the people. It is designed along monarchic line that imposes, excludes rather than includes. Now for a system change, this is deemed to be too hard, and thus through non-decision-making it allows a dysfunctional system of government to emerge and if left unaddressed to persist. The changes which are not mandated by the people like a new Prime Minister cannot become the prerogative of a 'coup' from factions in a political party. Once the people have decided through an election event about who represents a constituent, the agenda to be prosecuted by each constituency would also have been established, and then it is delivered through parliament. Warring factions within a party would have to develop inputs to the agenda in a transparent manner. These factions in a SE framework would be treated in a similar manner to any individual who is a citizen of the country. Pressure groups and lobbyists with undue influence would not force their views upon the country as a whole, with SE their motives and agendas would be transparent to the public at large, and it would be openly debated amongst the free press and academia. Scope creep from the people's mandate to support such pressure groups or lobbyist would be apparent to all citizens given use of SE processes that are traceable from requirements to specifications (policy/legislations). With SE the inputs from the powerful would be transparent.

Elected members links in promoting and delivering to such vested interest, the privileged, would become public knowledge with the public deciding at an election event about the effectiveness of the member as a representative for maintaining office. Such transparent debate is anticipated to ensure elected representatives are focused on delivery of clear outcomes for their constituents, for example an increase in crime would not be tolerated by the public as only a need for more police but rather as a need for addressing the causes of the criminality increase. Decision-making would become accountable and those seeking office would no longer be allowed to vote along party lines or pressure groups' agendas, a police station. As to the public at large, such behaviours would become exposed and transparency result with appropriate outcome from action by the public at the ballot box. Members seeking public office will need to understand and focus on what its electorate wants and needs and then justify through their decision making how they delivered those outcome or contributed to its delivery during each term in office. Representatives will have clear goals for delivery and thus set agendas during their mandate.

New ideas promoted by the office of the representative (a CI) would create a new dynamism that focuses on effective decisions of the people and for the people in parliament (as FCI). This is a goal often mentioned by every elected representative but rather poorly achieved in practice as more often than not these members do not understand what the people really need or want. A process flow described below allows such needs and wants to emerge for each constituency (CI) in Australia (as FCI). Over time these needs and wants data would grow, and if that data is captured, it would show patterns that would allow elected members to deliver by anticipating the people's needs and wants. Then using forecasting, when derived from these data, it is possible that even the future needs and wants of the people or agenda items (FCI) can be anticipated (political intelligence) to some degree.

Uninformed decision making has costs which are revealed down the track. Decisions made have a life cycle. SE makes the costs transparent as it has a life cycle approach to designing solutions, also the citizens must agree to the designed solution (agenda) which then a parliament legislates and implements in a transparent manner and which the governor-general as the primary FCI ratifies. Cost of legislation creates at time unnecessary red tape that can constrain as opportunities are wasted during its life cycle. These legislations can also be identified and alternatives sought as the SE process mentioned earlier provides a feedback process. Liberalism was meant to provide the space such that the people could have a private sphere to manage their affairs, and it was meant to create more freedom. The role of government is to empower its citizens so they can use their efforts freely to create the good life. The first step is to understand the people's needs and wants through direct inputs from them as the SE process Requirements does. The current conflictual nature of parliamentary debate cannot be allowed to continue as being along party agenda lines it encourages resources waste. Meaningful debate about a common Australian agenda (FCI) allows for a progressive government. SE can create both the technology and framework for improving the current system. In a democracy, politicians as CIs are to serve the people, so they must accept that they rule and they are ruled in turn by the people (FCI) through an agenda.

5.1 Cost of Bad Regulation

Regulations implementation have consequences that cannot be predicted when they are legislated, decisions as well as non-decisions has cost over the life cycle of the regulation. In a fast-changing technological environment, inadequate decisions can be detrimental for innovation, and it makes obsolescence the norm. Over regulation for example through strict standards in SE had to be relinquished to enable technology to provide innovative solutions. Lessons learnt from SE was to facilitate through guidelines which meant it had to relinquish the degree of control for a standards that were for a low technology paradigm, a knee jerk response to a problem in society tends to be regulation. Regulation is coercive and it creates more growth for litigation rather than long-term solution. Rather than using a SE framework to investigate various solutions and improve or design out the problem, elected representatives exercise powers to coerce through regulation. It is a pattern of behaviour that creates a society that is burdened to the point of too much red tape such that innovation is discouraged. Good regulations are rare, given the tendency for conflict between the two main parties bent on self-promotion to rule. Good outcomes are not feasible given a lack of a set agenda to which the public expects both parties to deliver. As a result of this need for conflict to win or disrupt each other, both business and society are subject to bad legislation that squanders scarce resources from taxpayers by different ideologues prevailing at the time and in those parties. Policy making even with good intent in this sense is deficient for good outcomes, wasteful.

As an example of profiteering, there is Australia's social security policy mentioned by Bob Brown that when it was applied led to alleged fraud. It was a social policy meant as a safety net but seemingly five doctors including specialists exploited it as they conspired 'to defraud Medibank and the social security system' (Brown 2007: 55). This case was reported on 2 April 1978, and it would appear the incompetence of the system was revealed as the case of fraud turned into a political farce. Brown a practitioner in politics reported numerous cases of decisions making in government; they describe the way government made policy to meet their own objectives. Some in academia even go so far as labelling the Australian taxation system design as 'perverse' for redistributing income from the poor to the rich (Brown 2007: 106). For a nation that aspires for fairness this may be problematic. Policy making in the way it is debated, which is confrontational, does not meet the people's objectives. It would appear that its application leaves room for improvements from the outcomes of the policies in practice, alleged fraud. Over the life cycle of a solution, the SE process is an applied approach designed to ensuring needs meet outcomes. The solution actually works.

Now for example it is due to a powerful few that some politicians cater to, climate change is assumed to be 'fake'. If the system in place remains, this contentious issue about damage to the environment from humans is one over which the parties can argue for decades to come. The insidious creeping of this idea as fake can be alluded from an undermining of scientific or science in general from a minority bent on resisting what is undeniable to some on a scientific basis. Resistance to agree provides media coverage to all those engaged in the conflict, and the longer it festers the more coverage to those protagonists. Issues are created to become electoral platforms as they tend to polarise public attention. However, it makes the debate a confusing one as rather than enlightening the public at large, the reasoning is lost superseded by alarmist propaganda, positive and negative, and this derails a true understanding of the issue.

Such stance against believing in climate change from those in power to rule ends in risk taking behaviours by a few to the detriment of the future generations. It is detrimental given the evidence from the data collected by scientists that shows a pattern of warming temperatures. There is tendency perhaps a conservative one, for sticking to obsolete ideas. This is normal as some politicians are averse to risks from change, it affects their vote banks. However, such mindsets for those who rule have consequences as now shown by the free market mantra conservatives tend to believe in, which critical citizens are questioning for its failings. The mantra from a belief that somehow the market is altruistic and will cater for the good of society perhaps reflects the type of myopic organisations that political parties have morphed into.

It is a paradigm which is proving itself wrong as organisations in a free market are profit driven for shareholders, those profits which may come at a cost to society like pollution. Coal powered station must deliver a return on its investment to its shareholders despite better technology being available to replace it. After making big claims for outsourcing of energy policy to a free market mantra, today Australia is closer according to commentators to a third world country with blackouts during peak period. Threat of blackouts can be used to coerce society from the bad decisions made earlier by those who allowed this to become a norm for political gain. With electric cars, as the technology matures, becoming a new norm for the future such market mantras show the limited and flawed thinking in parliament. Politicians normally take an adverse position to their opponents as this creates media coverage, and a market approach is a good cover for their bad decision making when they fail to adequately address the problem they or their party contribute in creating. Costs of the free market energy paradigm are on society given they are subjected to the decision making by a few in parliament who did not reveal the risk it involved. Outsourcing certain key capabilities society needs like electricity has risks over the life cycle of that decision.

A short-sighted approach that the market will provide despite lessons learnt from New Zealand which had massive blackouts during the peak of winter did not get noticed by the policy makers in Australia. Now the downside of electricity outsourcing has created due to non-decision making in Australia blackouts in two states. Even though Australia produces gas, its supply or security of supply for example has become an issue as the free market policy fails each and every Australian including decimating its manufacturing sector requiring gas from the market-driven approach where prices spiral out of control. Gas companies with a free market mantra seek higher returns from their overseas markets to the detriment of the people of Australia. They argue for higher prices from Australians and thus make the manufacturing sector uncompetitive, a collateral damage. Those elected to power caught in their own idea of short-term perceived efficiencies of a free market puts the whole Australian community at risk given a lack of long-term risk assessment or systemic competency in their decision making. The market is working for shareholders benefit and that is a profiteering organisation bent on optimising shareholders profit. Bad decisions put the Australian community at risk as the system is not holistic in its thinking. Also, as there is not set agenda from the people, there are many distractions for those elected to govern specially from the minor parties who distort the parliamentary debates. Often the system given its minority party promotion, for plurality perhaps, delivers senators who are people with very extreme agendas. These agendas impose costs to the tax-payers as these individuals have the power to disrupt decision making in the current political system.

One of the minor parties that the electoral system facilitated for gaining senate seats attacked minorities to provide it with a migration policy platform at election. Its propaganda, for example, by the leader claimed 'that 70 percent of migrants were Asians when the actual figures was about 37 percent', and it implied that Australia's population would double by 2050 when the official projection from the ABS was 28 million in that year (Brown 2007: 364); this party was rubbished for its ludicrous policies by the Sydney Morning Herald as polluting 'political debate'. However, the party through strategic manoeuvring for preferences elected more senators in the 2016 election with a senator with 19 direct votes as mentioned above. The electoral system improvements with preferential and proportional voting for allowing minor parties to emerge may not have anticipated poisoned political debates as an outcome from such minor parties in their bids to get elected. Such debate creates tension in a multicultural society and stresses the political system, from polluting the debates. Without a defined agenda from the people, if there is enough support for negativity, such behaviours will increase from the design failure of the system. A static bureaucracy mired with chaotic oppositional decision making in parliament is ineffective to improve the system.

Organisations' changes being shaped by decisions which may appear to be for improvements when they are implemented are inadequate for as new technology is created these become obsolete. The public sector, for example, with its performance measure of meeting targets rather than doing found that if 'many of ...[those targets] have not been set very intelligently, [it] is a sure-fire way of not improving public services' (Gray and Jenkins 2004: 269), performance measurement may even hinder change if the wrong thing is being measured. A recommendation for improvement was that new visions for the collective should be forthcoming seeking with it 'innovative forms of user involvement and decentralised, democratic accountability' from a partnership to be created (Gray and Jenkins 2004: 270). It would appear that measures sought to improve without end user involvement may have adverse results that are not anticipated when they are operationalised. Other scholars have disparaged the public service as lacking vision from their lack of competency mentioned earlier (Crisp 1983: 87–88); however, arguably if the system constrains by being defective, then the system actively contributes to enabling the current state of things which is bad decision-making. Like in other countries with a representational democracy, systemic weaknesses allow scare mongering, and this represents a paucity of ideas when better systems could be designed like through a SE upgrade.

As Wright puts it, if one is within a given paradigm every solution that one explores is constrained by the framework, one is thinking within (Wright 1994). That is the way a politician thinks will dictate the type of decision-making or solution that he or she will think is feasible. Thus, society is ruled by a few individuals that limit its progress from being subjected to the limits imposed through these powerful individuals. If the system is designed to perpetuate the flaw, then it becomes difficult to remove these enterprising parties and individuals from power. In the 1960s, Dhal predicted in a sense that the US system had become dysfunctional due to the entrepreneurial nature of the two major blocks bent on fighting each other rather than working towards people-oriented goals (Dahl 1995). This is despite technologies that exist today which should have allowed for better solutions or systemic change. A systemic failure would have been averted if according to Dhal creative solutions were forthcoming from the political leaders to engage the public in novel ways and improve on their own ideas for the country.

In fact Dhal goes on to remark that too much depends on the US President in his country for the role to perhaps be able to deliver given the dysfunctional Congress where individual political entrepreneurs do their best to follow their own agenda (Dahl 1995: 391). Though the Australian system has borrowed from both the USA and UK models, it is noted that the system does not operate with a people's generated agenda, and this is despite technology being available to allow that and yet no e-democracy policy seems to be forthcoming (Moss and Coleman 2014). Australian political system is no exception to the above issues of inadequate decision making. Volatility is becoming a norm in the Australian political system; leadership musical chairs are signs of systemic stress. The stress is becoming evident with quick overthrow of elected leaders as distractions for these enterprising individuals whose factions take turns to prevail. Subsequently, these displays of internecine warfare further reduced trust of politicians as Australians are forced to watch the madness emerging from these flash changes of leaders which they as the citizens elected to run the country. Elected leaders are changed in a flash action by factional ideologues.

Privatisation ideologies can be dangerous when combined with the pressure of decision making when in government. Negative news can distort the message or idea being debated with news oligarch having undue power to shape and manipulate outcomes. In a system like Australia which is highly connected negative news can be detrimental. Mood of voters seems to be a major factor for leadership decisions in the two main parties (Crisp 1983: 53). The private enterprise pursuing capitalism has had 'extraordinary successes' when in conflict with government. The outcome is that private rights have overcome through a re-interpretation of the Constitution nationalisation policies from government (Crisp 1983: 54). A need for a forum is required where the ideas can germinate and be researched prior to being tabled in parliament. Bureaucrats are mainly in charge of most policy making in government, and this bureaucratisation tends to have its own culture and also offers significant resistance to change in Australia. The chaotic relationships between a powerful permanent bureaucracy, news oligarchs and the transient politicians undermine the needs of the society as change for the long term rarely takes place in this system. Democracy needs another pillar in the current context.

The pillars of Australian democracy are an independent judiciary, the elected members representing society, and a free press. Now given the use of technology to improve both society and individuals' lives, a fourth element as another pillar is suggested; it is a people's agenda developed in a systematic manner. There is a need for independent input from the academia whereby ongoing research is the plank on which society's needs are met through a people's agenda. Schools and universities already provide a key function to society, which is that of educating its members. In Australia adopting and improving upon the SE process is therefore a natural fit for this functional area in society that already engages with members of society to improve their knowledge and skills, to train and also to create new. Technology is evolving so fast that the SE framework needs to be entrusted to the field responsible for both its creation and dissemination through skilling current and future generations, and it is a role that places of learning can assume. The following example shows how new technology can quickly make some decision makings obsolete.

An elected state member was approached concerning the behaviour of an electricity supply company in Australia. Solar panel on a house produces electricity and the argument was that it seemed unethical that while the producer, homeowner, produced the electricity the retailer of that electricity would charge the consumer (homeowner), for example 14 cents per kilowatt for electricity consumed at off-peak rates and then purchase the electricity produced by the homeowner at 11 cents. These retailer rates would then increase at peak time to nearly 4 times of the off-peak rates with no change to the homeowner producer rates. The elected state's representative for the constituency when contacted via email sent the issue to the Minister in Charge. An initial response in bureaucratese was the result through the member's office which technically did not even address the issue. The issue was raised again via email, to clarify the lack of transparency between what was being produced and consumed by the homeowner and the exploitation that was in progress. A second time the office of the Minister was contacted through the local representative and the bureaucrat in charge was informed to respond to the query. The bureaucrat requested data assuming that the homeowner did not have such data handy perhaps. Data of production and consumption was then provided to the bureaucrat from a digital application that the solar electricity company had installed. This data was sent to the bureaucrat who had promised that it would be used to address the issue if there was data (evidence). Response from the bureaucrat was tardy.

After a few months, the homeowner followed up and was informed that nothing could be done and that some policy was in the pipeline that would ultimately address the issue. The homeowner was also informed that the retailer could set any price (free market) that they wanted for the electricity that was being produced from the solar panels. The retailer now was also the supplier of electricity to the homeowner and also a major producer of electricity in the state and Australia reporting large profits and strong share prices. The homeowner, through a friend, was shown an electricity bill from the Sydney area where the solar production and consumption were clearly listed but this was not provided in his case by his supplier. His electricity retailer even though it was amongst one of the top producers and retailers of electricity in Australia did not provide such transparency. Furthermore, the homeowner was also informed that the issue alluded to which was electricity pricing was a federal government issue and not a state government issue. Now, the office of the federal member had previously been contacted through the federal member and the office had advised that electricity pricing was a state issue.

Such messy policy making is not uncommon, and it exists even when there are consumer protection laws. A second example is for obsolescence in understanding technology. A brand new car was purchased and even though the consumer protection tribunal was approached about a defect from the software in the car discovered within 3 months of purchase, the consumer could not get a refund. This is despite the car, the latest C250 model purchased was for $87,000 promising to a consumer a unique driving experience, but had a faulty software issue that could be potentially dangerous for those driving the car. There was no repair to fix the problem as per its warranty. The onus from the retailer's product failure was placed on the consumer asked to prove how the engine managed by a software was failing, or in simple terms the consumer became a test driver for the C250 software failure modes. This condition was imposed by both the retailer and manufacturer despite the retailer noting from the car's own report that it showed that there was an engine failure which was being captured by the engine management system at each failure events. These failures which were random could not be simulated.

The appropriate government office when contacted was shown the report and it explained that a microsecond glitch occurred as was shown in the report. It would appear that the person who was the bureaucrat did not even understand that software fails in less than a microsecond. The second line of defence from the bureaucrat was that such failures had not been reported. Now with more and more technology being integrated the skill set of those in the function of decision making allows such exploitation of the consumers as there is no requirement for the manufacturer to list transparently such failures reported for C250 cars it manufactured. Seemingly the consumer protection law in Australian was devised to protect but yet in both the above two instances failed to do so. There was no fix to the software problem which was sporadically shutting down one of the four cylinders of the car engine and it had recurred several times over a period of 3 years. Despite various troubleshooting and promises of software updates, the problem remained for that brand new vehicle purchased with a warranty. A protracted length of no solution to the engine problem culminated with the warranty period about to run out, the car was ultimately accepted by the retailer for a token price of $33,000 in lieu of a refund.

While for the electricity issue the bureaucrat was hiding behind legislation at different levels and not addressing the contentious issue of excessive free market power, in the case of the car failure it was a lack of knowledge (obsolescence) of how software fails and that when it does the failure is catastrophic. There is no dragged out record of its failure reported, software either works or it fails. In both instances, the decision-making system allows the customers to be exploited despite laws in place to deter such exploitative behaviours from the market. The market is empowered from its extensive resources to do what it wants and also freed (free market mantra) to do it. Technology (like political intelligence) is pervasive, neither positive nor negative. However, in the mentioned examples, the skill for managing these technology are perhaps beyond those of the decision makers. The result is costs for end users who are being limited from the decision maker's (bureaucrat's function, FCI) level of know-how (intelligence) or scope of action. Technology (political intelligence like AI) could assist the decision maker (empower) to improve. However, this functional improvement assumes that the political will exists to create a technology to assist good decision making for reducing costs to end users, the goal. Costs that could be safety related for the consumers of the car with a faulty software that operates the engine. An engine whose failure to power the car at the wrong moment would contribute to bad decisions at high speeds on busy roads. Regulations are failing to keep up as the people in the decision-making role are not skilled or perhaps lag to understand the technology they are dealing with. It makes for a case for a SE life cycle approach to regulation implementation and update, a life cycle demonstrated in the previous chapters which has a 360° feedback for ongoing improvement.

5.2 *Citizens in Search for a New Democratic Institution*

The state collects massive amounts of information on citizens with those who are more vulnerable like welfare recipients being most at risk. This is a form of inequality especially for liberal-oriented political system where due to the state's coercive powers one must remain wary of the state's intent in collecting information from the impact of policies that target a party's vote banks. To adopt a system paradigm, in the USA a large nation-state, Easton seemed to have been prophetic, when he suggested that universities conducting research should be the reformative driver for new thinking about political systems as the collection of facts has become problematic and the crude tools for eliciting and collecting information required to be refined (Easton 1957: 115). He was in need of an institution that can be up to the task to design a large system for better government which would cover the entire continent like the US nation-state. Despite improvements in technology e-engagement and e-participation to collect facts for an e-democracy, a digital political system design is still a challenge yet to be resolved even in modern democracies like Australia due to a lack of an appropriate framework (OECD 2003). The government for some scholars still lacks a coherent e-democracy policy for public engagement in policy-making as flagged earlier for the UK political system (Moss and Coleman 2014) and as reported by some politicians they are concerned that such engagement might open 'the floodgates to direct democracy' (Coleman and Gotze 2001: 15). The nation-state of Australia is no exception where its laws are used to defer reform, for example privacy laws when used as a crutch to support the status quo given direct democracy is a threat for some politicians.

Politicians seem fearful about change, and this fear is for their positions as elected members, politics is seen as a career. Representation is fundamentally a job. It is not a service to society even though the politician's rhetoric may be otherwise. In Australia, this elected members' fear is translated through the removal of their leaders as a stopgap measure to maintain the party in power when polling seemingly indicates that these leaders cannot win an election for them, or save the old furniture as the media tends to say. A state of affairs where excessive concerns about polling has resulted in five leadership changes of sitting Prime Ministers (primary functional role or key FCI) over a decade. Polling in Australia is seemingly a common excuse used to undermine the people's will in choosing who their elected leaders are.

Polling is an indirect measure at best and is used with media assistance, in Australia, to terminate the tenure of the incumbents in office many of whom are legitimately elected by the people as the preferred leads during the election term. Now negative news impact is not new as mentioned earlier by Norris when measuring democratic deficit. However, the extent of use of polling, this crude device for measuring perceived populism distorts the political system towards power struggles rather than policy outcomes or good deliverables by government. Under those circumstances, the elected team can at best maintain a status quo rather than meaningful change in anticipation of improving the future of the country. Leaders become fearful of loss of their position rather than focused on improvements for the country. This preoccupation with their seat results in that good ideas rarely emerge or long-term visions of the country remain illusory as the constant goal being sought is for the elected member to maintain and retain office or job permanency.

Under these operating conditions, an alternative must be found to secure a process for the emergence of good ideas in a context where changes that are meaningful to citizens are identified and prosecuted by the elected during their tenure. The SE process framework adopted in the context of an existing political system would allow a new paradigm to facilitate and nurture on an ongoing basis the interface between the people and the electorate office that represent them. Academia through universities is an independent arbiter which can also improve on SE itself and the tools that SE would require to invent and innovate using technology for this interface activity, like AI. Given the size of Australia and its complexity, every university (FCI) can fulfil this interface role by providing a service in terms of developing the key themes for the region they are located or tasked for in terms for their relevance to those constituencies (CIs). Universities in Australia are the hubs of cutting edge knowledge creation and technology, thereby they can provide a key support to the communities they are already serving. Now, this is assuming the political will exists in Australia to allow this reform.

Such an interface would however reduce the influence of the political parties to some extent. It is a move that may be resisted as it disempowers a few making policy, the traditional gatekeepers to power, factions and pressure groups. It shifts policy making to a more pluralistic rather than partisan-based approach. The use of SE as framework also shapes pressure groups and factions within parties as every elected member become accountable to their electorate with transparency reducing the opportunity for corrupt practice in government. Policy making through transparent and repeatable SE processes that are independently recorded and aggregated would deliver better government accountability in decision making. The system increases pluralism to the individual level (CI) for an input in decision making. In an information era, technology can be created to develop the software algorithms necessary for the data aggregation into meaningful policies. As technology improves (AI), this may allow for some level of automation.

Universities are already in the business of providing education for the many in different fields and thereby seem an ideal player to adopt and adapt SE to become a key resource for the community. Political systems design as an area of study would allow cross-disciplinary engagement to solve political issues using an engineering framework to design democracy. The universities (FCIs) geographic location allows them to be close to the communities (CIs), and they serve and build both the capability and capacity through growth to design better systems that serve the communities for designing the policies that their representative as selected members by the community takes to parliament. A distancing of the elected member from the policy agenda in this way provides a degree of independence with both transparency and accountability from SE techniques of Requirements and Configuration Management. The elected member becomes de facto a contractor or people's employee who is selected as a preferred candidate, and whose performance about what to deliver in parliament is then clear to its constituents and importantly themselves as legitimated representatives. SE designs to reduce democratic deficit as policies are transparent to their outcomes.

More importantly, it allows informed debates as universities in terms of learning institutions also drive improvements to both service delivery by existing institutions and also the training of the people for those institutions required by the community for both technology innovations and SE. As with growth of technology, it is anticipated these institutions would be improving the process of decision making by the people for policy formulation. An attempt in this direction was the case of the Fribourg University in Switzerland mentioned earlier which developed the Smartvote tool for the Swiss context to improve voting in a voting market. Australia may require one or more technologies to meet its needs based on the requirements of each hubs (universities) as the technology would require customisation based on context, as an example country areas electorate and mega-cities electorate would perhaps have different solutions given the size of each electorate. Each electorate in Australia aims to have an average of 100,000 voters, in country areas this means thousands of square kilometres per electorate. A degree of customisation is necessary given the context; these however are adapted by the hubs serving those areas. Universities and educational institutions serve certain areas or catchments; voting, for example, is conducted in schools around the country. Schools or set of schools treated as Configuration Items (CIs) feed inputs to the FCIs universities that delivers to the parliament as the ultimate FCI. Each electorate is a FCI and it is made from a set of CIs (individuals). Each FCI (electorate) generates their own Agendas to make the common Agenda in parliament.

Also, a key aspect is that it assists the genuine politician with the necessary legitimacy to debate clearly and openly in parliament what their community wants. The parliamentary debate which is open to the public and news media is then reported upon and from the way the members vote their accountability are visible to every Australian. It would be hard for elected members to follow their own or a party's agenda given a set agenda from their electorates. Scope creep is visible and can be questioned for accountability. SE links inputs to requirements to needs (specifications) these links are transparent. Unmet specifications (agenda items) are visible through technology and can be questioned by the electorate. The political system transforms, it upgrades to become a purposive one where SE processes deliver through the universities the specifications that are required by the people. In so doing the data that is gathered by the universities becomes a means for analysis and synthesis of current and future needs and wants of the people for the people. Both decision making and non-decision-making can be debated openly through a free press which adds to the debate in new ways.

Collusion between elected members to undermine the people will become visible through such an arrangement given there will be different roles for different entities within the system thus increasing the checks and balances for those in office when delegated the privilege to rule by the people or constituents. As the system learns and improves these input, hubs become a new democratic institution to both drive change and support ongoing change. Easton when initially developing system theory for political science may have been aspiring to such a role from institutions of learning but he was ahead of his time as the politics within the US universities themselves in those early days forced him to move to a university where he felt supported for his radical views (Gunnell 2013). SE would require a university or universities to run a pilot study for SE implementation into politics. However, in lieu of only technical systems designs which are SE's focus, the challenge in universities would be a political system design or the e-democracy that democratises.

5.3 SE Interface for Australia

SE is not unknown in Australia even though it may be new to politicians and political representatives in general. SE is new to politics as it has been mainly used in the Defence sector with some application now moving into civilian use as mentioned earlier and this bodes well for Australia as there exists a level of expertise in the use of SE techniques. Those Australians having served or worked for the Defence engineering sector would be acquainted with SE practices. This familiarity of an existing expertise pool with SE bodes well for a country like Australia if its politicians are willing to use the SE techniques. However, SE is not just a technique; it is a way of thinking and doing. In designing the upgrade for the enhanced e-democracy that democratises SE requires new thinking from its politicians, citizens and academia alike. There is a learning curve about SE in its role as a facilitator to government and governing as SE itself undergoes continuous improvement in this new system.

SE is dynamic it undergoes continuous process improvement (USAF 2010). This implies that moving SE into politics requires a learning phase for both existing and new politicians, and more importantly as a facilitator to the public, the universities that embrace it. Universities have as their key function to allow the people to engage with SE's processes. Now previously we explored the potential role of an institution like National Development Unit in Mauritius which was the case for a small country with an existing interface for people's inputs to policy. The interface for Australia given its size and with each electorate size variances to meet the average 100,000 voters per representative is suited to the universities as FCIs of the country. Universities provide education for the whole of Australia, and it can be upgraded for taking on this challenge to integrate SE within the political system of Australia. This interface is such that the design increases the political system's democraticness which as Wright alludes requires new thinking—new thinking that the research aspects of universities can facilitate in their catchments (CIs).

A paradigm change on the scale of Australia requires a learning organisation. SE in its initial stages will require experimentation and thereby a pilot case launched by a university with close working relationship with Defence for example is recommended. Lessons learnt from this pilot study will allow further roll-out to other universities. These entities (FCIs) would also have the additional responsibility to train and educate personnel required for this function in the new emerging Australian political landscape where citizen's inputs would be transformed into specifications set for a common communal agenda for parliament to action. It is anticipated that the applied aspect of use of SE in Defence would assist this paradigm shift as lessons learnt from technology use can be shared by these two entities. Academia has also been accused by media commentators for delivering training that is not required by industry. A synergy between the two (academia and Defence) allows SE to be improved to deliver outcomes using technology, and as mentioned earlier using technology Australian Defence had engaged to shape policy making (OECD 2003).

SE interface with the Australian political system would create new terminology for example that will be unique for democratisation. The technology that will emerge, if automated to some degree, will need to be created. This may require new partnerships with the private sector for finding and enhancing technological capability for the effective operation of the technology framework as recommended by the OECD (2003). When reporting on e-democracy experiments which have failed to operationalise one of it included Australia in 2000, the report from lessons learnt identified a need for a framework for effective implementation of ICTs technology. SE creates, implements and upgrades technology in a systemic manner. It also allows others to learn from how it has embedded technology within its framework where both obsolete and new technologies co-exist with the System of Systems (SoS) techniques. But as argued earlier once past its infancy stage (pilot stage), the processes and technology mature, and new political systems emerge from the upgrade of the dysfunctional ones currently in place.

Australia has already made a significant investment in providing broadband or the Internet's backbone through the roll-out of the NBN network. This is a start and in anticipation of the Internet of Things (IoT) becoming a reality, significant disruptions is expected in the way things are conducted both for existing businesses and the government sectors. New ideas like artificial intelligence (AI) are emerging to allow better information for decision making. Technological disruptions can be accommodated using a SE life cycle approach which manages technology from birth to death (obsolescence). Technology provides first mover advantage if within Australian politicians' ranks the political will to engage early exists. Contact with elected members at the federal level at both Member of Parliament or the lower house and Senator or upper house level, and a NSW State level elected member, and a prospective federal Member of Parliament indicates there is interest in a project to improve citizens' engagement through the university.

With the existing facilities like schooling from kindergarten to schools (primary/secondary) and finally university with its research facilities and the NBN in place as a key infrastructure, Australia is resourced to be a first mover towards democratisation using technology. Australia also aspires to be leader, an exemplar democracy in the region, with a role for the military in peacekeeping efforts in the region. Democratically oriented countries tend not to go to war but rather try to work out through peaceful negotiations for outcomes that are mutually beneficial to each other. The development of the SE framework will assist Australia's aspiration to lead by example for promoting peace and then share its lessons learnt from technological democratisation or an e-democracy in the region and around the world. Australia has already got a significant global foot print through its trade and international relations and using SE it can harness technology for a true democracy.

A lead role in improving democracy and designing a fairer system for all would align with Australia's national and international objective. As an OECD and UN member, it can then share its expertise in the e-democracy design area to other countries whose people intend to develop a system that is fair and egalitarian, a democracy where everyone's voice matters. The significant problem of refugees fleeing political repression may then become a thing of the past as countries open themselves up to better systems of governance and allow every member of their society to contribute equally to the decision making in those democratically oriented countries. Australia's current expenses of billions of dollars on refugee policy, which is for jailing the illegal refugees arriving by boat, can be redirected to e-democracy aid for other countries. A potential role of Defence will be to inform and educate and proactively anticipate ways and means that technology can be used to deliver better government to its peacekeeping objectives. With such initiatives feasible from ongoing e-democracy research, trust for Australians in their government and trust for the people of other countries with shared values will improve. This will reduce the effect of fear mongering in Australia with politicians rhetoric aimed at scare tactics for winning power rather than improving the good life of Australians. Democracy means ruling and being ruled; it is not meant to be a job for life, it is about serving the people.

Australia's aspiration to peacekeeping will be enhanced as it proactively assists those in need to improve to better governing systems with the knowledge gained in developing its system e-democracy. Significant efforts spent in deterrence from acquiring war machines by each country could be redirected towards true peacekeeping that is a government that is fair to all its community members or a fairer society in the same ethos as Australian national values. Like-minded countries do not go to war, as like-minded societies join together to govern by dialog rather than force.

As the IoT becomes reality further automation will become possible for the system e-democracy. Cross nation-states' discourse becomes feasible and political decisions like climate change which affects every living thing on the planet becomes possible. Removing the middle man (politician) from the risk of decision making which is tainted by vested interests of the few to one where it is anticipated a sensible majoritarian view emerges is where SE provides a potential for innovative ideas to tackle global threats to human existence. Rather than rhetoric of a few that are obsolete based on their vested interest to protect their seats or a jobs focus, initiatives on what benefits all the members of the human community that is the good ideas get facilitated by SE. With SE these ideas then have a better chance to emerge for implementation. Normally, every human being wants to leave a good legacy for their children. The voiceless majority protecting advertently or inadvertently the vested interest of a few makes a mockery of nation-states that call themselves democratic.

As mentioned earlier, the Australian political landscape is riddled with controversies which are symptomatic of a system in stress. Leadership, and as one commentator infers a vision that inspires Australia is necessary, for the OECD this leadership for e-democracy must come initially from political leaders (OECD 2003). In Australia, this leadership is sorely lacking in the current system of political party operation where factions and pressure groups have a significant clout to derail even good ideas. The culture in representative system of politics is one of confrontation or conflict based outcomes. It would appear that politics has morphed to become a professional job rather than a service to the community. Organised political parties are professionalising bent on winning at any cost, or the professional politician is a lucrative job which some want for life. Promises from these professionals are made that are then undelivered or under-delivered and then through smart political rhetoric deferred or deviated using justifications like it is about budgets surplus. The interparty fights are complicated further with political party infighting such that the political system has ground to a halt in improving itself to better serve the people. Innovation has died in politics as every party is too timid to take the leadership of transformative change. This timidness may not have been the intent of the people who originally designed the system that smashed together a set of penal colonies to form Australia. It was intended to be a fair society, an Australia with a social democracy intent. A new discourse for a new era for better government and meaningful governing is required where

SE upgrades the stressed Australian political system to one which uses technology or is an e-democracy that democratises.

6 A Case for Change to E-Democracy System

Political leadership to bring effective change to the political system is missing, and with the current design, the chaos in the system continues. During the Turnbull government crisis, two elected members of federal parliament were approached about the idea of improvements to the Australian political system from using the university to engage with public needs and wants. One was a Senate member who had been in the role of Cabinet Secretary to Turnbull and a Minister, the other was a lower house member in the Labor party. Contact was through using email, this was followed for the Senator with phone conversations with staff members of his office, while for the second the lower house member, being a local member, it was supplemented with a face-to-face interaction of around 43 minutes during the elected member's yearly community meeting event. Both members from the two main parties expressed their support for improvements. Now such views are often voiced and heard from other elected members on national TV albeit in different words. It is a common theme that they, the elected, represent the people. However, what this actually means in practice is again subject to the various interpretations of actions of that elected member making those public comments. An assumption is often made that elected members know what the Australian public desires in terms of delivery from their elected members in office. As various commentators point out very effectively during live TV debates, these elected members may profess to represent members of the public but they quite often have no clue as to what the Australian public needs and wants. This is not entirely the fault of the members as elected members are operating within a political system that was not designed to deliver agenda-setting from the public for the public. The representative system has been designed for winning votes and the rhetoric is about securing those 50% + 1 votes. However, if these members had a genuine interest for a people's agenda as new technology came on board they could have actively sought ways and means to improve the peoples' engagement for an agenda. Though some of these elected members may intend to improve, it would appear that a framework for change like SE is missing as the default goal is and remains to win government, we explain this further below.

The two main parties comment that they have their own polling that they conduct in various seats, the marginal seats as defined by the electoral office. This polling is to gauge their prospects of winning government at the next election. This is perhaps also a means to gauge the public's mood for change of the incumbent in office. Winning is a focus on power not the public needs and wants that these parties are supposed to deliver. In its current representative form, the people's voice was never part of the representative system of democracy that emerged in Australia. The Australian political system design is not for a traditional democracy (Hindess 2002). Usually, the debates about issues are hijacked through strident pressure groups and with the aid of media; it becomes a frenzied battle of opinions rather than facts or evidence for seeking public support. The idea is to sway the public opinion. It is not to use the publics' ideas in a pluralistic manner to transform society. Rather at times fears are promoted to convince for a vote, as votes are lucrative, and at every election there is a voting market to exploit.

The Australian tax payer through the electoral office pays for the number of votes acquired by those contesting an election event, the 2019 rate per vote is 275.642 cents (Australian Electoral Commission 2019). Now, instead of providing funds to political parties for the number of electoral votes these government funds could be diverted to the institutions involved in the development of policies. A voting market is about vote maximisation for each brand (parties) not about policy that matters to the public. Vote maximisation is a paradigm for power to rule rather than ideas to lead government through agenda-setting from the people by the people. In Australia, as votes to candidates are paid by the electoral office, it becomes a good business every four or three years to seek representation. A vote bank exists that the two main parties tap into for their benefits in the business of representation, now representation is a lucrative business when for example 19 direct votes can win a well remunerated senate seat, a systemic reform is required to look beyond voting at elections.

A change to provide resources to the university policy-making process then is about a shift of focus on empowerment to rule but with a communal agenda for ruling. The core ingredient that seems absent from the political system is a well-developed system to engage with the public to understand and then cater to the public needs and wants. This problem is not new and David Easton had identified that such a process of public inputs is both necessary and essential to any political system as the stress from unmet demands as mentioned earlier leads to system change.

A stressed system becomes volatile with the outcome at times for disruptions like revolutions or protests. It could appear ludicrous from what has been said before that even though Australia has got both the means and the technology to improve and some experiment like the Defence policy making that was attempted yet this did not get traction into the larger political system. There was no mention in the report though that the policy making used a SE framework (OECD 2003). There was no legislation passed in parliament or debate to make this form of public participation in policy making a feature of an electoral agenda, instead polling and the resulting perceived performance from polling are hot political topics. The reputation of Australia as an exemplar democracy from its volatile leadership changes that can be perceived as intraparty coups does not help its political image in the world.

So even though new legislations are hotly debated in parliament and there may be good intent like assistance to Australian immediate neighbours about peacekeeping efforts a short-sightedness of improving the Australian political system itself may have jeopardised its democratic image as a good political system. The Turnbull government and prior to that 4 other leadership coups may have led the previous Foreign Affairs Minister to comment that Australia is perceived as the coup capital of the world. This representational system may seem erratic given what we have mentioned earlier as Australia's aspiration to be an exemplar democracy. An Australian political system which itself is now perceived to be unstable and dysfunctional will fail to bring change for peace efforts in other nation-states. The Australian political system needs an upgrade.

The issues are many but they are rooted to the lack of good design when the system was created and its erratic evolution to democratise still keeps the people's voice out. To deliver a social democracy Australia did set itself the goal for its own political system to become a better one through copying the best of the two systems at the time, a federal and Westminster systems. One would assume it was easier to copy what was deemed at the time to be a good system of government rather than design through an iterative process like SE, and the best political system for a democracy that Australia may yet become. An iterative process is dynamic with ongoing improvements to the political system. So the question is how to transition from what is now termed a chaotic system to one where the ruled are in turn rulers, a democracy *a la* people rather than a political system imported to impose a worldview which is not even changing to suit the current era where technology allows reform. A majority of Australians commonly use or do have access to electronic devices to communicate like computers or smartphones. This bodes well for implementing change through technology.

With the two main political party intent to vie for power for the sake of power rather than people's rule, the problem of Australian representative system as technologically inadequate remains. Its political *staticism* means that changes cannot be implemented given the system in place as mentioned earlier. So could the existing system be improved to allow firstly a modernisation of the system using technology and then stopping the elected representatives becoming agents of chaos in pursuit of personal agendas, this is a trend in the existing political system. A claim of every elected member is that they represent their *mandants* or constituents. To give some credence to the members' claims to be elected or aspiring to be elected, a fundamental need exists to clearly define what that mandate is that these elected members represent. It becomes imperative for the system to be improved so as to articulate what the public wants and needs or its people's mandate. The reported experiment by Defence in policy making has not found any further inroads in the mainstream political parties for supporting a people's driven agenda. This is despite the infrastructure that currently exists like connectivity infrastructure from NBN and the technology like ICTs that can provide a means to this end with the use of smartphones and computers by Australians.

A problem is that the bureaucratic system that develops policy in Australia is in need for overhaul. To this aim one option is to use the universities in Australia as an institution to drive change to the political system. These institutions (or FCIs) are in the business of knowledge creation and delivery in the geopolitical locations in the states within Australia. Through their online activities, like online courses these universities also have a presence that technologically is even external to Australia, so a connectivity network is well established for these FCIs. SE, explained earlier, is an evolutionary design framework and the universities (FCIs) as knowledge management systems can conduct further research and development to improve and enhance the political system (e-democracy) design using the SE methodology. SE which itself is a dynamic framework is subject to change as new technology becomes available like AI. A SE developed e-democracy system from the various universities engagement would allow further growth of democratisation ideas to continuously improve political system's design. Ideas to use newer technology for parliamentary systems that gets created by these institutions is through their existing activities of research and development, it creates further progress. Research and development allow continuous improvement of e-democracy system designs which these FCIs contribute to. An issue is the resources for funding this activity.

Many elected members claim expenses for official work-related trips which as the media reports are subject to abuse at times. Technology, as an example, can now allow virtual meetings. Technology so applied improves taxpayers' money as trips are then to be used for actual work-related activities that existing communication technology could not support. Lately, some elected members expenses that are claimed as work related is becoming a contentious issue. There are media reports that some members who may be stretching their claims by mixing their personal business with their work-related trips. Senators are allowed to 'hire of charter aircraft and other vehicles for travel within and for service of the electorate' (Owen 2017), like a senator's whale watching trip with her daughter.

Most of these representational trips are information gathering or disseminating ones. Each electorate covering various areas could be customised around a university and its resources. This entity or FCI would provide the support for the activities that such a venture like elected members requirement of information gathering or dissemination from their electorates. Communication technology would make such trips unnecessary. Savings offset from use of technology will allow further development of the existing communication technology by the universities from their R&D—Research and Development. It will also make the role of representative clear with a potential outsourcing of mandate's development from people's inputs to universities such that in future an elected members would then focus on effective delivery of their mandate.

In an upgraded political system, there is through independent information gathering a potential for a changed role of the governor-general to a FCI. This role change occurs with the benefit of a defined agenda from use of SE Requirements process, described earlier. The role takes on the responsibility of the performance to a people's agenda. Performance is through the role's managing of scope creep and accountability, and this is from the configuration control of the agenda. The development of the agenda in Australia is by the educational institutions and technology use, software as a service assists these development and management processes, but once the agenda is agreed at an election it is Configuration Management that controls the agenda. This configuration oversight then becomes the responsibility of the office of the governor-general as the configuration controller of the people's agenda. The existing role already has oversight over the parliamentary workings and when it is fully independent of the parliamentary institution with the SE political system upgrade to a system e-democracy this FCI can enforce the people's agenda. The role becomes a feedback for accepting the outcome from the parliamentary debates on the defined mandate given by the people. The governor is the actor with oversight responsibility of the people's will. A people's will that is then exercised by the authority vested in this office. Issues that may arise would be transparent if the life cycle of the decision making is tracked through a workflow tool, a flow process that is mentioned in Fig. 1. The inputs from citizens can be tracked with a workflow tool as is mentioned in Appendix 3, automation links inputs to outcomes.

The key role of the office is to ensure scope creep does not occur from the people's mandate during a team's tenure. In a sense, the function of the office becomes the primary FCI for the Australian political system or system e-democracy. Even the judges who interpret legislation are appointed by the office, the office is uniquely placed to deliver reforms agreed to by the people who task the e-democracy actors or elected representatives who must deliver. Accountability of the actors is from their debates for or against the people's agenda. The task becomes one about delivery rather than bad decision making through the powers vested unto the parliamentary actors. The media can also report its views and opinion to add to the debates. The electorates are informed both by the media and by the governor-general's office as to whether the public interest is being met by those elected to office. Based on their performance, an elected person's contract is then renewed for another term from their official performance in prosecuting a people's agenda during their tenure. The office also has the power to ensure that vested interests of the powerful, lobbyist and pressure groups are transparent in the decisions made and those that are not made. Power concentration can be curbed to equalise making the system fair for all. Figure 1 depicts the suggested process for a people's agenda for Australia involving universities as requested by some Australian politicians. This flow can be further developed using SE techniques and technology.

The suggested flow chart, Fig. 1, through a data capture tool mentioned in Appendix 3 allows the development of people's needs and wants in a transparent manner. The specification or policies when developed by

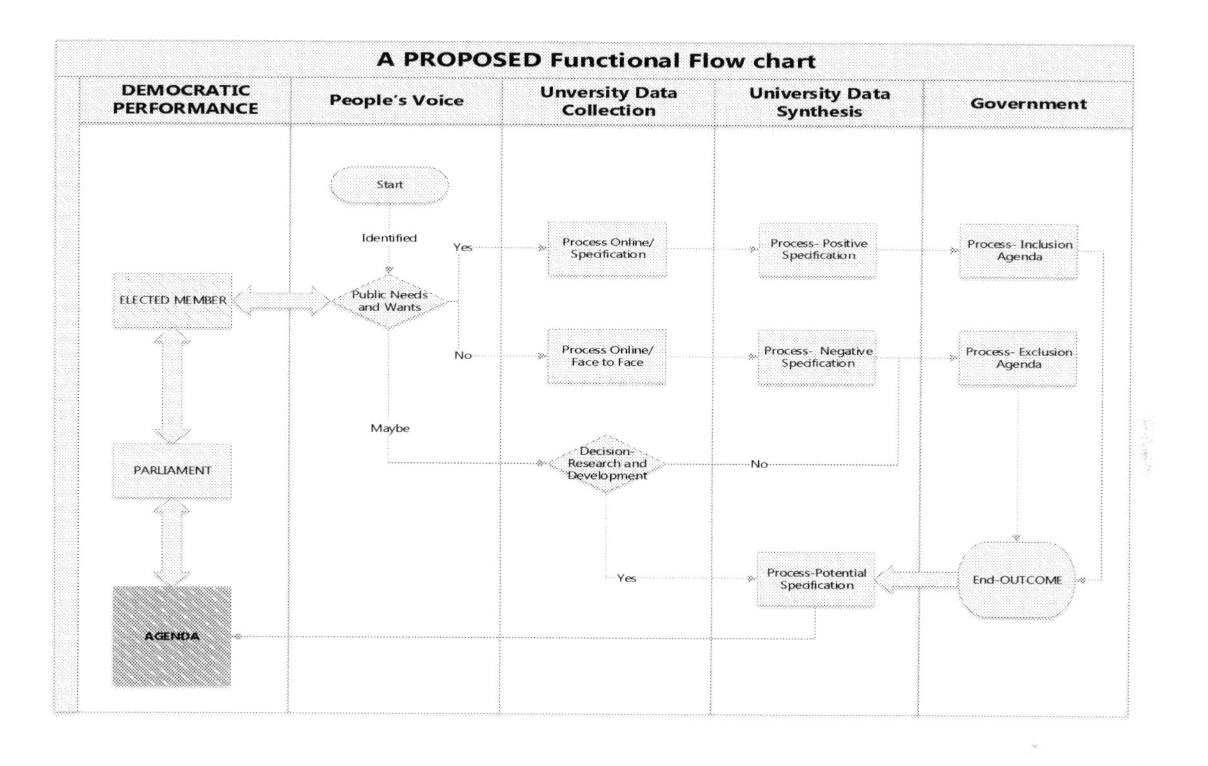

Fig. 1 A flow for data capture and synthesis for policy making

the university allow the elected member to champion what their electorate needs and wants. Such a workflow tool would link from people's needs and wants the various outcomes in parliament. Agenda the FCI, to be debated in parliament, becomes a debate of issues or ideas from the people and for the people. In a matured form the tool allows growth for predicting and also for new ideas to emerge to anticipate and fix issues within the communities making the electorates of Australia (CIs). Each CI (electorate) contributes to the FCI (Agenda).

An Agenda makes elected members become the voice of their communities that is elevated in parliament which then becomes a means to deliver for the common good agreed during each electoral cycle. A common good which can be synthesised by the universities and openly debated through the existing as well as new forums that such an activity would generate. Data accumulated over time would allow elected members to become more informed to their electorate needs and wants rather than partisan politics or pressure groups with narrow agendas to promote their existing advantages (vote banks) rather than the good of society at large. Importantly due to the increased participation from the potential that this channel represents the new ideas can emerge and allow changes to be considered for public debate which in the current system is ineffective, leading to wastage. Issues like long wait time for public at hospitals can become more prominent as it is common issue to all. To the public with an interest on how this issue is being managed, they can openly require changes for a proposed system using software tools to provide inputs about how the system should operate.

Now the software tools to develop the agenda can be improved upon as the system learns and new technology comes on board to allow improvements. This dynamism ensures that the system undergoes continuous improvements and progresses over time. Automation, it is anticipated, would create new processes as well as make obsolete redundant ones. New technology like AI (political intelligence) could be created to improve and SE can upgrade the political system as such technology becomes available through research and development. This allows a cutting edge technology to democratise to maintain a democracy as an evolutionary e-democracy system. Universities as knowledge systems would be actively developing what is required to maintain and upgrade the political system so the e-democracy is not obsolete.

Fixing the political system design in turn would then drive e-engagement. The universities involved to develop for example better health system will actively investigate improvement either through technological use or other means that would become subject to research for effective system designs. The SE framework, it is anticipated generates many such issues, for example like jobs and the educational gap for those jobs would become debates that matter and get actively explored by parliamentarian for solutions. Youth unemployment and regional underdevelopment would become visible and new solutions researched and implemented in a continuous cycle of systemic improvements.

A dynamism of the right kind would emerge with the SE framework rather than the existing tendency of throwing money at the problem as a means to grow the sector into what may be a system which is no longer adequate or meant to deliver efficiently in the technological context it is operating within. For example, a health system which is obsolete in an environment where technology would allow for different means of health delivery becomes things to research. The new system may require new technology and not populist ideas like more hospitals based on manipulating for votes in an electorate (pork barrelling). Rather with SE, it is about solutions for effective delivery of better health. Better health may not need more hospitals but rather better technological solutions to reduce existing health problems, a preventative solution in many cases as the intent is to design out the problem. If designing out the problem fails, not being cost-effective for example, then other controls are implemented to mitigate it until new technology (medical solution through research) is created to fix the problem.

The SE approach allows for embedding systems within systems, and thereby health becomes part of the larger system like education or society with overlaps and integration is with using technology as facilitating the various system to deliver better outcomes for the people. A new era for cooperation and partnering using technology would allow the elected members to focus on delivery rather than rhetoric, as it is often the case. Mostly, the voting public at elections are promised what cannot and would not be delivered. A SE flow explained above is the start of the reform that Australia must address if its political system is to deliver a people's agenda in an e-democracy. Each elected member would then be focused on the delivery of outcomes required by the public rather than perpetuating myth like a budget surplus by sales of assets to temporarily bloat budget figures or government spending more than it can tax to generate the revenues it requires to meet some, wasteful at times, perceived public need or want (pork barrelling). The business of government would become more focused on service delivery and negotiation between each electorate for common outcomes for the people of Australia. Resources allocation becomes transparent and the people decide on the government agenda, and if workflow tools are part of the technology used, prioritisation is visible to all.

The system's upgrade with SE and technology is that the person aspiring to represent the people then actually does represent the public will for they are elected with a defined agenda to deliver, from the public, for the public and to the public. In this task, they would be assisted by a governor-general's office with oversight over the business of government and by the universities with their activities of research and development, along with the training and skilling of the resources required to the effective delivery of a political system incorporating a SE paradigm. As new technologies emerge from these institutions, these would be trialled and rolled off to deliver an ever efficient political system that would meet and exceed the public needs. For example, like better quality education or better health or empowering of individuals so as to move them out of dependency from the dole system to become effective citizens contributing to a better society for themselves and others. Lessons learnt can then be shared through Australia's aid programmes to other countries around the world, for example for politicians in Mauritius discussed earlier. Politicians who are pro-e-democracy but yet averse of being the lead venture in an e-democracy system implementation given the risks of the unknown, but then they would readily embrace a system used and operated in Australia to adopt and adapt for their own purposes (ID No. 2 at Appendix 1). If the political will exists, Australia has the technological means to implement changes to enhance democracy, and become a world leader (exemplar) in this area of political system designs for improving a political system's democraticness or an e-democracy system that democratises.

7 Recommendation for the Australian E-Democracy System

In the Australian representative political system of government, parliamentary work is wasteful, as response to initiatives is wasteful conflicts. Conflict is aimed to oppose from 'tactics to harass the government' of the day in its work (Solomon 1978: 33). The modern political system from opportunistic decision making aimed for conflict is flawed from its systemic design. It creates tactical waste of people's resources (taxpayers' revenue), and it is waste that a modern democracy could control through the people and also eliminate by generating a people's agenda through an e-democracy design. Australia has both the resources and technological infrastructure to design an e-democracy using a SE upgrade. Since federation in 1901 where the Constitution for the legal system to follow was drawn its political system has faced many challenges (Peel and Twomey 2011; Macintyre 2016). The political system is meant to have an opponent even when technology though can make everyone a proponent of the people's agenda. The bicameral 150 member House of Representative (lower house) and 76 member Senate or upper house bent on conflict is becoming dysfunctional to bring about some of the big decisions that would position Australia to lead globally as it aspires in its immediate region and beyond (peacekeeping, for example). To generate conflict means that it cannot for example come to an agreement about climate change policy and its associated changes that it needs itself to adapt to for the climate change challenges that it predicts to affect Australia's future. This illogical climate change stance is despite the ravages that bush fires and drought wreak both on the farming community one of its key economic pillar, and there is the weather which constantly reminds every Australian of the cost of non-action. Non-action through rhetoric that is clever to discredit facts.

Rampant disagreement for short-term political gain tends to undermine a debate based on reason, and this is even when scientific reports are presented. These reports are then politicised from competing well-resourced opinion leaders such that no changes are made. Opinion leaders who are elected to serve for the public good, however, when their opinions at times are effectively combined with commentators bent to create a controversy, they perpetuate conflict as a means to maintain their and their party's relevance. The representative system of democracy with party agenda requires conflict, and therefore, the system actively promotes behaviours in the political system that is detrimental of the publics' interest. Leadership from progressive parties are necessary for a change to a new paradigm which allows a better governing system to emerge in Australia. With the failure of the electoral reforms, this book starts the SE and technology conversation in two political systems, Mauritius a small nation-state with limited resources, and Australia a well-resourced nation-state which through *staticism* is wasting itself as a potential e-democracy leader. It is time Australia uses technology to select the right people's representatives for a people's agenda; this will reduce bad legislation which is never adequate to deliver good solutions. Despite the threat of litigation, the powerful always have a way to circumvent these legislations' intent.

Australian legislation is seemingly passed without a holistic perspective about the impact of these legislations on existing businesses and society. The costs are forced on the people rather than mitigating the impact of these changes enforced by legislating. With SE thinking, assisting society to adapt through planned transition for solutions that works for all is feasible. Some policies for example zero asbestos create dilemma for organisations like Defence, where asbestos being a product that occurs naturally makes implementation of the policy aiming for zero a risky compliance issue. Due to its natural existence products that are imported or transported may have small amounts of contamination making zero a goal that is next to impossible. Also instrumentation used to measure for asbestos itself may have margin of errors which then leads to further complications for zero as a target for compliance. The asbestos elimination intent requires an implementation plan that reduces costs on society affected by the problem, this not only about litigation or its threat. The immediate growth from the legislation banning the product is the litigation industry growth rather than effective safety through an elimination programme being designed and then implemented to reduce the impact of the danger of its presence. The elimination programme design with SE creates the right industry as ultimately all asbestos gets removed. Through design methodology, engineering would investigate how to eliminate the problem and for making society safe.

Emotive debates undoubtedly with good intents perhaps lead to impractical decisions in parliament. When making the claim for pragmatism, it implies that solutions have a SE life cycle approach. That is solutions that are brought about independently by knowledge institutions that are further from the debate. Given the political status quo (*staticism*) though, applied politics is fast becoming mired into factionalist and individualist views and debates. Thus, a knowledge-based institution is required so as to develop those innovative political ideas and debates that are community oriented and are ready to be tabled in parliament for each elected members. There is a need to deliver an agenda to each member for those representing the communities. In the advanced and complex Australian society, the universities FCIs and its educational CIs involvement as an element to develop the policies sourced from the people is becoming necessary. SE techniques when applied in politics allow universities as established institutions to source and validate the needs and wants of the community, refer Fig. 1.

These needs and wants can be made visible to each and every and openly debated by both the media in terms of its role to inform, free speech and also transparency for the parliamentary debates in terms of which member voted for or against the needs and wants brought forward by elected members. Ideas from the community developed and verified by the universities are agenda for debate. Parliament then becomes a forum of debate for good ideas and not partisan politics. Partisan politics currently tends to distort and obscure the excessive powers wielded by some pressure groups that are well resourced to obstruct for their own interests the silent majority through support they provide to elected candidates. Using a transparent SE process, ideas that are defeated in parliament due to the fact that it was deemed unsound or needed further developmental work are visible. These may re-emerge as those who have championed it initially can elevate these ideas again in new amended forms through the feedback SE loops mentioned earlier. Thus, improved participations from the public or e-engagement will generate its own systemic changes as the e-democracy system is no longer constrained by the elected members' personal views. Or worse, hiding behind elected members decisions are views from vested interests, transparency with SE processes deters such behaviours. Technology can curb such behaviours as with a public agenda those representatives selected by the people are those who can best represent the interest of the community (defend a communal agenda).

Technology like the workflow tool mentioned in Appendix 3 can be used to provide transparency of the citizen's input through its life cycle from input to outcome. Thus in the SE process, technological tools assist in generating trust amongst the public and more importantly allow change to be made in parliament with the universities providing the support and maintenance of the system as it grows and learns. It is common practice in Australia, like in other countries, to provide universities with financial grants to conduct research. This proposed SE form of policy development creates new engagement using existing infrastructure amongst one of the neglected elements of society the incoming generation of thinkers and knowledge workers that educational institutions generate to create and design political systems. In this design process, the public also engages through the university as with SE people are the centre of design. This SE people's facilitation then provides new momentum to upgrade and reform a political system that is no longer relevant to modern-day Australia.

The synergy from the new e-democracy relationships between those who govern on behalf of the public and those in the business of creating new knowledge will balance the often over bureaucratic processes. The processes are technologically obsolete given government institutions as matured organisations have operated for more than a century in the current form of representative system without innovatively engaging the public. The power conflict between government bureaucracy and the elected members governing Australia frustrates attempts to bring change, this tension in governing agenda is well known and reported by commentators. Conflict also impedes progress.

When legislations get passed, with SE processes, it would have the benefit of pilot projects trialled through the knowledge creating arms of society (FCIs) and be monitored for the life cycle of the legislation. These trials when using a SE systems approach are holistic. Holistic given SE would explore the impact of the legislation on citizens (costs for example) and ensure that they are mitigated and appropriately rolled out. Both direct and indirect costs of any SE designed legislation would have been assessed and evaluated with appropriate measures to accompany its transition for its life cycle into society. The other benefit of the SE approach is that educational institutions are often accused of being out of touch about what they teach and what is required by industry. With data inputs from the people, these institutions become at the forefront of what society needs and actually wants and through their research can even anticipate future needs for new industries, like electric cars instead of petrol-driven ones given climate change issues. Grants could be provided to solve such challenges rather than legislation to ban petrol cars.

Working closely with government and industry, SE could bridge this perceived gap and provide higher return on investment for those universities grants that get provided by government. Some of the e-democracy grants did dry up when there is no outcome (Parycek et al. 2012; Parycek and Edelmann 2013), as shown from those speculators exaggerating e-democracy claims from technology discussed earlier. The adoption of SE creates designs of tools from technology to be applicable to solve the volatility problem of ruling in Australia and its political *staticism*. SE designs to create the appropriate e-democracy solution, customised in each context, for the present and the future. Each customised design is an evolutionary e-democracy solution (dynamic continuously improving) where the people's voice is ever present in governing and government. As the agenda is predefined, then instead of electing a party the focus of the people is in selecting the right representative. For the people, the government delivery is selected through appropriate representation, a performance issue for both the selected individual at local level and the selected team or teams at state and federal level. Technology with an agenda inputs and outputs allows people's representative selection of the best in lieu of the forced election from organised parties. Technology with SE would force political party reforms from vote entrepreneurs to service focus to its community, enhanced representation.

In Australia, the question is where will the collective political will come from to make SE part of the governing system to support effective decision making, will universities take the lead or the political agents (senators and representatives) genuinely seeking reform or yet the critical citizens of Australia. A system for a bygone era (obsolete) is not something that one wants to leave behind even if its processes are digitised, especially given the opportunity that SE provides for designing a democracy in the true sense. Maybe like federation in the years preceding 1901, the governors may assist someone like the Premier of NSW, as a leader, in his struggle to convince the rest to initiate the process of reform (Wise 1913), only time will tell. This book provides those willing to change with a roadmap which as suggested can be further improved in our collective journey to better government and most importantly governing with our inputs as the people in decision makings that affect us.

Using technology the SE framework, CIs and FCIs are used to source people's requirements and translate it into outcomes. The transparent process which a workflow tool at Appendix 3 allows traceability of inputs to outcome is to improve trust. It strengthens the governance, as it reduces social pressure and promotes appropriate values (less corruption and an education that improves civic values). Through the concept of SoS, the number of variants in each system can be bounded which then allows for continuous improvements of models. Each model FCI as an applied version that is bounded evolves, it gets researched for improvement and the applied version is then made more effective. Democracy is converted from its philosophising to its actualising in each context. Democracy, the concept, has a life cycle approach with traceability to its outcomes from SE, using technology the people's rule emerges through agenda control. Basically, the SE system approach delivers, and it is holistic, efficient and effective to successfully upgrade the current system of representative democracy. Requirements sourced from the people increases the quality of engagement with government through appropriate SE tools which are used to facilitate citizens' requirements extraction, and these requirements are then refined into specifications (Agenda) that are processed as outcomes for a democracy that democratises.

A metadata data library reduces e-terminology proliferation; this was a failing of initial e-democracy initiatives, as with SE it is controlled and managed through a database. Western countries collect enormous amount of data and Australia is no stranger to such pattern of behaviour. Importantly, SE's use creates a data library for a democratic entity. This can grow to bring an engineering structure for democratic tools development. Despite some claims that democracy is from Greece ancestry, democratic type decision making is very old and precedes its coinage from the Greeks according to some authors (Isakhan and Stockwell 2012). SE is not about establishing when democracy came into being rather it is about upgrading the system towards democratic decision making using digital technology like software as a service. The people decide using technology which the SE framework designs and applies.

In Australia, this democratic decision making is feasible given firstly many digital processes is already used by both government and the private sector from the existence of a key enabling backbone, the NBN. This NBN broadband network's existence is a capability which potentially allows every citizen to connect in Australia. Most smartphones can use free Wi-Fi access at shopping centres and Australian libraries also provide resources for accessing the web. These can be further improved upon to provide the public with more access points as the process is rolled out on the internet. If in Australia people mattered, then the political imagination must lead it to reimagine how to shape the affairs of government such that people mattered, thus democracy through SE is a design goal worth striving for. True democracy is when the members of Australian society as a political community exercise their imagination and creativity to have inputs on issues so they can select between viable options that shape their daily lives. In the system e-democracy with the SE process, the actors, selected representatives pursue in a transparent and accountable manner in parliament, the legitimate communal aspirations and goals from their electors options.

Australia has been improving its political system. It was apparent even during the empire days that changes were necessary to stop a revolution like in the USA. The mass of critical citizens grows seeking better political systems. Australia as a political system started its representative model with a federation process in the 1880s, but with technology that it has available today, for political system design, development and upgrade of the political changes can go further with an engineering framework like SE embedded in institutions like universities. Akin to the field of aerospace where technology consumed for defence keeps being upgraded to create better and more capable system for Defence, in politics, SE allows when embraced, better and more capable tools created for a political system to democratise to an e-democracy that democratises. Technology has risks, for digital ones they are hacking, and these technological risks are those that a SE design must address with research from the universities.

Options to upgrade the existing systems using the SE framework allow a genuine democracy as an outcome, as with SE people are its central consideration. This book starts the conversation and recommends to those Australians willing to change the existing system to one that is progressive with a means (initial roadmap) that allows the people to decide about what matters in their daily lives. The book is a framework for new thinking about political systems and their designs. It would indeed be disappointing that in an era where so much could be done so little is actually done for the current generations of Australian and even those to come. It behoves on all of us to seek ways and means to support improvements to our political systems (create political intelligence technology similar to AI). For Australia's future, reform through technology is necessary so that the current accidental and dysfunctional Australian system of another era that creates musical chairs Prime Ministers for example, are improved upon through a SE upgrade of its political system.

A SE upgrade leaves the legacy of a colonial system to one that it designs as an e-democracy that democratises. It is an evolving e-democracy designed such that its use of technology facilitates the changes to create, improve and maintain the political system of now and the future or a system e-democracy. SE is the framework used to successfully achieve a system e-democracy for Australia. Provided that the political will exists, any nation-state can use SE's e-democracy design framework for their own successful democratic political system upgrades no matter what their size. SE customises the solution to one that suits best to each country in the context of its citizens' needs and wants. SE puts the citizens at the centre of the design process, as the upgrade is driven by the citizen, from the citizen and for the citizen. The SE designed solution is not about elections and electing, it is about the citizen. As described above, Australia has all the necessary elements for the people to select from aspiring representatives the right citizens or team for the delivery of ongoing reform that puts citizens first by enabling through technology their inputs and outputs in government decision making.

In March 2019, the federal election showed that the same old tactics worked to deliver government to the existing ruling party. The author's engagement with two elected federal members and one unsuccessful Liberal party candidate during and after the election shows that a rhetoric for change exists rather than actual move to change, hence the recommendation that the strategy for Australia is through its institutions for learning or its universities. Through these universities, SE has a significant role to play to assist in upgrading the political system.

References

ABS. 2018. Population Projections. Australian Bureau of Statistics. http://www.abs.gov.au/ausstats/abs@.nsf/Lookup/by%20Subject/1301.0~2012~Main%20Features~Population%20projections~48. Consulted 28 December 2018.

AEC. 2019. *Electorate Redistributions*. Canberra: Australian Electoral Commission. https://www.aec.gov.au/Electorates/Redistributions/index.htm. Consulted 14 February 2019.

ALP. 2019. *Will You Support the National Integrity Commission?* Melbourne, VIC, Australia: ALP. https://www.alp.org.au/petitions/support-the-national-integrity-commission/ Consulted 2 February 2019.

Australian Electoral Commission. 2019. *Electoral Public Funding*. Canberra: Australian Electoral Commission. https://www.aec.gov.au/parties_and_representatives/public_funding/Current_Funding_Rate.htm. Consulted 14 February 2019.

Birch, A.H. 1972. *Representation*. Key Concepts in Political Science. London: Macmillan Press.

Brissenden, M., and B. Andersen. 2018. Liberal Party Elders Lash Tony Abbott for Acts of Revenge on Turnbull's Government. ABC. https://www.abc.net.au/news/2018-08-27/liberal-party-elders-lash-tony-abbott-for-acts-of-revenge-on-tu/10166590. Consulted 20 November 2018.

Brown, R.J. 2007. *Governing Australia: A Century of Politics, Policies and People*. Nelson Bay, NSW: R.J. Brown.

Coleman, S., and J. Gotze. 2001. *Bowling Together: Online Public Engagement in Policy Deliberation*. London: Hansard Society.

Conifer, D. 2018. Senator Fraser Anning Gives Controversial Maiden Speech Calling for Muslim Immigration Ban. ABC News. https://www.abc.net.au/news/2018-08-14/fraser-anning-maiden-speech-immigration-solution/10120270. Consulted 25 February 2019.

Coorey, P. 2018. Payback, Reward and Healing: Scott Morrison Unveils a New Ministry. *Australian Financial Review*. https://www.afr.com/news/scott-morrison-announces-new-ministry-20180825-h14ie2. Consulted 27 August 2018.

Craig, J. 1993. *Australian Politics: A Source Book*. Sydney: Harcourt Brace.

Crisp, L.F. 1983. *Australian National Government*, 5th ed. Melbourne: Longmans.

Dahl, R.A. 1995. Justifying Democracy. *Society* 32 (3): 386–392.

Dahl, R.A. 2005. What Political Institutions Does Large-Scale Democracy Require? *Political Science Quarterly* 120 (2): 187–197.

Doran, M. 2018. *Labor's Federal ICAC Proposal Knocked Back by Government, as Coalition Argues Lack of Detail*. Canberra: ABC. https://www.abc.net.au/news/2018-05-23/labor-federal-icac-proposal-knocked-back-by-government/9791674. Consulted 25 February 2019.

Easton, D. 1957. Traditional and Behaviorial Research in American Political Science. *Administrative Science Quarterly* 2 (1): 110–115.

Gray, A., and B. Jenkins. 2004. Government and Administration: Too Much Checking, Not Enough Doing? *Parliamentary Affairs* 57: 269–287.

Gare, D., and D. Ritter. 2008. *Making Australian History: Perspectives on the Past Since 1788*. South Melbourne: Thomson Learning Australia.

Gunnell, J.G. 2013. The Reconstitution of Political Theory: David Easton, Behavioralism, and the Long Road to System. *Journal of the History of the Behavioral Sciences* 49 (2): 190–210.

Held, D. 1983. *States and Societies*. New York: New York University Press.

Held, D. 1993. *Prospects for Democracy: North, South, East, West*. Cambridge, UK: Polity Press.

Held, D. 1995. *Democracy and the Global Order: From the Modern State to Cosmopolitan Governance*. Cambridge: Polity Press.

Held, D. 2006. *Models of Democracy*, 3rd ed. Stanford, CA: Stanford University Press.

Hindess, B. 2002. Deficit by Design. *Australian Journal of Public Administration* 61 (1): 30–38.

Hirst, J.B. 2014. *Australian history in 7 questions*. In Collingwood, Victoria: Black Inc. http://ezproxy.newcastle.edu.au/login?url=https://ebookcentral.proquest.com/lib/newcastle/detail.action?docID=1887446.

Hughes, C.A. 1968. *Readings in Australian Government*. St Lucia: University of Queensland Press.

Isakhan, B., and S. Stockwell. 2012. *The Edinburgh Companion to the History of Democracy*. Edinburgh: Edinburgh University Press.

Jupp, J. 2018. *An Immigrant Nation Seeks Cohesion: Australia from 1788*. London: Anthem Press.

Kant, I. 2000. *The Critique of Pure Reason*. South Bend, IN: Infomotions.

Levi, M., and L. Stoker. 2000. Political Trust and Trustworthiness. *Annual Review of Political Science* 3 (1): 475–507.

Lovell, David, Ian McAllister, William Maley, and Chandran Kukathas. 1998.*The Australian Political System*, 2nd ed. Melbourne: Addison-Wesley.

Lukes, S. 2002. *Power a Radical View*, 2nd ed. New York: *New York University Press*.

Macintyre, S. 2016. *A Concise History of Australia*. Port Melbourne, VIC: Cambridge University Press.

Marsh, D., C. Lewis, and J. Chesters. 2014. The Australian Mining Tax and the Political Power of Business. *Australian Journal of Political Science* 49: 711–725.

Mckenna, M. 2017. Fraser Anning Is Pauline Hanson's New Low-Vote Senator. *The WeekEnd Australian*, November 11.

Moss, G., and S. Coleman. 2014. Deliberative Manoeuvres in the Digital Darkness: E-Democracy Policy in the UK. *The British Journal of Politics & International Relations* 16 (3): 410–427.

Norris, P. 2004. *Electoral Engineering: Voting Rules and Political Behavior*. Cambridge, UK: Cambridge University Press.

OECD. 2003. *Promise and Problems of E-Democracy-Challenges of Online Citizen Engagement*. Paris, France: OECD.

Owen, M. 2017. $4k Bill for Whale-Watching Trip. *The Australian*, 3 July.

Parliament of Australia. 2010. *Australia's Constitution—With Overview and Notes by the Australian Government Solicitor*. Canberra: Parliamentary Education Office and Australian Government Solicitor. https://www.aph.gov.au/About_Parliament/Senate/Powers_practice_n_procedures/Constitution. Consulted 25 February 2019.

Parliament of Australia. 2019. *Senators and Members*. Canberra: Parliament of Australia. https://www.aph.gov.au/Senators_and_Members/Parliamentarian_Search_Results?q. Consulted 11 January 2019.

Parycek, P., N. Edelmann, and M. Sachs. 2012. CeDEM12 Proceedings of the International Conference for E-Democracy and Open Government. Paper Presented to Conference for E-Democracy and Open Government, Danube University Krems, Austria.

Parycek, P., and N. Edelmann. 2013. CeDEM13 Proceedings of the International Conference for E-Democracy and Open Government. Paper Presented to Conference for E-Democracy and Open Government, Danube University Krems, Austria.

Peel, M., and C. Twomey. 2011. *A History of Australia*. Palgrave Essential Histories. New York: Palgrave Macmillan.

Qvortrup, M. 2007. *The Politics of Participation: From Athens to E-Democracy*. Manchester: Manchester University Press.

Smith, B. 2008. *Australia's Birthstain: The Startling Legacy of the Convict Era*. Sydney, NSW, Australia: Allen & Unwin.

Smith, R., A. Vromen, and I. Cook. 2012. *Contemporary Politics in Australia: Theories, Practices and Issues*. Port Melbourne, VIC: Cambridge University Press.

Solomon, D. 1978. *Inside the Australian Parliament*. Hornsby, NSW: Allen & Unwin.

The Guardian. 2018. Malcolm Turnbull Tells Q&A His Removal Was an Act of Madness. *The Guardian*. https://www.youtube.com/watch?v=b_dopFlIbc4. Consulted 25 February 2019.

Transparency International. 2018. Australia—Corruption Perception Index 2018. https://www.transparency.org/country/AUS. Consulted 2 February 2019.

USAF. 2010. *Air Force Systems Engineering Assessment Model*. Air Force Center for Systems Engineering, Air Force Institute of Technology: Secretary of Air Force, Acquisition. http://www.afit.edu/cse/. Consulted 8 September 2013.

Whitlock, G., and D. Carter. 1992. *Images of Australia: An Introductory Reader in Australian Studies*. St Lucia: University of Queensland Press.

Wise, B.R. 1913. *The Making of the Australian Commonwealth, 1889–1900: A Stage in the Growth of the Empire*. New York: Longmans, Green, and Company.

Wright, E.O. 1994. Political Power, Democracy, and Coupon Socialism. *Politics & Society* 22: 535.

Appendix 1: Interviews

Interview Process

Interpretative analysis unpacks meaning through a narrative form of explanation from political phenomena where beliefs are situated in webs of belief embedded within past and present traditions (Bevir 2006). The interpretative analyses of the relevant policy documents, newspapers, online web pages and observations of a key event like a general election in Mauritius were supplemented with a set of interviews using semi-structured questions (attached in Appendix 2). The interview of policy makers in accordance with the University of Newcastle policy was carried for the members listed below. A targeted initial set of 55 interviewees, all government members listed on the official government web page, were approached due to their key role in politics and policy making in Mauritius. These potential candidates were emailed for a request for an interview but there was no response. Then using a local contact, Mr. Hemraj Bungsraz, a number of potential interviewees were approached based on the criteria of senior public servants and current/previous Ministers or Members of Parliament. The data collection has been facilitated by Mr. Hemraj Bungsraz who contacted all of the candidates listed below to schedule interviews. Securing the interviews was through a 16-month preparatory activity preceding the actual interviews that were conducted by the author. Without Mr. Hemraj Bungsraz's ongoing efforts on the

S. Bungsraz, *Operationalising e-Democracy through a System Engineering Approach in Mauritius and Australia*,
https://doi.org/10.1007/978-981-15-1777-8

ground, the coordination and completion of the 15 interviews over a two-week 6–19 March 2016 period would not have been successful. Recordings and transcripts of the interviews were then analysed using an interpretative framework along the lines developed by Bevir and Rhodes (2002).

The purpose of the interviews was to gain background information on how political actors understood e-democracy. The interviewer used semi-structured questions (see Appendix 2) to prompt and elicit the interviewees' perspectives about e-democracy in the Republic of Mauritius. These answers were recorded and compiled to develop the transcript of the interviews. The analysis of the transcript allowed themes to emerge, and these were interpreted to provide the basis of the findings which were then discussed in the book. Furthermore, for a 360-degree feedback a reflective technique was used by the interviewer. Prior to the conclusion of each interview, the candidates were requested to reflect on the interview and then provide feedback about questions discussed with the interviewer. Email contact details were provided to each interviewee along with a copy of the interview script, Appendix 2. At the interview, each interviewee assured me that they would provide me with feedback on the interview. However, out of the 15 interviewed only one interviewee responded using the email provided. To shape my analysis, I also included informal conversations with the people like a senior adviser and Ministers past and present from the National Development Unit; field notes also recorded the context of the interviews and the conversations with Mauritians. From these interpretations, I developed the findings in chapter 8, Upgrading Mauritius a Legacy Political System, along with my recommendations.

List of interviewees

Id no.	*ROLE and description*
1	Private Parliamentary Secretary (PPS)—attached to the Prime Minister's Office—Junior Minister assisting the Prime Minister for national development of Rodrigues island, an **autonomous Constituency** of Mauritius—OPR Party—is one of the representative from Constituency Number 21 of Mauritius
2	Ex-Minister of Foreign Affairs in the previous government—Labor Party—and since winning a by-election in Constituency Number 18 in late 2017 is about to become a MP in 2018, once Parliament sits
3	PPS—attached to the Prime Minister's Office—Junior Minister reporting to the Prime Minister's Office in charge of national development—MSM Party—is one of the representative from Constituency Number 2 of Mauritius
4	Elected member of the National Assembly—MSM Party—is one of the representative from Constituency Number 10 of Mauritius

(continued)

(continued)

Id no.	*ROLE and description*
5	Ex-Private Parliamentary Secretary—attached to the ex-Vice Prime Minister's Office—Junior Minister assisting the Vice Prime Minister for national development of Mauritius in the previous government, mainly infrastructure projects—Labor Party—was one of the representative from Constituency Number 9 of Mauritius
6	Ex-Minister of Arts and Culture—Labor Party—was one of the representatives from the Constituency Number 7 of Mauritius
7	Ex-Vice Prime Minister and Minister of Infrastructure—Labor Party—was one of the representatives from Constituency Number 9 of Mauritius
8	Registrar (CEO) Medical Council Mauritius—a senior public servant dealing with government policy for medicine in Mauritius—interviewee shared that he was keen in politics and aim is to join the Labor Party
9	Minister of Cooperatives—MSM Party—is one of the representatives from Constituency Number 10 of Mauritius
10	Ex-Minister of Tertiary Education Science, Research and Technology—Labor Party—was one of the representatives from Constituency Number 10 of Mauritius
11	PPS—attached to the Prime Minister's Office—Junior Minister reporting to the Prime Minister's Office in charge of national development—MSM Party—is one of the representatives from Constituency Number 5 of Mauritius
12	Minister of Agro Industry and Food Security—MSM Party—is one of the representatives from Constituency Number 11 of Mauritius
13	Assistant Permanent Secretary, Ministry of Agro Industry and Food Security, a senior bureaucrat dealing with government policy in Mauritius—he shared that he was part of the committee to develop the ICT strategic policy documents for the government
14	PPS—attached to the Prime Minister's Office—Junior Minister assisting the Prime Minister for national development—MSM Party—is one of the representatives from Constituency Number 17 of Mauritius
15	PPS—attached to the Prime Minister's Office—Junior Minister assisting the Prime Minister for national development—MSM Party—is one of the representatives from Constituency Number 9 of Mauritius

References

Bevir, M. (ed.). 2006. *Interpretation and Method: Empirical Research Methods and the Interpretive Turn*. Armonk, NY: M. E. Sharpe.

Bevir, M., and R.A. Rhodes. 2002. Interpretive Theory. *Theory and Methods in Political Science* 1: 1.

Appendix 2: Interview Script

This schedule consists of demographic data and outlines the topic areas and themes that will be explored in a semi-structured interview. It is anticipated the interview will take a maximum of 1.5 hours and will reflect the areas of enquiry noted below. The interview will concentrate on the issues for discussion so as to gain an insight of the use of ICT to improve democracy.

S. Bungsraz, *Operationalising e-Democracy through a System Engineering Approach in Mauritius and Australia*,
https://doi.org/10.1007/978-981-15-1777-8

Participant Demographic- Data

1. Gender:

Male Female

2. What is your age group from the following: *(Circle one of the Age groups)*

18-30

30-40

40-60

60+

3. Qualification and Experience
    a. (a) What is your highest educational qualification? *(for example School Certificate, High School Certificate, BSc etc.)*
    b. Where did you do your studies? *(for example Mauritius- SSR Medical school, overseas- University of Newcastle)*
    c. What was your occupation prior to your current role?

Digital Democracy Themes

1. What is your understanding of Information and Communication Technologies (ICTs)?

2. Do you see ICTs having a role to play to enhance or improve the ways in which a political system operates (for example, to create an e-democracy)?

3. Could you describe what you understand by the idea of e-democracy?

4. Do you see a role for ICT innovations to contribute to the democratic political system in Mauritius?
    a. If not, could you elaborate why you think there might not be such a role?
    b. If so:
        i. What might that role (or roles) be?
        ii. What problems or barriers to the adoption or implementation of ICTs (in the political system in Mauritius) do you think might need to be addressed?

Appendix 3: Software as Political Service

A prototype software that can be customised to conduct the data capture and synthesis is provided by the firm listed below, see link. This prototype software allows the NDU (FCI) of Mauritius, for example, to provide transparency for issues being raised by citizens and to follow their inputs to the final outcome. The process flow allows an agenda item to be raised in the parliament and followed across its life cycle, evolving through various iterations that can be tracked and recorded until it, if it is not rejected before that, becomes a policy change. Transparency in decision making becomes the norm in the nation–state; software as political service enables the transparency for each citizen (CI).

Universities and politicians in Australia can use the software to deliver training to those aiming to become policy developers or facilitators or advocates (critical citizens) for agendas that impact their constituencies. Each constituency can provide, through their affiliated Universities as FCIs, a link to the software to make it accessible to the citizens who wish to raise issues with their elected members. Technology used in this way will allow citizens' ideas to emerge about improvements to the Australian system, and equally allow those policies to emerge that are applicable to actual rather than perceived needs. These policies once received and screened can become a routine part of the parliamentary debates so

S. Bungsraz, *Operationalising e-Democracy through a System Engineering Approach in Mauritius and Australia*,
https://doi.org/10.1007/978-981-15-1777-8

that appropriate laws are developed. More information about the demo of the political service software is available at http://www.customate.biz/contact.html.

References

ABS. 2018. Population Projections. Australian Bureau of Statistics. http://www.abs.gov.au/ausstats/abs@.nsf/Lookup/by%20Subject/1301.0~2012~Main%20Features~Population%20projections~48. Consulted 28 December 2018.

AEC. 2019. *Electorate Redistributions*. Canberra: Australian Electoral Commission. https://www.aec.gov.au/Electorates/Redistributions/index.htm. Consulted 14 February 2019.

Alexander, D. 2002. Democracy in the Information Age? *Representation* 39 (3): 209–214.

Allen, R.B. 1999. *African Studies: Slaves, Freedmen and Indentured Laborers in Colonial Mauritius*. African Studies Series 99. Cambridge, GB: Cambridge University Press.

Almond, G.A. 1988. The Return to the State. *The American Political Science Review* 82: 853–874.

ALP. 2019. *Will You Support the National Integrity Commission?* Melbourne, VIC, Australia: ALP. https://www.alp.org.au/petitions/support-the-national-integrity-commission/ Consulted 2 February 2019.

Antill, L., and T. Wood-Harper. 1985. *Systems Analysis: Made Simple Computerbooks*. London: Heinemann.

Australian Electoral Commission. 2019. *Electoral Public Funding*. Canberra: Australian Electoral Commission. https://www.aec.gov.au/parties_and_representatives/public_funding/Current_Funding_Rate.htm. Consulted 14 February 2019.

S. Bungsraz, *Operationalising e-Democracy through a System Engineering Approach in Mauritius and Australia*,
https://doi.org/10.1007/978-981-15-1777-8

Backus, M. 2001. EGovernance in Developing Countries. Research Report No. 3, April 3. International Institute for Communication and Development (IICD). https://scholar.google.com.au/scholar?hl=en&as_sdt=0%2C5&q=Backus+RESEARCH+REPORT+No.+3%2C+April+2001&btnG=.

Badcock, C.R. 2014. *Levi-Strauss (RLE Social Theory): Structuralism and Sociological Theory*. London, UK: Routledge.

Ball, T. 2007. Political Theory and Political Science: Can This Marriage Be Saved? *Theoria: A Journal of Social & Political Theory* 54 (113): 1–22.

Barber, B.R. 1997. The New Telecommunications Technology: Endless Frontier or the End of Democracy? *Constellations: An International Journal of Critical & Democratic Theory* 4 (2): 208–228.

Barry, B.M. 1991. *Essays in Political Theory*. Clarendon Paperbacks. New York: Clarendon Press.

Beetham, D. 2012. Defining and Identifying a Democratic Deficit. In *Imperfect Democracies the Democratic Deficit in Canada and the United States*, ed. R. Simeon and P.T. Lenard. Vancouver and Toronto: UBC Press.

Berlinski, D. 1976. *On Systems Analysis: An Essay Concerning the Limitations of Some Mathematical Methods in the Social, Political, and Biological Sciences*. Cambridge: MIT Press.

Bevir, M. (ed.). 2006. *Interpretation and Method: Empirical Research Methods and the Interpretive Turn*. Armonk, NY: M. E. Sharpe.

Bevir, M., and R.A. Rhodes. 2002. Interpretive Theory. *Theory and Methods in Political Science* 1: 1.

Birch, A.H. 1972. *Representation*. Key Concepts in Political Science. London: Macmillan Press.

Birch, A.H. 1993. *Concepts and Theories of Modern Democracy*, 2nd ed. London: Routledge.

Blanchard, B.S., and W.J. Fabrycky. 1998. *Systems Engineering and Analysis*, 3rd ed. Upper Saddle River, NJ: Prentice Hall.

Boyd, O.P. 2008. Differences in eDemocracy Parties' eParticipation Systems. *Information Polity* 13 (3): 167–188.

Bräutigam, D. 1997. Institutions, Economic Reform, and Democratic Consolidation in Mauritius. *Comparative Politics* 30 (1): 45–62.

Briggs, C., and M. Sampson. 2006. *Tying Requirements to Design Artifacts*. Orlando: INCOSE. http://onlinelibrary.wiley.com/doi/10.1002/j.2334-5837.2006.tb02795.x/full. Consulted 28 January 2018.

Brissenden, M., and B. Andersen. 2018. Liberal Party Elders Lash Tony Abbott for Acts of Revenge on Turnbull's Government. ABC. https://www.abc.net.au/news/2018-08-27/liberal-party-elders-lash-tony-abbott-for-acts-of-revenge-on-tu/10166590. Consulted 20 November 2018.

Brooks, R.T., and A.P. Sage. 2006. System of Systems Integration and Test. *Information Knowledge Systems Management* 5 (4): 261–280.

Brown, R.J. 2007. *Governing Australia: A Century of Politics, Policies and People*. Nelson Bay, NSW: R.J. Brown.

Buckley, W. 1968. *Modern Systems Research for the Behaviorial Scientist: A Sourcebook*. Chicago: Aldine Publishing Company.

Burke, M. 2012. A Decade of e-Government Research in Africa. *The African Journal of Information and Communication* 12: 1–25.

Cant, T., J. McCarthy, and R. Stanley. 2006. *Tools for Requirements Management: A Comparison of Teleologic DOORS and the HIVE*. Edinburgh, SA: Defence Science and Technology Organisation. https://www.researchgate.net/publication/27253971_Tools_for_requirements_management_a_comparison_of_telelogic_DOORS_and_the_HiVe. Consulted 28 January 2018.

Carammia, M., S. Princen, and A. Timmermans. 2016. From Summitry to EU Government: An Agenda Formation Perspective on the European Council. *Journal of Common Market Studies* 54 (4): 809–825.

Carroll, B.W., and T. Carroll. 1999. The Consolidation of Democracy in Mauritius. *Democratization* 6 (1): 179–197.

Carter, A. 2017. Reading Lessons: C.B. Macpherson's Immanent Critique. *University of Toronto Quarterly* 86 (2): 19–42.

Caws, P. 1965. *The Philosophy of Science: A Systematic Account*. Princeton, NJ: Van Nostrand.

Chadwick, A. 2006. *Internet Politics: States, Citizens, and New Communication Technologies*. Oxford: Oxford University Press.

Chalmers, A. 2013. *What Is This Thing Called Science?* St Lucia: University of Queensland Press. https://ebookcentral-proquest-com.ezproxy.newcastle.edu.au/lib/newcastle/reader.action?docID=1181566&query=. Consulted 28 January 2018.

Chen, P., R. Gibson, and K. Geiselhart. 2006. *Electronic Democracy? The Impact of New Communications Technology on Australian Democracy*. Canberra: Australian National University.

Clifford, D. 2010. *ISO/IEC 20000 an Introduction to the Global Standard for Service Management*. London: Ebook Library.

Coleman, S., and J.G. Blumler. 2009. *The Internet and Democratic Citizenship: Theory, Practice and Policy*. Communication, Society and Politics. Cambridge: Cambridge University Press.

Coleman, S., and J. Gotze. 2001. *Bowling Together: Online Public Engagement in Policy Deliberation*. London: Hansard Society.

Coleman, S., and G. Moss. 2012. Under Construction: The Field of Online Deliberation Research. *Journal of Information Technology & Politics* 9 (1): 1–15.

Conifer, D. 2018. Senator Fraser Anning Gives Controversial Maiden Speech Calling for Muslim Immigration Ban. ABC News. https://www.abc.net.au/news/2018-08-14/fraser-anning-maiden-speech-immigration-solution/10120270. Consulted 25 February 2019.

Coorey, P. 2018. Payback, Reward and Healing: Scott Morrison Unveils a New Ministry. *Australian Financial Review*. https://www.afr.com/news/scott-morrison-announces-new-ministry-20180825-h14ie2. Consulted 27 August 2018.

Corcoran, P.E. 1983. *Before Marx: Socialism and Communism in France, 1830–48*. London: Macmillan.

CountryWatch, I. 2017. *Mauritius Country Review*. Houston, TX: CountryWatch Incorporated. http://web.b.ebscohost.com.ezproxy.newcastle.edu.au/ehost/command/detail?vid=0&sid=83858dab-466d-43a1-8559-41e43a834c04%40sessionmgr101&bdata=JnNpdGU9ZWhvc3QtbGl2ZSZzY29wZT1zaXRl#jid=DWE&db=bsu. Consulted 28 January 2018.

Craig, J. 1993. *Australian Politics: A Source Book*. Sydney: Harcourt Brace.

Crespo, Rubén González, Oscar Sanjuán Martínez, José Manuel Saiz Alvarez, Juan Manuel Cueva Lovelle, B. Cristina Pelayo García-Bustelo, and Patricia Ordoñez de Pablos. 2013. Design of an Open Platform for Collective Voting through EDNI on the Internet. In *E-Procurement Management for Successful Electronic Government Systems*. Hershey: Information Science Reference (an imprint of IGI Global).

Crisp, L.F. 1983. *Australian National Government*, 5th ed. Melbourne: Longmans.

Croucher, R., and J. McIlroy. 2013. Mauritius 1938: The Origins of a Milestone in Colonial Trade Union Legislation. *Labor History* 54 (3): 223–239.

Curd, M., and J.A. Cover (eds.). 1998. *Philosophy of Science: The Central Issues*, 1st ed. New York: W. W. Norton.

Dahl, R.A. 1995. Justifying Democracy. *Society* 32 (3): 386–392.

Dahl, R.A. 2005. What Political Institutions Does Large-Scale Democracy Require? *Political Science Quarterly* 120 (2): 187–197.

Dahl, R.A., and R.E. Tufte. 1973. *Size and Democracy*. The Politics of the Smaller European Democracies. Stanford, CA: Stanford University Press.

Dahlberg, L., and E. Siapera (eds.). 2007. *Radical Democracy and the Internet: Interrogating Theory and Practice*. Basingstoke: Palgrave Macmillan.

Day, R.B., R. Beiner, and J. Masciulli (eds.). 1988. *Democratic Theory and Technological Society*. Armonk, NY: M. E. Sharpe.

Dewan, E.M. (ed.). 1969. *Cybernetics and the Management of Large Systems: Proceedings of the Second Annual Symposium of the American Society for Cybernetics*. New York: Spartan Books.

Doebelin, E.O. 1980. *System Modeling and Response: Theoretical and Experimental Approaches*. New York: Wiley.

Doran, M. 2018. *Labor's Federal ICAC Proposal Knocked Back by Government, as Coalition Argues Lack of Detail*. Canberra: ABC.

https://www.abc.net.au/news/2018-05-23/labor-federal-icac-proposal-knocked-back-by-government/9791674. Consulted 25 February 2019.
Drack, M., and D. Pouvreau. 2015. On the History of Ludwig von Bertalanffy's "General Systemology", and on Its Relationship to Cybernetics—Part III: Convergences and Divergences. *International Journal of General Systems* 44 (5): 523–571.
Dryzek, J.S., B. Honig, and A. Phillips (eds.). 2008. *The Oxford Handbook of Political Theory.* Oxford: Oxford University Press.
Dunn, J. 2005. *Setting the People Free: The Story of Democracy.* London: Atlantic.
Dunn, J., and I. Harris (eds.). 1997. *Aristotle Volume II.* Great Political Thinkers, 2. Cheltenham, UK and Lyme, NH: Edward Elgar.
Easton, D. 1957a. Traditional and Behavioral Research in American Political Science. *Administrative Science Quarterly* 2 (1): 110–115.
Easton, D. 1957b. Classification of Political Systems. *American Behavioral Scientist* 1 (2): 3–4.
Easton, D. 1965a. *A Systems Analysis of Political Life.* New York: Wiley.
Easton, D. 1965b. *A Framework for Political Analysis.* Prentice-Hall Contemporary Political Theory Series. Englewood Cliffs, NJ: Prentice-Hall.
Easton, D. 1966a. *Varieties of Political Theory.* Prentice-Hall Contemporary Political Theory Series. Englewood Cliffs, NJ: Prentice-Hall.
Easton, D. 1966b. *A Systems Approach to Political Life.* Lafayette, IN: Purdue University. http://www.eric.ed.gov/contentdelivery/servlet/ERICServlet?accno=ED013997. Consulted 28 January 2018.
Easton, D. 1971. *The Political System: An Inquiry into the State of Political Science*, 2nd ed. New York: Alfred A. Knopf.
Easton, D. 1972. Some Limits of Exchange Theory in Politics. *Sociological Inquiry* 42 (3–4): 129–148.
Easton, D. 1975. A Re-assessment of the Concept of Political Support. *British Journal of Political Science* 5 (4): 435–457.
Easton, D. 1981. The Political System Besieged by the State. *Political Theory* 9 (3): 303–325.
Easton, D. 1985. Political Science in the United States: Past and Present. *International Political Science Review* 6 (1): 133–152.
Easton, D., J.G. Gunnell, and M.B. Stein. 1995. Democracy as a Regime Type and the Development of Political Science. In *Regime and Discipline: Democracy and the Development of Political Science*, ed. D. Easton, J.G. Gunnell, and M.B. Stein. Ann Arbor: University of Michigan Press.
Easton, D., J.G. Gunnell, and L. Graziano (eds.). 2002. *The Development of Political Science: A Comparative Survey.* London and New York: Routledge.
Eisner, H. (2011). Systems Engineering: Building Successful Systems. *Synthesis Lectures on Engineering* 6 (2): 1–139.

Elliott, W.Y., and N.A. McDonald. 1965. *Western Political Heritage*. Englewood Cliffs, NJ: Prentice-Hall.

Estlund, D.M. 2002. *Democracy*. Blackwell Readings in Philosophy; 4. Malden, MA: Blackwell.

European Commission. 2018a. EU Membership. Communication Department of the European Commission. https://europa.eu/european-union/about-eu/countries_en#tab-0-1. Consulted 4 February 2018.

European Commission. 2018b. Policies, Information and Services. European Commission. https://ec.europa.eu/info/about-commissions-new-web-presence_en. Consulted 28 January 2018.

Faulk, S. et al. 1994. *Experience Applying the Core Method to the Lockheed C-130J Software Requirements*. Gaithersburg, MD: IEEE. http://ieeexplore.ieee.org.ezproxy.newcastle.edu.au/stamp/stamp.jsp?tp=&arnumber=318472. Consulted 28 January 2018.

Finger, M. and F.N. Sultana. 2012. *E-Governance: A Global Journey*, vol. 4. Amsterdam and Fairfax: IOS Press.

Freitas, E., and M. Bhintade. 2017. *Building Bots with Node.js*. Birmingham: Packt Publishing.

Furneaux, B., and M. Wade. 2017. Impediments to Information Systems Replacement: A Calculus of Discontinuance. *Journal of Management Information Systems* 34 (3): 902–932.

Gallie, W.B. 1955. Essentially Contested Concepts. *Proceedings of the Aristotelian Society* 56 (1955–1956): 167–198.

Galston, W.A. 1991. *Liberal Purposes: Goods, Virtues, and Diversity in the Liberal State*. Cambridge Studies in Philosophy and Public Policy. Cambridge and New York: Cambridge University Press.

Gare, D., and D. Ritter. 2008. *Making Australian History: Perspectives on the Past Since 1788*. South Melbourne: Thomson Learning Australia.

Garnsey, P., and R.I. Winton. 1997. *Political Theory*, vol. II, ed. J. Dunn and I. Harris. Cheltenham, UK and Lyme, NH: Edward Elgar.

Goos, G. et al. (eds.). 2002. Electronic Government. First International Conference, EGOV 2002. Aix-en-Provence, France: Springer.

Gordon, B.M. (ed.). 2011. *Artificial Intelligence: Approaches, Tools, and Applications*. Hauppauge, NY: Nova Science Publishers.

Government of Mauritius. 1968. Part 1—The Constitution GN 54 of 1968, March 12. Mauritius: Government Gazette. http://attorneygeneral.govmu.org/English/Documents/A-Z%20Acts/T/Page%201/THE%20CONSTITUTION,%20GN%2054%20of%201968.pdf. Consulted 28 January 2018.

Government of Mauritius. 2002. *Information and Communication Technologies Act—Act 44 of 2001*, 11 February. Mauritius: Government of Mauritius. http://mtci.govmu.org/English/Rules-Regulations-Policies/Acts/INFORMATION_AND_COMMUNICATION_TECHNOLOGIES.pdf. Consulted 23 September 2016.

Government of Mauritius. 2014. *Government of Mauritius—Consultation Paper on Electoral Reform—Modernising the Electoral System*. Mauritius: Government of Mauritius. http://www.gov.mu/English/Pages/default.aspx. Consulted 14 April 2014.

Government of Mauritius, M.o.F.a.E.D. 2017. Budget Speech 2016/2017. Government of Mauritius. http://budget.mof.govmu.org/budget2017/budgetspeech2016-17.pdf. Consulted 29 January 2017.

Government of Mauritius, M.o.F.a.E.D. 2018a. Summary of Expenditure by Votes 2016–2020. Government of Mauritius. http://budget.mof.govmu.org/budget2017-18/V_00_112017_18ExpbyVotes.pdf. Consulted 6 January 2018.

Government of Mauritius, M.o.F.a.E.D. 2018b. Statement of Government Operations 2016–2020. Government of Mauritius. http://budget.mof.govmu.org/budget2017-18/V_00_102017_18SGovOperations.pdf. Consulted 6 January 2018.

Government of Mauritius, M.o.T., Communication and Innovation. 2014. *National Information & Communication Technology Strategic Plan (NICTSP) 2011–2014: Towards i-Mauritius—Copy Requested Through Minister of Arts and Culture*. Mauritius: Government of Mauritius—Ministry of Technology, Communication and Innovation.

Government of Mauritius, M.o.T., Communication and Innovation. 2018a. Republic of Mauritius e-Government Strategy 2013–2017 l *Empowering Citizens l Collaborating with Business l Networked Government l*. Mauritius Government of Mauritius. http://mtci.govmu.org/English/Documents/eGovernment%20Strategy%20finalv1.pdf. Consulted 28 January 2018.

Government of Mauritius, M.o.T., Communication and Innovation. 2018b. *Expression of Interest for a Market Sounding Exercise For the Implementation of a Third International Gateway Through the Installation of a New Submarine Cable for Both Mauritius and Rodrigues*. Mauritius: Ministry of Technology, Communication and Innovation. http://www1.govmu.org/portal/sites/mfamission/pretoria/documents/bids/ict/Third_Submarine_Cable_EOI.pdf. Consulted 7 January 2018.

Government of Mauritius, M.o.T., *Communication* and Innovation. 2018c. *E-Ideas Online Service*. Mauritius: Ministry of Technology, Communication and Innovation. http://mtci.govmu.org/English/Pages/e-Ideas-Online-Service.aspx. Consulted 28 January 2018.

Government of Mauritius, N.D.U. 2016. *National Development Unit, Customer Charter—Obtained from Honourable Jean Francisco Francois, Private Parliamentary Secretary (Personal Copy)*. Mauritius: Republic of Mauritius, Prime Minister's Office.

Government of Mauritius, N.D.U. 2018. Citizens Advice Bureau, Web Page Accessed on 23 January 2017. National Development Unit, Prime Minister's Office: National Development Unit, Prime Minister's Office. http://ndu.govmu.org/English/Citizens%20Advice%20Bureau/Pages/default.aspx. Consulted 28 January 2018.

Government of Mauritius, P.M.O. 2018. List of Ministers. http://pmo.govmu.org/English/Documents/LIST%20OF%20MINISTERS%20as%20at%2024%20January%202017.pdf. Consulted 28 January 2018.

Grady, J.O. 2006. *System Requirements Analysis*. London, UK: Elsevier.

Gray, A., and B. Jenkins. 2004. Government and Administration: Too Much Checking, Not Enough Doing? *Parliamentary Affairs* 57: 269–287.

Greengard, S. 2015. *The Internet of Things*. The MIT Press Essential Knowledge Series. Cambridge: MIT Press.

Grönlund, Å. 2001. Democracy in an It-Framed Society. *Communications of the ACM* 44 (1): 22–26.

Grönlund, Å. 2004. State of the Art in e-Gov Research—A Survey. *Electronic Government* 3183: 178–185.

Grönlund, Å. 2008. Lost in Competition? The State of the Art in e-Government Research. In *Digital Government*, ed. H. Chen et al. Berlin and Heidelberg: Springer.

Grönlund, Å. 2011. Connecting e-Government to Real Government—The Failure of the UN Eparticipation Index. *Electronic Government* 6846: 26–37.

Grönlund, Å., and T.A. Horan. 2004. Introducing e-Government: History, Definitions, and Issues. *Communications of the Association for Information Systems* 15: 713–729.

Gunnell, J.G. 2013. The Reconstitution of Political Theory: David Easton, Behavioralism, and the Long Road to System. *Journal of the History of the Behavioral Sciences* 49 (2): 190–210.

Gutmann, A. 1980. *Liberal Equality*. Cambridge and New York: Cambridge University Press.

Hague, B.N., and B. Loader. 1999. *Digital Democracy: Discourse and Decision Making in the Information Age*. London and New York: Routledge.

Hall, J.M. 2013. The Rise of State Action in the Archaic Age. In *A Companion to Ancient Greek Government*, ed. H. Beck. Chichester, West Sussex: Wiley.

Held, D. 1983. *States and Societies*. New York: New York University Press.

Held, D. 1993. *Prospects for Democracy: North, South, East, West*. Cambridge, UK: Polity Press.

Held, D. 1995. *Democracy and the Global Order: From the Modern State to Cosmopolitan Governance*. Cambridge: Polity Press.

Held, D. 2006. *Models of Democracy*, 3rd ed. Stanford, CA: Stanford University Press.

Hilbert, M. 2009. The Maturing Concept of E-Democracy: From E-Voting and Online Consultations to Democratic Value Out of Jumbled Online Chatter. *Journal of Information Technology & Politics* 6 (2): 87–110.

Hindess, B. 2002. Deficit by Design. *Australian Journal of Public Administration* 61 (1): 30–38.

Hindman, M. 2009. *The Myth of Digital Democracy*. Princeton, NJ: Princeton University Press.

Hirst, J.B. 2014. *Australian history in 7 questions*. In Collingwood, Victoria: Black Inc. http://ezproxy.newcastle.edu.au/login?url=https://ebookcentral.proquest.com/lib/newcastle/detail.action?docID=1887446.

Hoff, J., I. Horrocks, and P. Tops. 2003. New Technology and the 'Crises' of Democracy. In *Democratic Governance and New Technology: Technologically Mediated Innovations in Political Practice in Western Europe*, ed. J. Hoff, I. Horrocks, and P. Tops. London and New York: Routledge.

Hughes, C.A. 1968. *Readings in Australian Government*. St Lucia: University of Queensland Press.

Huxley, G. 1997. On Aristotle's Best State. In *Aristotle Volume II*, ed. J. Dunn and I. Harris. Cheltenham, UK and Lyme, NH: Edward Elgar.

IDNO. 2016. *Policy Makers in Mauritius, List of Interviewees—Interviews from 6–19 March 2016, Appendix A of Book*. Nvivo.

IHS Global Inc. 2016. *Country Reports—Mauritius*. Lexington, MA: IHS Global Inc.

INCOSE, I.C.O.S.E. 2007. *Systems Engineering Vision 2020*. San Diego, CA: International Council on Systems Engineering.

Isakhan, B., and S. Stockwell. 2012. *The Edinburgh Companion to the History of Democracy*. Edinburgh: Edinburgh University Press.

Jupp, J. 2018. *An Immigrant Nation Seeks Cohesion: Australia from 1788*. London: Anthem Press.

Kant, I. 1991. *Kant: Political Writings*, 2nd ed. Cambridge Texts in the History of Political Thought. Cambridge, UK and New York: Cambridge University Press.

Kant, I. 2000. *The Critique of Pure Reason*. South Bend, IN: Infomotions.

Kaplan, J. 2016. *Artificial Intelligence: What Everyone Needs to Know*. New York: Oxford University Press.

Keane, J. 2010. *The Life and Death of Democracy*. New York: Simon & Schuster.

Kearney, R.C. 1990. Mauritius and the NIC Model Redux: Or, How Many Cases Make a Model? *The Journal of Developing Areas* 24 (2): 195–216.

Kearney, R.C. 1991. Mauritius: Managing Success by World Bank. *African Studies Review* 34 (3): 136–137.

Keman, H. 2014. Democratic Performance of Parties and Legitimacy in Europe. *West European Politics* 37 (2): 309–330.

Kerzner, H. 2017. Introduction to Scope Creep. In *Project Management Metrics, KPIs, and Dashboards—A Guide to Measuring and Monitoring Project Performance*, 2nd ed. Hoboken: Wiley.

Kincaid, H. 1996. *Philosophical Foundations of the Social Sciences: Analyzing Controversies in Social Research*. Cambridge, UK and New York, NY: Cambridge University Press.

Kneuer, M. 2016. E-Democracy: A New Challenge for Measuring Democracy. *International Political Science Review* 37 (5): 666–678.

Kossiakoff, Alexander, William N. Sweet, Samuel J. Seymour, and Steven M. Biemer. 2011. *Systems Engineering Principles and Practice*, 2nd ed. Hoboken: Wiley. http://app.knovel.com/hotlink/toc/id:kpSEPPE006/systems-engineering-principles/systems-engineering-principles. Consulted 26 November 2017.

Lacy, S. 2010. *Configuration Management*. Swindon and Biggleswade: British Computer Society, The Turpin Distribution Services Limited.

Ladyman, J. 2012. *Understanding Philosophy of Science*. Hoboken: Routledge.

Lambert, G. 2012. *In Search of a New Image of Thought: Gilles Deleuze and Philosophical Expressionism*. Minneapolis: University of Minnesota Press. http://ebookcentral.proquest.com/lib/newcastle/detail.action?docID=1047458. Consulted 26 February 2019.

Lanvin, B. et al. 2010. Promoting Information Societies in Complex Environments: An In-Depth Look at Spain's Plan Avanza. *The Global Information Technology Report 2009–2010*, 127–140. World Economic Forum.

Lee, J.S., and L.E. Miller. 1998. *CDMA Systems Engineering Handbook*. Boston, MA: Artech House.

Lee, C., K. Chang, and F.S. Berry. 2011. Testing the Development and Diffusion of E-Government and E-Democracy: A Global Perspective. *Public Administration Review* 71 (3): 444–454.

Levi, M., and L. Stoker. 2000. Political Trust and Trustworthiness. *Annual Review of Political Science* 3 (1): 475–507.

Lidén, G. 2015. Technology and Democracy: Validity in Measurements of E-Democracy. *Democratization* 22 (4): 1–16.

Locke, J. 1960. *Locke's Two Treatises of Government/A Critical Edition with an Introduction and Apparatus Criticus by Peter Laslett*, Rev ed. New York, NY: Cambridge University Press.

Lovell, David, Ian McAllister, William Maley, and Chandran Kukathas. 1998.*The Australian Political System*, 2nd ed. Melbourne: Addison-Wesley.

Luhmann, N. 1995. *Social Systems*. Stanford: Stanford University Press.

Lukes, S. 2002. *Power a Radical View*, 2nd ed. New York: New York University Press.

Macintyre, S. 2016. *A Concise History of Australia*. Port Melbourne, VIC: Cambridge University Press.

Manin, B. 1997. *The Principles of Representative Government*. Themes in the Social Sciences. Cambridge and New York: Cambridge University Press.

Marsh, D., C. Lewis, and J. Chesters. 2014. The Australian Mining Tax and the Political Power of Business. *Australian Journal of Political Science* 49: 711–725.

McCall, G., and K. Widerquist. 2017. *Prehistoric Myths in Modern Political Philosophy*. Edinburgh: Edinburgh University Press.

Mckenna, M. 2017. Fraser Anning Is Pauline Hanson's New Low-Vote Senator. *The WeekEnd Australian*, November 11.

McMahon, M. 2013. *Structural Functionalism*. Salem Press. http://ezproxy.newcastle.edu.au/login?url=http://search.ebscohost.com/login.aspx?direct=true&db=ers&AN=89185764&site=eds-live. Consulted 26 February 2019.

Mead, M. 1969. Cybernetics of Cybernetics. In *Purposive Systems: Proceedings of the First Annual Symposium of the American Society for Cybernetics*, ed. H. von Foerster. New York: Spartan Books.

Meier, A. 2012. *EDemocracy & EGovernment: Stages of a Democratic Knowledge Society*. Berlin: Springer.

Meisenhelder, T. 1997. The Developmental State in Mauritius. *The Journal of Modern African Studies* 35: 279–297.

Mennell, S., and J. Stone (eds.). 1980. *Alexis de Tocqueville on Democracy, Revolution, and Society: Selected Writings*. Heritage of Sociology. Chicago: University of Chicago Press.

Mesarović, M.D. 1964. *Views on General Systems Theory: Proceedings of the Second Systems Symposium at Case Institute of Technology*. Case Institute of Technology, Systems Research Center publications. New York: Wiley.

Mezey, M.L. 2008. *Representative Democracy: Legislators and their Constituents*. Lanham, MD: Rowman & Littlefield.

Miles, W.F. 1999. The Mauritius Enigma. *Journal of Democracy* 10: 91–104.

Miller, E.F. 1971. David Easton's Political Theory. *The Political Science Reviewer* 1: 184–235.

Morgan, G. 2016. New Actors and Old Solidarities: Institutional Change and Inequality Under a Neo-Liberal International Order. *Socio-Economic Review* 14: 201–225.

Moss, G., and S. Coleman. 2014. Deliberative Manoeuvres in the Digital Darkness: E-Democracy Policy in the UK. *The British Journal of Politics & International Relations* 16 (3): 410–427.

Mouffe, C. (ed.). 1996. *Deconstruction and Pragmatism/Simon Critchley, Jacques Derrida, Ernesto Laclau, and Richard Rorty*. London and New York: Routledge.

Mulder, B., and M. Hartog. 2013. Applied E-Democracy. In *Proceedings of the International Conference of E-Democracy and Open Government*, ed. P. Parycek and N. Edelmann. Krems an der Donau, Austria: Edition Donau-Universitat Krems.

Nassehi, A. 2005. Organizations as Decision Machines: Niklas Luhmann's Theory of Organized Social Systems. *The Sociological Review* 53: 178–191.

Nielsen, Claus Ballegaard, Peter Gorm Larsen, John Fitzgerald, Jim Woodcock, and Jan Peleska. 2015. Systems of Systems Engineering: Basic Concepts, Model-Based Techniques, and Research Directions. *ACM Computing Surveys* 48: 11–18.

Norris, P. 1997. Representation and the Democratic Deficit. *European Journal of Political Research* 32: 273–282.

Norris, P. 2001. *Digital Divide: Civic Engagement, Information Poverty, and the Internet Worldwide*. Port Melbourne, VIC, Australia: Cambridge University Press.

Norris, P. 2004. *Electoral Engineering: Voting Rules and Political Behavior*. Cambridge, UK: Cambridge University Press.

Norris, P. 2011. *Democratic Deficit: Critical Citizens Revisited*. New York: Cambridge University Press.

Norris, P. 2012. The Democratic Deficit Canada and the United States in Comparative Perspective. In *Imperfect Democracies: The Democratic Deficit in Canada and the United States*, ed. R. Simeon and P.T. Lenard. Vancouver and Toronto: University of British Columbia Press.

Norris, D.F., and C.G. Reddick. 2013. E-Democracy at the American Grassroots: Not Now Not Likely? *Information Polity: The International Journal of Government & Democracy in the Information Age* 18: 201–216.

Oates, S., D. Owen, and R.K. Gibson. 2006. *The Internet and Politics: Citizens, Voters and Activists*. London and New York: Routledge and Taylor & Francis Group.

OECD. 2003. *Promise and Problems of E-Democracy-Challenges of Online Citizen Engagement*. Paris, France: OECD.

OECD. 2005. *Report from OECD Forum 2005 to the OECD Ministerial Council Meeting*. Paris: OECD Publishing. http://www.oecd-ilibrary.org/docserver/download/0105101e.pdf?expires=1517024994&id=id&accname=guest&checksum=6E744C2E44301A5B640ADBF534EFC209. Consulted 27 January 2018.

OECD. 2010a. *The Development Dimension—ICTs for Development—Improving Policy Coherence*. Washington, DC: Organization for Economic Cooperation & Development, OECD iLibrary. http://www.oecd-ilibrary.org.ezproxy.newcastle.edu.au/docserver/download/0309091e.pdf?expires=1517024328&id=id&accname=ocid194270&checksum=61AB79A0E079DDDBBFE3B6E4568A2B0F. Consulted 27 January 2018.

OECD. 2010b. *Good Governance for Digital Policies—How to Get the Most Out of ICT—The Case Of Spain's Plan Avanza*. OECD Publishing. www.oecd.org/gov/egov/isstrategies. Consulted 27 January 2018.

OECD. 2013. *Reaping the Benefits of ICTS in Spain Strategic Study on Communication Infrastructure and Paperless Administration*. Paris: Organisation for Economic Co-operation and Development, OECD iLibrary. http://www.oecd-ilibrary.org.ezproxy.newcastle.edu.au/docserver/download/4212081e.pdf?expires=1517023432&id=id&accname=ocid194270&checksum=28A93F4721E71AADB84F3539BCB992DE. Consulted 27 January 2018.

OECD. 2015. *States of Fragility 2015 Meeting Post-2015 Ambitions*. Paris: OECD Publishing. http://dx.doi.org/10.1787/9789264227699-en. Consulted 27 January 2018.

OECD. 2016. *Development Aid at a Glance Statistic at a Glance 2.0 Africa*. OECD Publishing. http://www.oecd.org/dac/stats/documentupload/2%20Africa%20-%20Development%20Aid%20at%20a%20Glance%202016.pdf. Consulted 27 January 2018.

OECD, and H. Gaël. 2016. *The Internet of Things-Seizing the Benefits and Addressing the Challenges—2016 Ministerial Meeting on the Digital Economy*. OECD Publishing. http://dx.doi.org/10.1787/5jlwvzz8td0n-en. Consulted 27 January 2018.

Owen, M. 2017. $4k Bill for Whale-Watching Trip. *The Australian*, 3 July.

Päivärinta, T., and Ø. Sæbø. 2006. Models of E-Democracy. *Communications of the Association for Information Systems* 17 (37): 1–42.

Papacharissi, Z. 2002. The Virtual Sphere the Internet as a Public Sphere. *New Media & Society* 4: 9–27.

Parliament of Australia. 2010. *Australia's Constitution—With Overview and Notes by the Australian Government Solicitor*. Canberra: Parliamentary Education Office and Australian Government Solicitor. https://www.aph.gov.au/About_Parliament/Senate/Powers_practice_n_procedures/Constitution. Consulted 25 February 2019.

Parliament of Australia. 2019. *Senators and Members*. Canberra: Parliament of Australia. https://www.aph.gov.au/Senators_and_Members/Parliamentarian_Search_Results?q. Consulted 11 January 2019.

Parsons, T. 1956. Suggestions for a Sociological Approach to the Theory of Organizations–I. *Administrative Science Quarterly* 1: 63–85.

Parsons, T. 1969. *Politics and Social Structure*. New York: Free Press.

Parsons, T. 1971. *The System of Modern Societies*. Englewood Cliffs, NJ: Prentice-Hall.

Parycek, P., and N. Edelmann. 2013. CeDEM13 Proceedings of the International Conference for E-Democracy and Open Government. Paper Presented to Conference for E-Democracy and Open Government, Danube University Krems, Austria.

Parycek, P., N. Edelmann, and M. Sachs. 2012. CeDEM12 Proceedings of the International Conference for E-Democracy and Open Government. Paper Presented to Conference for E-Democracy and Open Government, Danube University Krems, Austria.

Parycek, P. et al. 2013. CeDEM13 Proceedings of the International Conference for E-Democracy and Open Government. Conference for E-Democracy and Open Governement, Danube University Krems, Austria, Danube University Krems.

Peel, M., and C. Twomey. 2011. *A History of Australia*. Palgrave Essential Histories. New York: Palgrave Macmillan.

Pennock, J.R. 1978. *Liberal Democracy: Its Merits and Prospects*. Westport, CT: Greenwood Press.

Pennock, J.R., and J.W. Chapman. 1977. *Human Nature in Politics*. Nomos. New York: New York University Press.

Pruulmann-Vengerfeldt, P. 2007. Participating in a Representative Democracy: Three Case Studies of Estonian Participatory Online Initiatives. In *Media Technologies for Democracy in an Enlarged Europe: The Intellectual Work of the 2007 European Media and Communication Doctoral Summer School*, ed. Nico Carpentier, Pille Pruulmann-Vengerfeldt, Kaarle Nordenstreng, Maren Hartmann, Peeter Vihalemm, Bart Cammaerts, and Hannu Nieminen. Tartu: Tartu University Press.

Qvortrup, M. 2007. *The Politics of Participation: From Athens to E-Democracy*. Manchester: Manchester University Press.

Reddick, C.G. 2005. Citizen Interaction with E-Government: From the Streets to Servers? *Government Information Quarterly* 22 (1): 38–57.

Rios Insua, D., and S. French. 2010. *E-Democracy*, vol. 5. Dordrecht, Heidelberg, London, and New York: Springer.

Rodríguez Bolívar, M.P. et al. 2010. Trends of E-Government Research: Contextualization and Research Opportunities. *The International Journal of Digital Accounting Research* 10 (16): 87–111.

Rosenberg, A. 2013. *Philosophy of Science a Contemporary Introduction*. Routledge Contemporary Introductions to Philosophy. Florence: Taylor & Francis.

Sainsbury, L. 2017. But the Soldier's Remains Were Gone: Thought Experiments in Children's Literature. *Children's Literature in Education* 48: 152–168.

Sandbrook, R. 2005. Origins of the Democratic Developmental State: Interrogating Mauritius. *Canadian Journal of African Studies/La Revue canadienne des études africaines* 39 (3): 549–581.

Scholl, H.J. 2006. Is E-Government Research a Flash in the Pan or Here for the Long Shot? In *Electronic Government*, ed. M.A. Wimmer, H.J. Scholl, A. Grönlund, and K.V. Andersen. Berlin and Heidelberg: Springer.

Scholl, H.J. 2013. Electronic Government Research: Topical Directions and Preferences. In *Electronic Government*, ed. M. Wimmer, M. Janssen, and H. Scholl. Berlin and Heidelberg: Springer.

Schumpeter, J.A. 2010. *Capitalism, Socialism and Democracy*, 1st ed. Florence: Taylor & Francis.

Schumpeter, J.A. 2013. *Capitalism, Socialism and Democracy*. Abingdon: Routledge.

Shaw, T. 2008. Max Weber on Democracy: Can the People Have Political Power in Modern States? *Constellations: An International Journal of Critical & Democratic Theory* 15: 33–45.

Silver, P., W. McLean, and P. Evans. 2013. *Structural Engineering for Architects a Handbook*. London, UK: Laurence King Publishing.

Skinner, Q. 2012. On the Liberty of the Ancients and the Moderns: A Reply to My Critics. *Journal of the History of Ideas* 73 (1): 127–146.

Smartvote. 2018. Smartvote Web Page. Smartvote. https://www.smartvote.ch/about/ideaEGOV. Consulted 5 February 2018.

Smith, B. 2008. *Australia's Birthstain: The Startling Legacy of the Convict Era*. Sydney, NSW, Australia: Allen & Unwin.

Smith, R. 2016. Confidence in Paper-Based and Electronic Voting Channels: Evidence from Australia. *Australian Journal of Political Science* 51: 68–85.

Smith, R., A. Vromen, and I. Cook. 2012. *Contemporary Politics in Australia: Theories, Practices and Issues*. Port Melbourne, VIC: Cambridge University Press.

Solomon, D. 1978. *Inside the Australian Parliament*. Hornsby, NSW: Allen & Unwin.

Stahl, B.C. 2008. *Information Systems Critical Perspectives*, vol. 2. Routledge Studies in Organization and Systems. Hoboken: Taylor & Francis.

Statistics Mauritius. 2017. Table 4—Population Enumerated at Each Census by Community and Sex, 1962–1972. Statistics Mauritius. http://statsmauritius.govmu.org/English/Pages/POPULATION–And-VITAL-STATISTICS.aspx. Consulted 16 November 2017.

Steele-Vivas, R.D. 1996. Creating a Smart Nation: Strategy, Policy, Intelligence, and Information. *Government Information Quarterly* 13 (2): 159–173.

Stokes, G., and A. Carter (eds.). 2001. *Democratic Theory Today: Challenges for the 21st Century*. Malden, MA: Blackwell Publishers.

The Guardian. 2018. Malcolm Turnbull Tells Q&A His Removal Was an Act of Madness. *The Guardian*. https://www.youtube.com/watch?v=b_dopFlIbc4. Consulted 25 February 2019.

Tobias, O., S. Håkan, and D. Peter. 2003. An Information Society for Everyone? *International Communication Gazette (Formerly Gazette)* 65 (4): 347.

Transparency International. 2018. Australia—Corruption Perception Index 2018. https://www.transparency.org/country/AUS. Consulted 2 February 2019.

Transparency Mauritius. 2013. *Corruption Perception Index 2013*. Port Louis, Mauritius: Transparency Mauritius. http://www.transparencymauritius.org/corruption-perception-index/corruption-perception-index-2013/. Consulted 25 December 2016.

Transparency Mauritius. 2015. *Corruption Perception Index 2015*. Port Louis, Mauritius: Transparency Mauritius Organisation. https://www.transparencymauritius.org/wp-content/uploads/2016/01/CPI-2015.pdf. Consulted 25 December 2016.

Transparency Mauritius. 2017. *Corruption Perception Index 2016*. Port Louis, Mauritius: Transparency Mauritius Organisation.

https://www.transparencymauritius.org/wp-content/uploads/2017/06/CPI-2016-25-01-2017.pdf. Consulted 30 October 2017.

Ubaldi, B. 2013. *Open Government Data: Towards Empirical Analysis of Open Government Data Initiatives*. OECD Working Papers on Public Governance, 1–61. OECD Publishing.

United Nations. 2005. *UN Global E-Government Readiness Report from E-Government to E-Inclusion, UNPAN/2005/14*. New York: United Nations.

USAF. 2010. *Air Force Systems Engineering Assessment Model*. Air Force Center for Systems Engineering, Air Force Institute of Technology: Secretary of Air Force, Acquisition. http://www.afit.edu/cse/. Consulted 8 September 2013.

USAF. 2013. *SMC Systems Engineering Primer and Handbook*. Space and Missile System Center. EverySpec. http://everyspec.com/search_result.php?cx=partner-pub-0685247861072675%3A94rti-pv850&cof=FORID%3A10&ie=ISO-8859-1&q=SMC+Systems+Engineering+Primer+and+Handbook&sa=Search&siteurl=everyspec.com%2F&ref=&ss=20497j31595539j48. Consulted 8 February 2018.

U. S. Department of Defense. 1985. *Military Standard Specification Practices MIL-STD-490a*. Andrews Air Force Base, Washington, DC. http://everyspec.com/MIL-STD/MIL-STD-0300-0499/MIL-STD-490A_10378/. Consulted 28 January 2018.

U. S. Department of Defense. 1992. *Military Standard Configuration Management MIL-STD-973*. Falls Church, VA. http://everyspec.com/MIL-STD/MIL-STD-0900-1099/MIL_STD_973_1146/. Consulted 28 January 2018.

U. S. Department of Defense. 1993. *Military Standard Systems Engineering MIL-STD-499b*. EverySpec. http://everyspec.com/MIL-STD/MIL-STD-0300-0499/MIL-STD-499B_DRAFT_24AUG1993_21855/. Consulted 28 January 2018.

U. S. Department of Defense. 1997. *Military Handbook Configuration Management Guidance MIL-HDBK-61*. Falls Church, VA. http://everyspec.com/MIL-HDBK/MIL-HDBK-0001-0099/MIL-HDBK-61_11531/. Consulted 28 January 2018.

U.S. Department of Defense. 2003. *Department of Defense Standard Practice Defense and Program-Unique Specifications and Format MIL-STD-961e*. Fort Belvoir, VA: Defense Standardization Program Office. http://everyspec.com/MIL-STD/MIL-STD-0900-1099/MIL-STD-961E_11343/. Consulted 28 January 2018.

U.S. Department of Defense. 2008. *Systems Engineering Guide for Systems of Systems*. Office of the Deputy Under Secretary of Defense for Acquisition and Technology, Systems and Software Engineering. Systems Engineering Guide for Systems of Systems, Version 1.0. Washington, DC: ODUSD(A&T). http://www.everyspec.com/. Consulted 28 January 2018.

van der Hof, S., and M.M. Groothuis. 2011. *Innovating Government Normative, Policy and Technological Dimensions of Modern Government.* The Hague and The Netherlands: T.M.C. Asser Press.

von Bertalanffy, L. 1950. An Outline of General System Theory. *The British Journal for the Philosophy of Science* 1: 134–165.

von Foerster, H. 1969. *Purposive Systems: Proceedings of the First Annual Symposium of the American Society for Cybernetics.* New York: Spartan Books.

Von Mises, R. 1951. *Positivism: A Study in Human Understanding.* Cambridge: Harvard University Press.

Wallas, G. 1962. *Human Nature in Politics.* Lincoln: University Nebraska Press.

Ward, D. 2002. *European Union Democratic Deficit and the Public Sphere.* Amsterdam, The Netherlands: IOS Press.

Webster, F. 2007. *Theories of the Information Society.* International Library of Sociology. Hoboken: Taylor & Francis.

Weinberger, J. 1988. Liberal Democracy and the Problem of Technology. In *Democratic Theory and Technological Society*, ed. R.B. Day, R. Beiner, and J. Masciulli. Armonk, NY: M. E. Sharpe.

White, S., and N. Nevitte. 2012. Citizens Expectations and Democratic Performance the Sources and Consequences of Democratic Deficits from the Bottom Up. In *Imperfect Democracies the Democratic Deficit in Canada and the United States*, ed. R. Simeon and P.T. Lenard. Vancouver and Toronto: University of British Columbia Press.

Whitlock, G., and D. Carter. 1992. *Images of Australia: An Introductory Reader in Australian Studies.* St Lucia: University of Queensland Press.

Wilhelm, A.G. 2000. *Democracy in the Digital Age.* New York: Routledge.

Willmore, L. (2003). Universal pensions in Mauritius: Lessons for the Rest of Us, ST/ESA/2003/DP. 32. DESA Discussion Paper No. 32. United Nations DESA Discussion Paper Series. Division for Public Economics and Public Administration, Room DC2-1446, United Nations, New York, NY, 10017, United Nations.

Williams, G., and D. Hume. 2010. *People Power: The History and the Referendums in Australia.* Sydney: University of New South Wales Press.

Wise, B.R. 1913. *The Making of the Australian Commonwealth, 1889–1900: A Stage in the Growth of the Empire.* New York: Longmans, Green, and Company.

Wright, E.O. 1994. Political Power, Democracy, and Coupon Socialism. *Politics & Society* 22: 535.

Wyatt, S. 2007. Technological Determinism Is Dead: Long Live Technological Determinism. In *The Handbook of Science and Technology Studies*, ed. E.J. Hackett, O. Amsterdamska, M. Lynch, J. Wajcman. Cambridge: MIT Press.

Youngblood-Coleman, D. 2014. *Mauritius: 2014 Country Review*. Houston, TX: CountryWatch Incorporated. http://ezproxy.newcastle.edu.au/login?url=http://search.ebscohost.com/login.aspx?direct=true&db=bth&AN=99017947&site=eds-live. Consulted 8 January 2018.

Youngblood-Coleman, D. 2016. *Mauritius: 2016 Country Review*. Houston, TX: CountryWatch Incorporated. http://web.b.ebscohost.com.ezproxy.newcastle.edu.au/ehost/pdfviewer/pdfviewer?vid=1&sid=ddbc9060-af26-4b18-8af7-09fd6888144c%40sessionmgr120. Consulted 8 January 2018.

Zweifel, T.D. 2002. *Democratic Deficit? Institutions and Regulation in the European Union, Switzerland, and the United States*. Lanham, MD: Lexington Books.

Printed in the United States
By Bookmasters